Springer Theses

Recognizing Outstanding Ph.D. Research

Aims and Scope

The series "Springer Theses" brings together a selection of the very best Ph.D. theses from around the world and across the physical sciences. Each volume has been selected for its scientific excellence by an internationally renowned institute or department. For greater accessibility to non-specialists, the published versions include an extended introduction, as well as a foreword by the student's supervisor explaining the special relevance of the work for the field. The series as a whole provides a valuable resource both for newcomers to the research fields, and for other scientists seeking detailed background information on special questions. Finally, it serves as an accredited documentation of the valuable contributions made by today's younger generation of scientists.

How to Apply

Theses prepared at internationally leading universities and institutes may be nominated for publication by supervisors with the endorsement of the head of department. Each such department is entitled to nominate one thesis per year. Enquiries and requests for a nomination form should be sent, together with the thesis, to Ms. Nathalie Berg (see link on right).

Eligibility Criteria

To qualify for consideration, theses should fulfil the following criteria:

- They must be written in good English.
- The topic should fall within the confines of Chemistry, Physics, Earth Sciences, Engineering and related interdisciplinary fields such as Materials, Nanoscience, Chemical Engineering, Complex Systems and Biophysics.
- The work reported in the thesis must represent a significant scientific advance.
- A maximum of 30% of the text may be a verbatim reproduction from the author's previous publications.
- They must have been examined and passed during the 12 months prior to nomination.
- The theses should have a clearly defined structure including an introduction accessible to new PhD students and scientists not expert in the relevant field.
- After acceptance, theses should be supplemented by a "Supervisor's Foreword" outlining the wider scientific relevance of the work

Indexed by zbMATH.

Matthias Fabry

Detailed Structure and Evolution Modeling of the Tightest Massive Binary Stars

Springer

Matthias Fabry
Dept of Astrophys and Planetary Science
Villanova University
Villanova, PA, USA

ISSN 2190-5053 ISSN 2190-5061 (electronic)
Springer Theses
ISBN 978-3-032-28787-8 ISBN 978-3-032-28788-5 (eBook)
https://doi.org/10.1007/978-3-032-28788-5

This Springer imprint is published by the registered company Springer Nature Switzerland AG
The registered company address is: Gewerbestrasse 11, 6330 Cham, Switzerland

to the curious

Foreword

Stars are the dominant sources of visible light in the Universe and are responsible for producing most of the chemical elements, including those essential for life. A precise understanding of their inner workings and final fate is thus essential as stars provide the input physics to many of our modern cosmic models, from the reionization of the Universe, to the evolution of the galaxies, the chemical enrichment of the Universe, and the rates and properties of gravitational wave signals.

In the imagination of most people, as likely already in Aristotle's mind, stars are depicted as perfect gaseous spheres (circles, in two-dimensional projections such as children's drawings). Reality, however, is more complex. Rotation alone deforms stellar shapes, with differences between polar and equatorial radii reaching up to 50% in the most extreme cases. In close binary systems, the mutual gravitational pull and centrifugal forces associated with orbital motion further reshape their surfaces: semi-detached systems resemble (massive) droplets, while, in the most extreme contact configurations, the two stars share a join surface with a characteristic dumbbell like geometry, "peanut stars" as Matthias Fabry likes to call them.

Although such deformations have been observed for more than half a century, theoretical models have long lagged behind in their ability to self-consistently describe these extreme configurations. Processes such as tidal deformation, rotational mixing, mass transfer, and the establishment of contact structures profoundly alter the stars' internal structure, surface properties, and evolutionary fates. Capturing these effects quantitatively remains one of the central challenges of modern stellar astrophysics, and it is especially relevant for massive stars given the widespread presence of close binaries.

In this context, it is a pleasure to introduce the doctoral work of Matthias Fabry, which provides a significant contribution to the theory of stellar structure and evolution in close binary systems. Matthias's PhD work was awarded the Springer PhD Thesis Prize, a distinction that acknowledges its scientific merit and represents the culmination of several years of research at the KU Leuven Institute of Astronomy.

When a star is deformed, the assumption of spherical symmetry, which is commonly used to reduce the three-dimensional hydrodynamical problem of stellar structure to a one-dimensional system of equations, breaks down. In such cases, one-dimensional prescriptions become, at best, inaccurate and, at worst, irrelevant. Matthias Fabry's thesis contributes to address these issues at the most fundamental level. He develops new physical prescriptions that enable one-dimensional stellar evolution models to incorporate rotation, tidal deformation, and energy transfer. These processes are critical for understanding the tightest binaries yet absent from standard evolutionary calculations. The resulting models quantify how deformation modifies stellar radii, temperatures, and luminosities, and the modified equations can be incorporated into detailed binary evolution computations to explore the evolution and fate of such systems, both individually and at the population level.

Another major contribution of this thesis is its physically motivated treatment of energy transfer in massive contact systems. Observationally, massive contact binaries exhibit nearly identical surface temperatures despite frequently having strongly unequal masses, an empirical fact that contradicts traditional stellar-structure models. By introducing a parameterized mechanism that reproduces the expected thermodynamical outcome of energy exchange across equipotential layers, Matthias's work investigates a new pathway toward resolving this discrepancy. A large grid of binary models shows that energy transfer alone does not fully reproduce the observed mass-ratio distribution of contact systems, but it significantly shifts theoretical predictions in the correct direction and discusses which additional processes must also be considered.

The thesis further includes an observational analysis of 9 Sagittarii, a long-period, very massive O-type binary in the Lagoon Nebula. Through a combination of interferometric astrometry, spectral disentangling, and atmosphere modeling, Matthias obtains the most precise dynamical masses to date for the components of this system. As a wide and effectively non-interacting binary, 9 Sgr offers a rare and valuable benchmark for calibrating evolutionary tracks in the upper Hertzsprung–Russell diagram. The analysis not only resolves previous inconsistencies in the literature but also demonstrates the synergy between state-of-the-art observational techniques and advanced theoretical modeling.

Taken together, the developments in this thesis significantly advance our ability to model the tightest massive binaries and their evolved descendants. Their importance extends well beyond classical stellar astrophysics: close massive binaries play a central role in several proposed formation channels of the binary black holes now observed through gravitational waves. As this field progresses toward increasingly large and precise datasets, the need for accurate, physically grounded models becomes ever more pressing. Finally, I would like to underline the qualities of the researcher behind this work. Together with Dr. Pablo Marchant (KU Leuven/UGhent, Belgium), it has been our privilege to guide Matthias Fabry in his early scientific years. Throughout his doctoral studies, Matthias demonstrated scientific maturity, a critical mindset, and the ability to combine theoretical developments, numerical computations, and state-of-the-art observations. His drive to understand the fundamental physical processes, and his collaborative spirit made him a valued member of our KU Leuven research group, as well as of the broader stellar astrophysics community. The recognition conferred by the Springer PhD Thesis Prize is therefore very well deserved and I am delighted to recommend this thesis to the reader. This thesis does not only describe substantial scientific achievement by Matthias but also provides an instructive review of stellar physics. In particular, Chap. 1 will delight newcomers and expert readers alike.

KU Leuven, Leuven
March 2026

Hugues Sana

Preface

Massive stars are responsible for producing heavy elements in the universe, and scatter them into their cosmic neighborhoods as they explode as supernovae. These processes pave the way for the formation of a new generation of stars that have the possibility to form planets out of the formed dust, potentially harboring life.

With masses over eight times that of our Sun, massive stars lead bright and fast lives, taking only millions of years from birth to death, compared to billions for solar-type stars. Furthermore, much more so than the low-mass stars, massive stars come predominantly in pairs, called binary systems. Throughout their evolution, stars in binary systems interact, transferring mass between each other, which has major implications for their ultimate fate. This is one of the reasons that makes studying massive binary stars so interesting.

For tight binary systems in particular, a whole list of extra effects come into play. Some binaries orbit so close that they deform through tidal interaction (the same process that sloshes the oceans on Earth), and in the extreme case, the stars come into contact, forming "peanut stars." Massive contact binaries do not necessarily immediately merge into one, as they are rotating around each other very quickly, and they can be stable for millions of years. What is more is that, by observing samples of contact binaries, they seem to be stable while having unequal masses. The mass-ratio distribution is fairly uniform between mass ratios of around one half and unity. Unfortunately, no binary models that simulate the life of these stars have been able to reproduce this constraint, and predict that contact systems should overwhelmingly appear as equal-mass binaries.

When modeling the life of stars, today, we use computers to solve the set of differential equations that govern their evolution. We include several effects that modify the structure of the stars, such as rotation, internal mixing, stellar winds, as well as the binary-star process of mass transfer. Up until today, tidal deformation and the process of energy transfer in contact binaries have not yet been accounted for in detailed models of binary evolution. Instead, the state-of-the-art models are spherically symmetric, or at most deformed cylindrically due to rotation.

In this thesis, we develop methods to represent tidal deformation of close binaries in one-dimensional models of stars, and study its effects. From numerical integrations of the Roche geometry, we calculate the deformation corrections needed to adapt the stellar-structure equations and their boundary conditions. With these new methods, we compute stellar models including the effect of tidal deformation by a companion, and conclude that a 5% variation can arise compared to models that are spherically symmetric. Single-star rotation, causing axisymmetric deformation, is a good first-order approximation to this effect. Thus, to model close binary stars (whether in contact or not), rotational and/or tidal deformation needs to be taken into account to make accurate predictions about observed systems. This is important in particular for eclipsing binaries because they can have their surface conditions (such as their radius) measured with a precision better than 3%. For example, if non-rotating models would be used to fit the radius of a strongly deformed eclipsing star, systematically higher masses would be found, as non-rotating models are around 5% smaller than deformed ones (at the same mass and age).

We also investigate the effect of energy transfer in massive contact binaries. Motivated by the observation that the temperatures of both components are similar, despite their differing mass, and the theoretical argument that otherwise the stars would be hydrodynamically unstable, there is consensus that energy must be flowing in the common layers of a contact binary. In one model, the energy transfer is compared to how a refrigerator cools its inside by moving heat to a radiator, but the precise structure of the thermodynamical flow in contact binaries is an unsolved topic. In this thesis, instead of attempting to model the flow, we model its expected outcome by implementing an energy source or sink in the interacting stellar models. The result is that the common layers of the models are shellular, which means that the properties of temperature, density, and pressure are equal across equipotential surfaces of the binary. It also makes the surface temperatures of both components identical, which is close to what is observed.

Finally, we apply the developed methodologies to a large population of massive contact-binary models. In particular, we investigate if our model of energy transfer is sufficient to solve the discrepancy between the observed mass-ratio distribution of contact binaries versus the modeled one. Our models with energy transfer do not fully solve the issue. While it does have a positive impact on the amount of contact binaries predicted with unequal masses, the vast majority of the models still quickly evolve to equal-mass binaries. From the modeling side, we conclude that other processes, such as internal mixing and the rejuvenation process of mass-gaining stars should be investigated in tandem with energy transfer. From the observational point of view, it should be investigated what the true birth distributions of short-period binaries are, because our models indicate binaries are most likely not born with initial periods as short as one day. Both angles should be pursued to resolve this open question.

Contact binaries represent a fascinating phase in massive binary-star evolution. They are a unique laboratory to test the structure and evolution of stars, and play an important role in the formation of gravitational-wave sources, which are being observed for almost ten years now with LIGO and VIRGO. With the methodology developed in this thesis, further modeling studies of peanut stars can be undertaken to uncover the life of the tightest binary stars.

Villanova, PA, USA Matthias Fabry

Acknowledgments What a journey.

In the first half of my PhD, the pandemic required all interactions with students and colleagues to move to Zoom (2/10; would not recommend). Despite the unbelievable situation, once it receded, I had the pleasure to meet many cool people abroad, both at conferences and during visits. Special thanks go to Silvia, Tomer, and the API members for hosting me in Amsterdam, and to Norbert and the Bonn stellar group for letting me stay for four months! Also big thanks to the `MESA` devs, TAs, and students at the Santa Barbara summer school for your inexhaustible enthusiasm for stars.

Back home, I had a most wonderful time at the Institute of Astronomy here in Leuven. It really is a vibrant place, with lots of sports, receptions, and an especially tight-knit group of PhD students and post-docs. Special thanks go to:

The volleyball group (Robin, Karan, Nico, Olivier, Luka, and many, many more…) for the many Friday nights on the sand (and at Spuye or Metafoor afterwards…).

The 2023 Uitje team (too many to list here, THANKS everyone!) for organizing three days in the Ardennes full of activities.

The UK/Ireland folks (Abi, Gareth, Calum, Emily) who introduced me to Six Nations Rugby at the Irish Pub.

The SkIvS adventurers (Nich, Luc, JP, Silke, …) whom I was able to join twice, taking in the sights, showing off the *drip*, and *shredding the gnar* on the French alps.

My fellow San Francisco travelers (Mathias, Joey, Vincent, Annachiara) for sitting through my very first live baseball game with me, walking through the Californian Redwoods, driving down highway 1, and otherwise making my first trip to the United States one to never forget.

Clio, Anita, Monique, and the System team who all play invaluable support roles in the institute.

Jesus and Saskia for the support on my three trips to La Palma. Observing on the mountain is truly an unforgettable experience!

And finally, *massive* thanks to the massive-star group. I consider myself blessed to have been able to meet every single one of you. Thanks Abi, Calum, Gareth, Laurent, Annachiara, Kunal, Soetkin, Tinne, Emma, Alan, Frank, Jaime, Julia, Michael, Cyprien, Philippe, Maddalena, Tomer, Reinhold, Dominic, and Karan for the many fruitful discussions, interesting science, and the fun on the team Uitjes! I deliberately left Hugues and Pablo out of this list. That's because, of course, you two deserve separate paragraphs.

Merci mille fois, Hugues, d'être un super mentor. I consider starting a PhD with you one of the best decisions I have ever made in my life (9/10; would recommend). Even before completing my work on your project, hell, even before finishing the first three months, you immediately let me write my own project, which resulted in the bulk of this thesis. You kick-started my academic career (which I hope to continue for many years to come) and supported me in my personal and professional development. I especially like your no-nonsense attitude, and making it extremely clear what to do and also (maybe more importantly) what *not* to do. If ever I need an opinion on anything even remotely related to astrophysics, I know who to turn to.

Muchas gracias, Pablo, por mostrarme las estrellas. While I learned what stars are in my courses, it was you who taught me the fine details and showed me the inner workings of `MESA`, not to mention the myriad effects of binary processes. You have provided the majority of the scientific motivation for this thesis, which I will never take for granted. Alongside Hugues, you have supported me throughout the past years, and thanks to you, I have become the scientist I am today. You are a continuous inspiration to me and I will always look to you and your work as the gold standard for research in theoretical stellar astrophysics. I hope we can work together for many more years. Thanks also for being a cool guy and a good friend, and 3D printing some of the designs I made over the years.

I'd be remiss not thanking all my other friends. Thanks to my pals at Golf Park Tervuren (and formerly DMGC), for all the fun rounds on the many Sunday mornings. I also have good memories from our interclub competition campaigns! Y'all made it a really, really fun club I feel at home at. The little nine-hole course I learned to play on will always have a special place in my heart. I'll definitely visit in the future!

Thanks to the people of Guikui (who came up with this name anyway?), for dealing with me during our bachelor degree. You are a bit crazy sometimes, but that's OK, and exactly why I like you. I was maybe not the most outgoing in this group during my PhD, but I still want to thank you since you helped me getting through some of the toughest times of my life.

Big thanks to my three partners of the *Lucky Basterds*, Bart, Jordi, and Lennart. By my estimation we played board games together for over a thousand hours, adventuring in Gloomhaven, Frosthaven, the 7th continent, Aeon Trespass, Scythe, and elsewhere. It takes a special kind of geek to be willing to do this, and I'm happy to have found them in you guys. I think it's all worth it though, right? Remember *tick tock*, the (glorious) deus ex machina d10 crit roll in Aeon, me trying desperately to make expeditions work in Odin, and, of course, the terrible *semi-cooperative* shenanigans in Gloom/Frosthaven. On to 10,000?

For some more random acknowledgments: Thanks to *deadmau5* for slapping some chords together that I enjoy,[1] Frank Herbert for writing *Dune*, the Miller brothers and Cyan Worlds for creating *Myst* and the *D'ni* universe, James SA Corey for writing *the Expanse*, Mike Pondsmith for creating *Cyberpunk* upon which both CDPR's *2077* and *Netrunner* are based (*not* random: huge thanks to the Belgian Netrunner community, but especially Ruben for playing many games with me Tuesday nights in Demo-Spel!), Asobo studios for creating the best flight simulator to date, with which I can feel just a bit closer to my late grandfather, and Jon van Caneghem for developing *Heroes of Might and Magic III*, a simply awesome and timeless game. These people will probably never read this, but still, thanks, you kept

[1] like "some chords" – deadmau5

me busy during most of my off-time in the past years, and that's equally important I think.

Ik heb de belangrijksten voor het laatst gehouden. Dank u meter, oma en opa om interesse te blijven tonen in mijn sterrenkundige avonturen. Wat ik bestudeer ligt misschien wat ver van jullie bed, maar toch laten jullie mij maar doen, met mijn hoofd in de ruimte. Dank u peter om mijn *science-nonkel* te zijn. Klinkt als iets uit een Amerikaanse sitcom, maar ik heb het geluk een échte te hebben! Dankzij jou weet ik wat mogelijk is als academicus, en ook bij wie ik terecht kan voor persoonlijk advies. Ten laatste, dank u mama, papa, en zus voor de ongeziene steun. Dank u om ten minste alsof te doen geïnteresseerd te zijn in wat ik doe, en te blijven luisteren naar mijn gepalaver over sterren, pindanoten, en zwarte gaten. De voorbije jaren waren niet altijd de gemakkelijkste, maar desondanks heb ik dit boekje kunnen schrijven, en dit was niet mogelijk zonder jullie hulp gedurende mijn hele leven. Ik ben hier extreem fier op, en ik hoop jullie ook.

Heverlee, Leuven — Matthias Fabry
May 2024

Competing Interests The author has no competing interests to declare that are relevant to the content of this manuscript.

Contents

List of Acronyms

AGB	Asymptotic giant branch (star)
BC	Boundary condition
BH	Black hole
CDF	Cumulative distribution function
CHE	Chemically homogeneous evolution
CE	Common envelope
ET	Energy transfer
GRB	Gamma-ray burst
GW	Gravitational wave
MESA	Modules for Experiments in Stellar Astrophysics (code)
MLT	Mixing length theory
MT	Mass transfer
HMXB	High-mass X-ray binary
HRD	Hertzsprung-Russell diagram
MB	Magnetic braking
MC	Monte Carlo
MCMC	Markov chain Monte Carlo (method)
NIR	Near infrared
PMS	Pre-main sequence
RGB	Red giant branch (star)
RL	Roche lobe
RLOF	Roche-lobe overflow
RV	Radial velocity
S/N	Signal-to-noise ratio
SB	Spectroscopic binary, SB1 or SB2 for single- or double-lined
SN	Supernova
WR	Wolf-Rayet (star)
ZAMS	Zero-age main sequence
ZKL	von Zeipel-Kozai-Lidov (oscillations)

List of Figures

Chapter 1
Introduction

"Space, the final frontier..."
— Captain James T. Kirk

When you look up on a clear night in or around Leuven, you can see a thousand or so objects on the sky. If it is above the horizon, the Moon will be most prominent, and it presents itself as a crescent or a bright disk depending on its phase. The next brightest objects are the planets, such as Jupiter, Venus, Mars and Saturn. Even with a modest telescope or good binoculars, you can see Saturn's rings, or the big red spot of Jupiter. Mercury can only be seen when it is reflecting a good amount of sunlight back to the Earth, while for Uranus you will probably need to squint your eyes in order to see it. Neptune is invisible to the naked eye in any but the very darkest skies, for which the Belgian ones anno 2024 definitely do not qualify.

Next in line are objects outside our own solar system, like stars, galaxies and nebulae. In contrast to the planets, these objects are up to thousands of light years away in the case of stars or nebulae in our own Galaxy to millions or even billions of light years for other galaxies. The brightest star in the sky is Sirius, the 'Dog star', and can be seen toward the south in wintertime. It is so bright it bears significant importance in many societies and cultures throughout history. It was used by the Egyptians to predict the flooding of the Nile (Clagett 1989), while many Polynesian and other seafaring nations used it (and other bright stars) for navigation purposes (Bateson 1959). In modern day Leuven some of the bright stars you will be able to spot easily include Polaris, the pole star, Betelgeuse and Rigel in the constellation of Orion, Castor and Pollux in Gemini (twins), and others like Vega, Deneb and Altair. When it comes to seeing galaxies, nebulae and star clusters, in Leuven we are similarly limited to bright objects like the Pleiades star cluster, also named 'the Seven Sisters,' the Andromeda galaxy, or the Orion Nebula.

Unfortunately, the wonders of modern technology limit our ability to see the night sky clearly. The use of artificial light to illuminate our streets, monuments, buildings and soccer pitches sends large amounts of light into the atmosphere that reflects back and hides fainter stars and galaxies, an effect called light pollution. Today, Flanders hosts some of the most light-polluted skies in western Europe. Better nearby stargazing sites are located in rural France or Spain and, with a nearly

M. Fabry, *Detailed Structure and Evolution Modeling of the Tightest Massive Binary Stars*, Springer Theses, https://doi.org/10.1007/978-3-032-28788-5_1

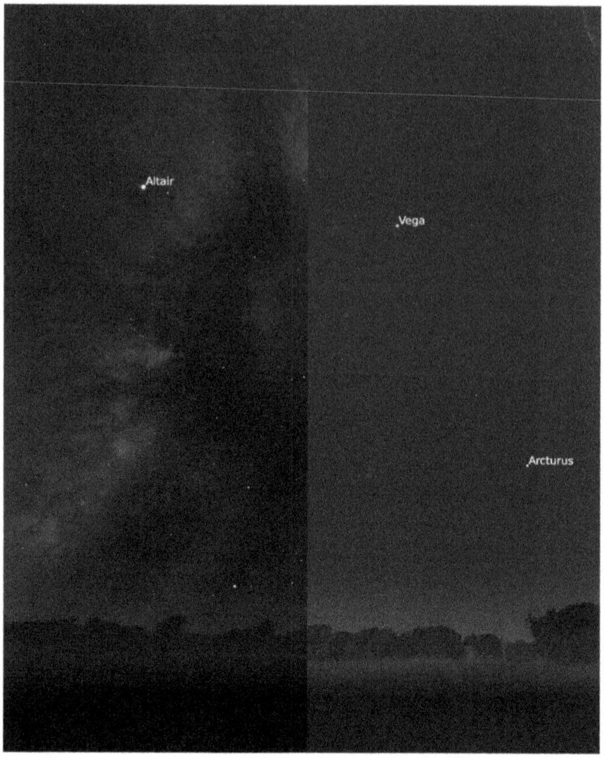

Fig. 1.1 Simulated view of the southwestern Leuven sky with, on the left, light pollution disabled, while on the right the typical Flemish light-pollution level enabled. In a dark sky, the Milky Way is easily visible, while strong light pollution washes most sources away, apart from some of the brightest stars. *Software credit: Stellarium (Zotti et al. 2020)*

perfect sky, the Scottish highlands (although once there, you do have some terrain to contend with, which might block the view of the horizon). If we go back in history to before the industrial revolution, when the skies were (relatively speaking) perfectly dark, the number of stars one could see increases from around a thousand to nearly five thousand. Figure 1.1 offers a simulated view of the Flemish southwestern sky in summer with, on the left, light pollution removed. You can see that the Milky Way and some of its structures, along with many stars, were clearly visible in those pre-industrialized times. The right hand side of Fig. 1.1 shows that, today, only the very brightest stars are visible.

Given the sky was much darker in the past, it should not be surprising then that humans studied the heavens since ancient times. There exist Mesopotamian, Chinese, Indian, as well as Aztec records of astronomical observations dated from well before the Common Era (Needham 1959; Ashfaque 1977; Edmonson 1988; Hunger and Pingree 1999). The Greek astronomer Hipparchos would improve on Babylonian measurements and accurately predict the motion of the planets in the second century BCE. Western astronomy further boomed during the Greek and Roman period in which the Ptolemaic system was established in the second century CE, with Earth at its center and the heavenly spheres above. Although geocentrism would prevail for at least 1700 years, Aristarchus of Samos already proposed the idea that the Earth and all other planets actually orbited the Sun, and he supported Anaxogoras' conception of the Sun being star that is close to Earth, as opposed to fixed on the celestial dome (Curd 2019). Aristarchus' theory was not accepted, unfortunately, because no parallax of the stars was observed. If the Earth was moving within the celestial sphere, which is located just beyond that of Saturn, the Greeks realized the apparent distances between stars should change as we viewed them under slightly different angles. Little did they know that the stars are *so* far away that it would take the invention of telescopes and other precise equipment to measure the tiny effect.[1]

The middle ages were a dark time for the sciences in Europe, but Islamic civilizations in the Middle East continued to develop new tools such as the astrolabe with which they conducted surveys of stars, noting their positions, color and brightness (Saliba 1994). It would take until the sixteenth century in the Renaissance period when Polish polymath Nicolaus Copernicus revolutionized astronomy by again proposing the heliocentric model of the heavens, which placed the Earth in an orbit around the Sun.[2] Not much later it was Johannes Kepler that discovered the planets moved in ellipses around the Sun, rather than perfect circles, after which Isaac Newton made the connection with gravitational forces in his treatise called the Principia Mathematica (Newton 1687).

In modern times, astronomy and astrophysics have become near synonyms, where, rather than just describing celestial bodies (e.g., their motion), we try to determine the physical processes happening in and around such bodies. There are many subfields in astrophysics, such as high-energy and extragalactic astrophysics, solar physics, asteroid dynamics, planetary and atmospheric physics, astrochemistry, and also astrobiology. The subject of this thesis, stellar astrophysics, has the goal to understand the birth, structure, evolution, and death of stars, as well as the impact they have on their neighborhood. This is a daunting task, but luckily many people from all across the world contributed to this field in the past, allowing us, today, to build on previous results.

In particular, in this thesis, we investigate the structure and evolution of stars that orbit very close to one another in binary systems. Sometimes they are so close to each other that they come into contact. In the following sections, we give an overview of how we expect a single star to evolve (Sect. 1.1). We then move to how binary-star systems were discovered, and introduce how they can change stellar evolution (Sects. 1.2 and 1.3). In Sect. 1.4 we give special attention to the contact binaries, while Sect. 1.5 introduces the youngest branch in astrophysics: gravitational-wave astronomy.

[1] Henderson (1839) is usually credited with the first good parallax determination, of both stars in α Centauri.

[2] It is unclear to what extent Copernicus was influenced or aware of Aristarchus' work. According to Gingerich (1985), all evidence points to Copernicus independently arguing for a heliocentric solar system.

1.1 Single Stars

The historical introduction in this section is based mainly on Longair (2013).

The study of the structure and evolution of single stars started in the second half of the nineteenth century. At that time, the theoretical study of stars was sparked by large forward leaps in observational aspects of stellar astrophysics. The advent of photography allowed astronomers to take images of stars as opposed to relying on the human eye to determine the brightness or color of an object. This way, many stars and star clusters were photographed and classified at the Harvard College Observatory. Of historical importance is that, back then at Harvard, nearly all of the analysis of photographic plates and the calculations of positions, color and brightness of stars were done by women. The director of the so-called 'Harvard Computers,' Edward Charles Pickering, reportedly hired over eighty women to do astronomical analysis (Von Mertens 2024). While paying them less than his male contemporaries, he did pave the way for women to have a prominent role in astronomy and astrophysics (Rossiter 1980). Many of these women, like Williamina Fleming, Mary Anna Draper, Antonia Maury, and Hienrietta Swan Leavitt, made significant scientific contributions (e.g., Maury and Pickering 1897; Leavitt and Pickering 1912; Fleming 1917).

The most basic information that can be collected from celestial objects is the amount of light that they send to the Earth, and thus the amount of light that falls on photographic plates. The brightness of stars is commonly described using their apparent magnitude, which is a logarithmic scale comparing their apparent brightness to that of the star Vega. Somewhat confusingly, but to agree with Ptolemy's original designations of 'first magnitude' for bright stars to 'sixth magnitude' for barely discernible stars, the modern magnitude scale is inverted: fainter stars have higher magnitudes, while brighter stars have lower magnitudes, going below zero for stars brighter than Vega. The absolute (or intrinsic) magnitude of an object can be determined from the apparent magnitude of the star and the distance to that star. Distances were found from the size of the wobble the star exhibits with respect to far away galaxies as the Earth moves around the Sun, called the parallax. Color indices can be calculated by comparing the brightness of objects when photographed through different filters.

Apart from the brightness and color, further information of a celestial object can be obtained from its spectrum. Spectroscopy of stars was pioneered by Joseph Fraunhofer who, in 1817, used high-quality prisms to split sunlight into its components and thus created what he called a "color-image" (Fraunhofer 1817). On the image he saw dark bands, which we now call absorption lines. They have their origin in the quantized energy levels of atoms, and would later be explained by the quantum revolution of the 1920s. Only in the late nineteenth century were spectra of stars numerous enough to build a classification system. They were ordered in a sequence according to what lines appeared, and in what strength. The letter sequence OBAFGKM for the main spectral types, with numerical subtypes 0–9, was created

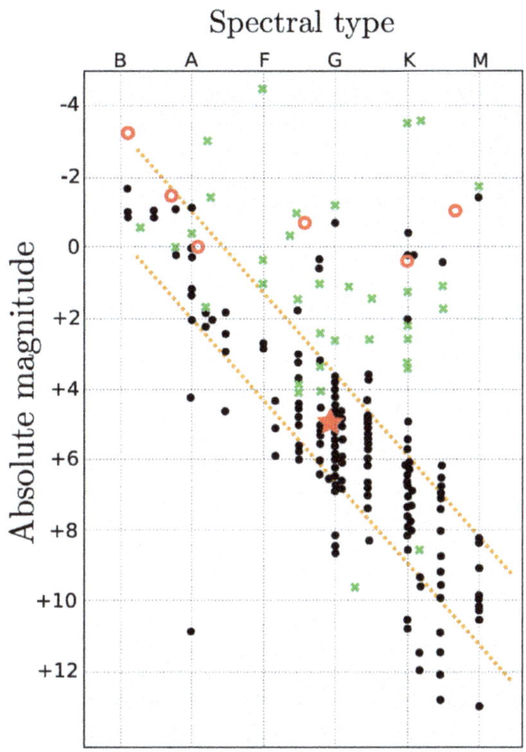

Fig. 1.2 Reproduction of the magnitude vs. spectral type plot of Russell (1914). The red star indicates the position of the Sun. Black dots have good parallax measurements, and thus a good absolute magnitude estimate, while green crosses have poor parallax and magnitude estimations. Open circles are averages from numerous stars of high brightness (around 120 in total) of which the parallax is very poorly constrained. Between the orange lines lies the main sequence. *Figure adapted from Marchant (2018)*

by Annie Jump Cannon, also at the Harvard Observatory (Cannon and Pickering 1901). It is still in use today, with some additions. The subtypes allow for more gradual classification, so that, for example, an B9 star is nearly a A0 star. Spectral types O and B (called early types) show lines of helium, while A and F-type spectra are strong in lines of hydrogen. In late types, G, K, and M, mostly metal[3] lines are seen, primarily ionized calcium in G types, neutral metals in K types, and molecular titanium-oxide bands in M types. With the knowledge that such lines are related to element transitions, it was quickly realized that this sequence closely matches a temperature scale. The hottest stars have spectral types O and B with surface temperatures reaching over 50,000 K, while the coolest stars are of types K and M, where the temperatures go down to around 3000 K. The Sun is of spectral type G2, and has a surface temperature of around 5800 K.

Ejnar Hertzsprung and Henry Norris Russell independently took the measurements of the Harvard team and plotted the absolute magnitude versus the spectral type of each star (Hertzsprung 1911; Russell 1914). Figure 1.2 shows Russell's diagram of the stars most nearby to the Sun. Of note is that most stars fall in a

[3] In astrophysics, a "metal" refers to any element heavier than helium. This undoubtedly confuses chemists and material scientists as, for example, carbon is chemically a non-metal.

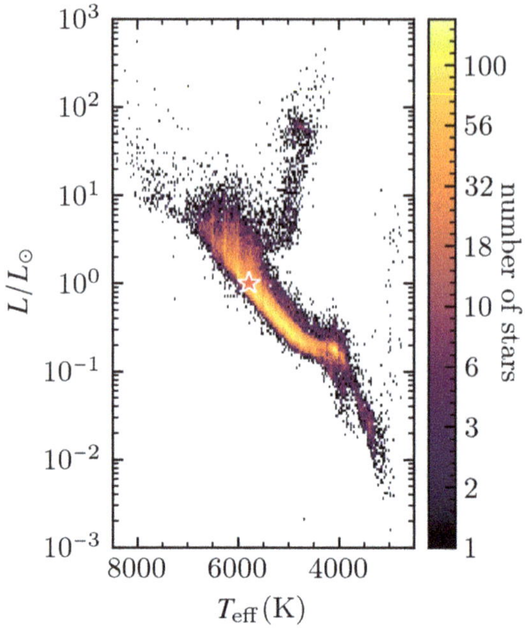

Fig. 1.3 HRD of the 50,000 stars closest to the Sun of magnitude 12 or brighter, as observed by the Gaia mission. As in Russell's first diagram, a clear main sequence is present, as well as a giant branch. The location of the Sun is marked with the red star

sequence from the top left to the bottom right, meaning hot stars of spectral types OBA are also bright, while cooler stars of types GKM are dim. Hertzsprung called this the main sequence, a name that stuck until today, and is marked with the orange lines. There are also a fair number of stars in the upper right corner of the diagram, which is called the giant branch and contains rather cool, but very bright stars. The location of stars on this diagram relates to their evolutionary state, although that would take decades of theoretical work to figure out.

Today, we still make similar diagrams to those of Hertzsprung and Russell, but we changed the axes to represent more physical properties of the stars. Instead of spectral type, we present directly its effective temperature. This is the temperature of the photosphere, which is the layer near the surface of the star were most of the light escapes the stellar material. Since the hot OB types were plotted left, this results in the axis being reversed from the intuitive direction, i.e., increasing to the left. The left side of the diagram is also called the "blue" side, while the right is the "red" side, because of the typical colors stars have at those locations. On the vertical axis we put luminosity, a measure of how much energy is radiated from the stellar surface per second. Figure 1.3 shows the Hertzsprung-Russell diagram, or HRD for short, of the 50,000 nearest stars to the Sun of (apparent) magnitude brighter than 12, using data from the Gaia spacecraft (Gaia Collaboration et al. 2016). Just as the original diagrams, we observe two clear features: a main sequence from the top-left to the bottom-right, were the vast majority of stars reside, and a giant branch going up to relatively cool but very luminous stars.

By the late nineteenth century, it was realized that stars are large balls of gas that radiate energy in the form of light. However, it was yet unknown where this energy came from. Lord Kelvin along with Hermann von Helmholtz thought it was the slow contraction of the Sun that liberated gravitational potential energy required for it to shine. With modern measurements of the solar luminosity, $L_{\odot} = 3.828 \times 10^{26}$ W, we can estimate the time in which the Sun, with mass $M_{\odot} = 1.988 \times 10^{30}$ kg, would have radiated away gravitational potential energy by contracting to its current radius, $R_{\odot} = 6.957 \times 10^{8}$ m:

$$\tau_{\mathrm{KH}} = \frac{GM_{\odot}^{2}}{R_{\odot}} \cdot \frac{1}{L_{\odot}} \approx 3.1 \times 10^{7}\ \mathrm{yr}. \quad (1.1)$$

The so-called Kelvin-Helmholtz or thermal timescale of the Sun is thus around thirty million years.[4] However, from meteoric rocks, we have evidence that the solar system is older than four billion years, much more than the thermal timescale of the Sun. It seems then that the available amount of gravitational potential energy alone cannot explain its energy output. Either there must be another energy reservoir not yet discovered, or the Sun must have been many orders of magnitude dimmer in the past than it is today.

At the start of the twentieth century, atomic theory took a large leap forward by the initial quantum mechanical insights from Max Planck (Planck 1901). Niels Bohr then proposed the shell model of the atom, with the electrons moving around the nucleus in discrete energy levels (Bohr 1913). The connection between spectral lines and atomic energy levels of specific elements was then made, because it could be shown that the wavelengths of the spectral lines of hydrogen exactly matched those predicted by Bohr's model. The model was later generalized to other elements by Erwin Schrödinger, who wrote down the wave equations of quantum mechanics during the quantum revolution of the 1920s (Schrödinger 1926). While the spectra of many elements were known empirically by that time, with quantum mechanics the wavelengths of the spectral lines could be calculated.

During that same decade, Cecilia Payne-Gaposchkin conducted groundbreaking work when, in 1925, she established that the Sun consisted mostly of hydrogen and helium (Payne 1925). The crucial insight she provided was that the variation of stellar spectra, going from type O to type M, was not due to different abundances of elements, but rather due to different ionization levels of the atomic species, for which she used the theory of Indian physicist Meghnad Saha. Her thesis was called "the most brilliant PhD thesis ever written is astronomy," by Otto Struve in 1960 (Soter and deGrasse Tyson 2001). Since stars in different environments exhibit mostly the same lines (even across different spectral types), it was concluded that the abundance of hydrogen and helium must be high across the whole galaxy (and

[4] Another definition of this timescale has a factor 1/2 in front, giving $\tau_{\mathrm{KH}} \approx 1.5 \times 10^{7}$ yr. The order of magnitude is what is important here, however, not the specific value.

indeed the universe). This fact was the final piece of information needed to answer the energy question of stars.

One year after Payne-Gaposchkin's thesis, it was realized by Sir Arthur Stanley Eddington that the Sun, being mostly hydrogen, possesses an enormous reservoir of energy in nuclear form (Eddington 1926). If the nuclei of four hydrogen atoms are fused together to form a helium nucleus, and helium is lighter than the original four hydrogen particles, Albert Einstein's famous equation $E = mc^2$, with $c \approx 3 \times 10^8\,\mathrm{m\,s^{-1}}$ the speed of light, says that even a tiny amount of mass can be converted into a large amount of energy. Francis Aston had experimentally measured the masses of hydrogen and helium, and found a hydrogen atom weighed 1.008 atomic units versus helium's 4 (Aston 1925). The atomic mass unit, m_u, is defined as one twelfth of the mass of carbon-12, which comes to $m_\mathrm{u} \approx 1.66 \times 10^{-27}\,\mathrm{kg}$. The net reaction

$$4\,{}^{1}\mathrm{H} \longrightarrow {}^{4}\mathrm{He} \tag{1.2}$$

then releases an amount of energy of about

$$\Delta E = \Delta m c^2 = \left(4 m_{^1\mathrm{H}} - m_{^4\mathrm{He}}\right) c^2 = (4.032 - 4)\, m_\mathrm{u} c^2 \approx 4.8 \times 10^{-12}\,\mathrm{J}. \tag{1.3}$$

This seems like a small number, but if we assume for a moment that the Sun is entirely made out of hydrogen, it could sustain the luminosity it has today for

$$\tau_\mathrm{nuc} = \frac{\Delta E}{L_\odot} \cdot \frac{N_{^1\mathrm{H}}}{4} = \frac{\Delta m c^2 M_\odot}{4 L_\odot m_{^1\mathrm{H}}} \approx 1.2 \times 10^{11}\,\mathrm{yr}. \tag{1.4}$$

This number is around ten times bigger than the estimated age of the universe! If the Sun has an age of several billion years, only a few percent of the mass of the Sun needs to be actually available as fusible hydrogen. This is indeed the case as only the hydrogen in the very center of the Sun is hot enough to be fused. The actual nuclear timescale of the Sun is then computed with a factor of 0.1 multiplying the estimate in Eq. (1.4), so that $\tau_{\mathrm{nuc},\odot} \approx 1.2 \times 10^{10}\,\mathrm{yr}$. The Sun thus has plenty of nuclear fuel in its tank to explain its brightness and age. This also means that almost all energy on Earth (which we use to generate electricity, or drive our cars), including fossil fuels, hydro, and wind power, can be traced back to nuclear fusion in the Sun.[5] With this

[5] Exceptions to this are tidal energy, which originates from the gravitational potential energy of the Earth, Moon, and Sun; nuclear power, which comes mostly from uranium made in neutron star collisions of a previous generation of stars; and geothermal energy, coming from the heat of Earth's interior. The latter, in turn, is split in two sources: gravitational from its formation and settling of the core, and again nuclear, this time from the radioactive decay of naturally occurring radioisotopes within the Earth. The fraction that the nuclear component contributes to geothermal energy is still debated, and ranges from 50% (Gando et al. 2011) to 80% (Turcotte and Schubert 2014), depending on the model.

final realization, it was Eddington who completed the set of equations that govern the life of a star, the general form of which has remained unchanged since.

Further advances were made thanks to new quantum-mechanical insights. More accurate theories on the opacity of stellar material (i.e., the degree with which photons interact with atoms) were developed (Kramers 1923; Rosseland 1924). Additionally, the precise rates at which nuclear reactions take place and the amount of energy released as a result of them, depending on the temperature and density, could be computed from new nuclear theories (Atkinson and Houtermans 1929; Atkinson 1936; Bethe and Critchfield 1938; Bethe 1939). As a consequence of the exclusion principle put forward by Wolfgang Pauli (Pauli 1925), which states that particles with half-integer spin (such as electrons) cannot occupy the same quantum state, electron degeneracy pressure was discovered, and applied to astrophysics by Fowler (1926) in the form of an equation of state independent of temperature. Electron degeneracy would be instrumental in understanding the late stages of stellar evolution. Most notably, Subrahmanyan Chandrasekhar computed, using the relativistic version of electron degeneracy, that there is a maximal mass (roughly $1.4\,M_{\odot}$) to an object having this equation of state (Chandrasekhar 1931a,b). Despite a rather public dispute between the well established Eddington, who rejected the theory of the up-and-coming student Chandrasekhar (Mena and Peres 2018), this idea proved to be correct, and the limiting mass of a white dwarf (see Sect. 1.1.4) is now known as the Chandrasekhar limit.

Until the invention of the multipurpose digital computer, all calculations involving stellar structure and evolution had to be done by hand, which necessitated that the equations were solvable analytically. This often required making multiple simplifications and approximations of the physical conditions, for example by stating that the gas of the whole star could be described by a single polytrope. A polytrope is the solution of the Lane-Emden equation (Lane 1870; Ritter 1898; Emden 1907), which was the then-prevailing equation describing stellar structure. It assumed a single mode of energy transport inside the star, and could not account for energy generation in the core. Furthermore, Eddington had to initially consider fully homogeneous stars (with a constant gas-to-photon pressure, a source of debate) to keep the equations manageable, meaning he could only compute models for fully-mixed, uniform stars.

The homogeneous stellar models of Eddington were not able to simultaneously explain the two populations in the HRD. You have the giant stars who were orders of magnitude more luminous than the main-sequence stars (also called dwarf stars) at roughly the same surface temperature. In 1938, Ernst Öpik first considered stellar models that were not homogeneous by entertaining the idea that the interior of stars is generally *not* well mixed (Öpik 1938). This meant that the core of a star could run out of hydrogen to burn and become inert. He predicted the core would rapidly contract and release gravitational energy instead, and as a result the envelope, that is still hydrogen rich, would expand. This process, i.e., the depletion of hydrogen in the core and the accompanying structural changes of the star, would naturally produce the giant stars in the HRD.

Once computers entered the scene in the 1950s, stellar models could be computed using numerical techniques. In those days, machines with the computational power of today's pocket calculators had the size of whole classrooms, but the physical processes occurring in stars could be programmed (with magnetic tapes or punch cards!),[6] and detailed stellar-evolution models were produced as early as 1952. As computers became more and more part of the standard equipment of the astrophysicist, innovative techniques in the implementation of the differential-equation solver were developed, most notably with the contribution by Louis Henyey (Henyey et al. 1959, 1964). The so-called Henyey method is still used today for calculating stellar-evolution models, and it uses a trial solution and a generalized Newton-Raphson root-finding algorithm to iteratively approach a better solution to the differential equations.

As atomic and nuclear data kept improving—driven in part by research on nuclear weapons during World War II and the Cold War—and as computers kept getting faster exponentially, stellar-model calculation saw substantial progress decade after decade. Rather than computing single models of one particular mass and composition, whole sets, or populations, of stellar models could be explored. This enabled to study the evolution of clusters of stars, where the assumption can be made that all stars are born more or less at the same time and more or less with the same composition. Alternatively, having a lot of computing power allows to study the impact of varying certain parameters that describe physical processes, such as the efficiency of internal mixing, the rate at which certain nuclear reactions occur, or the strength of mass loss of the star during its life (see Sect. 1.1.5).

Today, although supercomputers are used to calculate grids of tens or even hundreds of thousands of stellar models, astrophysicists do not need supercomputers hosted at academic or governmental institutions to produce detailed models. Most stellar-evolution codes are designed to run on basically any computer, and it will take only on the order of hours to complete the integration of, for example, the life of the Sun, taking around ten thousand integration steps. This is in stark contrast with the situation of yesteryear, as Martin Schwarzschild noted in 1958:

> A person can perform more than twenty integration steps per day ... so that for a typical single integration of, say, forty steps, less that two days is needed.
> — Schwarzschild (1958).

If one would attempt to compute the stellar models contained in this thesis without the help of modern computers, instead relying on manual integration like in the times of Schwarzschild, this thesis would require an estimated 10,000 years to complete. It is fairly certain that no funding agency would want to back a grant for that long. This goes to show how powerful computers have become in just seventy years or so. Progress in computer science does not stand still however, and, with machine-learning technologies finding applications in the natural

[6] Anecdotally, Rudolf Kippenhahn recalls that, from the patterns in the flashing lights and the noises the computer made, he and his collaborators could infer which specific part of the star it was calculating at that moment (Gingerich 1978).

sciences, including astrophysics, maybe the field of stellar structure and evolution, in particular the computation of stellar models, will again be revolutionized soon.

1.1.1 The Equations

The study of stellar structure and evolution revolves around the solution of a set of partial differential equations, each one encoding a fundamental physical principle. They are:

(i) Conservation of mass,
(ii) Conservation of momentum,
(iii) Conservation of energy,
(iv) Transport of energy, and
(v) Changes in composition.

Under general conditions, the set of equations associated to these principles is not analytically solvable since we are dealing with coupled, non-linear, partial differential equations in three spatial dimensions plus a time dimension. However, by observing that stars are, generally speaking, round balls, we can make use of spherical symmetry to cast the equations in the following form:

$$\frac{\partial M}{\partial r} = 4\pi r^2 \rho, \tag{1.5a}$$

$$\frac{\partial P}{\partial r} = -\rho \frac{\partial \Phi}{\partial r} = -\rho \frac{GM}{r^2}, \tag{1.5b}$$

$$\frac{\partial L}{\partial r} = 4\pi r^2 \rho \varepsilon, \tag{1.5c}$$

$$\frac{\partial T}{\partial r} = -\rho \frac{GM}{r^2} \frac{T}{P} \nabla, \tag{1.5d}$$

$$\frac{\partial X_i}{\partial t} = \sum_j r_{ji} - \sum_k r_{ik} - \frac{1}{\rho r^2} \frac{\partial}{\partial r} \left(\rho r^2 D \frac{\partial X_i}{\partial r} \right). \tag{1.5e}$$

We now have only one spatial dimension, r, and the time, t, left as independent variables. In these equations, several functions are defined. We have the set consisting of $M = M(r, t)$, which is the mass at time t enclosed in a sphere of radius r, and $P = P(r, t)$, $T = T(r, t)$ and $L = L(r, t)$, which are, respectively, the pressure, temperature and luminosity of the stellar material at the radius r and time t. These four functions specify the *structure* of the star at each point in time. We have also $\rho = \rho(P, T, X_i)$, the density of the material, which is computed via an equation of state, and the potential $\Phi = \Phi(M, r)$, which is the gravitational potential associated with the mass configuration of the star, and can be solved from Poisson's equation, $\nabla^2 \Phi = 4\pi G \rho$. For spherically symmetric matter distributions, we have $\Phi(r) = -GM(r)/r$, as used above. The function $\varepsilon = \varepsilon(\rho, T, P, X_i)$ encodes energy generation by nuclear reactions as well as thermodynamic heating or cooling by compression or expansion, and ∇ is the logarithmic derivative of temperature to pressure required to transport the luminosity $L(r)$ through the star. The last equation with $\partial X_i/\partial t$ is actually a set of multiple equations that describes, for each chemical isotope i, the transmutation of i into different isotopes k at a rate of r_{ik}, while other isotopes j are changed into i at the rate r_{ji}. The third term in the equation represents internal mixing of the stellar material, occurring with diffusion coefficient D. A full solution of all equations, giving the functions $M(r, t)$, $P(r, t)$, $T(r, t)$, $L(r, t)$ and the elemental species $X_i(r, t)$, with appropriate initial and boundary conditions is then called a *stellar-evolution model.*

1.1.1.1 Equations of Continuity and Hydrostatic Equilibrium

The first two equations are results of classical physics and were well understood already at the start of the twentieth century. The first equation, Eq. (1.5a), called the continuity equation, simply states that the mass, dM, contained in a thin shell of radius r is the product of the density, ρ, of the shell, its area, $4\pi r^2$, and its thickness, dr. Equation (1.5b), asserts that forces arising from the gravitational potential, Φ, are balanced by the internal pressure gradient coming from the gas making up the star and the radiation it produces. It describes conservation of momentum, which, without acceleration terms reduces to the situation called *hydrostatic equilibrium.* In most of the life of a star, hydrostatic equilibrium is satisfied, or very nearly so. However, for example, when a massive star collapses at the end of its life, we speak of hydrodynamic evolution, and an acceleration term $-\rho\, \partial^2 r/\partial t^2$ must be added to the right hand side of Eq. (1.5b). Additionally, Eq. (1.5b) ignores the fact that, at the surface of the star, radiative forces can launch a stellar wind, but this effect is usually taken into account by incorporating mass loss in the models (see Sect. 1.1.5).

The first two equations alone can already tell us something about the pressure and density profile of a star, if we make some assumptions. Taking the derivative to r of the equation of hydrostatic equilibrium, and substituting $\partial M/\partial r$ with the continuity equation, leaves you with:

$$\frac{\mathrm{d}}{\mathrm{d}r}\left(\frac{r^2}{\rho}\frac{\mathrm{d}P}{\mathrm{d}r}\right) = -4\pi G r^2 \rho. \tag{1.6}$$

Assuming then that the functions for the density and pressure are of the form $\rho = \lambda\theta^n$, and $P = K\rho^{1+1/n}$, with λ, K, and n constants, we arrive at:

$$\left(\frac{(n+1)K}{4\pi G}\lambda^{1/n-1}\right)\frac{\mathrm{d}}{\mathrm{d}r}\left(r^2\frac{\mathrm{d}\theta}{\mathrm{d}r}\right) = -\theta^n. \tag{1.7}$$

We can make the equation fully dimensionless using

$$r = \alpha\xi, \tag{1.8a}$$

$$\alpha = \sqrt{\frac{(n+1)K}{4\pi G}\lambda^{1/n-1}}, \tag{1.8b}$$

to finally arrive at the Lane-Emden equation for polytropes:

$$\frac{1}{\xi^2}\frac{\mathrm{d}}{\mathrm{d}\xi}\left(\xi^2\frac{\mathrm{d}\theta}{\mathrm{d}\xi}\right) = -\theta^n. \tag{1.9}$$

For complete polytropes, i.e., when the entire star can be described by a single polytrope with index n, the constant λ is equal to the central density, $\lambda = \rho_c$. As mentioned in the historical introduction above, the Lane-Emden equation was studied extensively in the past (see, e.g., Chandrasekhar 1939), and analytic solutions for $n = 0$, 1, 5, and ∞ exist. For example, if $n = 1$, the solution for θ is:

$$\theta_1(\xi) = \frac{\sin\xi}{\xi}, \tag{1.10}$$

and we can calculate the density and pressure profile by substituting back Eq. (1.8). For other values of n, numerical techniques or power series expansions are needed to write down solutions. Eddington's famous "Standard Model" of the Sun used an $n = 3$ polytrope, which is also a good approximation for a relativistic, fully degenerate electron gas. The case $n = 1.5$ on the other hand well describes large, fully convective stars, like giants or pre-main-sequence stars. In modern times, however, as stellar models use more sophisticated equations of state, which can vary throughout the interior of stars, knowing the specific solutions of the Lane-Emden equation is of less importance as it was in the times before computer calculations were possible.

1.1.1.2 Equations of Energy Conservation and Transmutation

The third equation, Eq. (1.5c), is the equation of conservation of energy and is intimately connected with Eq. (1.5e) specifying the transmutation of elements into others. As discussed in the introductory paragraphs, it was found in the early twentieth century that the main source of stellar energy is nuclear. On the main sequence, stars fuse hydrogen into helium. In low-mass stars ($M \lesssim 2\,M_\odot$), this process is achieved mostly by direct assembly, also called the proton-proton or *pp* chain. Two hydrogen nuclei, ^{1}H, which consist of just a single proton, combine while one of them changes into a neutron, forming deuterium, and emit a positron, e^+, and a neutrino, ν_e:

$$^1\mathrm{H} + {}^1\mathrm{H} \longrightarrow {}^2\mathrm{H} + \mathrm{e}^+ + \nu_\mathrm{e}. \tag{1.11}$$

Adding then another proton forms ^{3}He and emits a photon, γ:

$$^2\mathrm{H} + {}^1\mathrm{H} \longrightarrow {}^3\mathrm{He} + \gamma. \tag{1.12}$$

Final assembly to ^{4}He then occurs by either combining two ^{3}He nuclei (*pp*-I) or, with ^{4}He already present, going via beryllium and lithium (*pp*-II) or boron (*pp*-III), as presented in Table 1.1. Each of these reactions releases an amount of energy, ε, to be used to hold up the star against its own gravity.

The second way to fuse hydrogen is through the CNO cycle, which is the dominant process in intermediate-mass ($2\,M_\odot \lesssim M \lesssim 8\,M_\odot$) and massive stars ($M \gtrsim 8\,M_\odot$). Instead of assembling a helium nucleus from scratch by starting from single protons, in this process we start with already present carbon and nitrogen nuclei, and add protons one at a time until a helium nucleus splits off. This happens in two distinct cycles, going through both stable and unstable isotopes of carbon, nitrogen, oxygen and fluorine as shown in Fig. 1.4. An important property of these cycles is that the proton capture of nitrogen,

$$^{14}\mathrm{N} + {}^1\mathrm{H} \longrightarrow {}^{15}\mathrm{O} + \gamma, \tag{1.13}$$

has the lowest nuclear cross section. This means it is the slowest reaction of the loop, meaning that the CNO cycle, while burning hydrogen to helium, converts a fraction of the carbon and oxygen into nitrogen. Observing an enhanced nitrogen abundance on the surface of a star is therefore a good indication that nuclear processed material of the core has been brought to the surface by internal mixing processes (see Sect. 1.1.5).

When a star leaves the main sequence as its hydrogen runs out in the core, the next nuclear fusion reaction available is the combination of three helium nuclei to a carbon nucleus, which is called the triple alpha process:

$$3\,{}^4\mathrm{He} \longrightarrow {}^{12}\mathrm{C} + \gamma. \tag{1.14}$$

This reaction is crucially dependent on the stability of the intermediary product beryllium:

$$^4\mathrm{He} + {}^4\mathrm{He} \rightleftarrows {}^8\mathrm{Be} \tag{1.15}$$

Table 1.1 Final assembly reactions in the *pp* chain of hydrogen fusion into helium

pp-I	*pp*-II	*pp*-III
$^3\mathrm{He} + {}^3\mathrm{He} \longrightarrow {}^4\mathrm{He} + 2\,{}^1\mathrm{H}$	$^3\mathrm{He} + {}^4\mathrm{He} \longrightarrow {}^7\mathrm{Be} + \gamma$	
	$^7\mathrm{Be} + \mathrm{e}^- \longrightarrow {}^7\mathrm{Li} + \nu_\mathrm{e}$	$^7\mathrm{Be} + {}^1\mathrm{H} \longrightarrow {}^8\mathrm{B} + \gamma$
	$^7\mathrm{Li} + {}^1\mathrm{H} \longrightarrow 2\,{}^4\mathrm{He}$	$^8\mathrm{B} \longrightarrow {}^8\mathrm{Be} + \mathrm{e}^+ + \nu_\mathrm{e}$
		$^8\mathrm{Be} \longrightarrow 2\,{}^4\mathrm{He}$

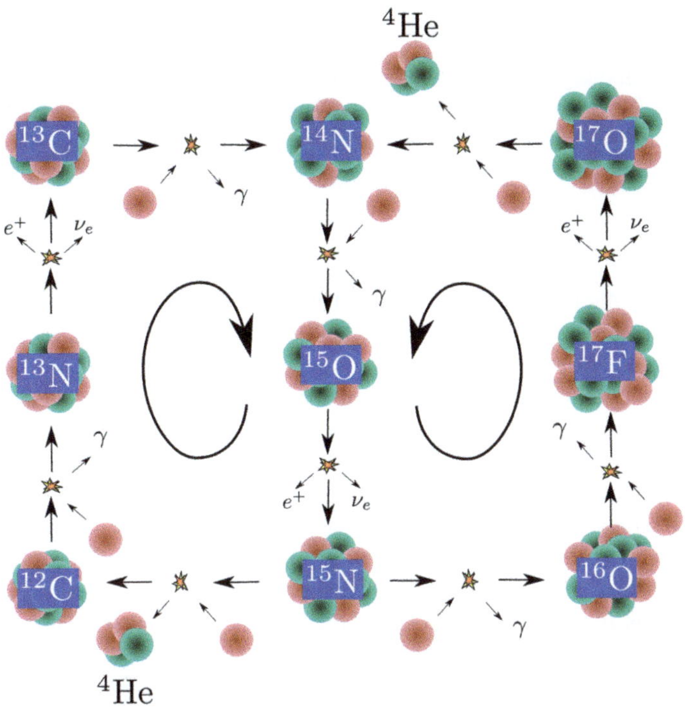

Fig. 1.4 CNO cycle of hydrogen burning into helium. Protons (red circles) are sequentially added until a helium nucleus splits off. In massive main-sequence stars this is the dominant process from which nuclear energy is liberated. *Adapted from Marchant (2018), reproduced with permission*

Beryllium-8 is unstable, and decays back to two ^{4}He nuclei with a half-life of only about 8.2×10^{-17} s. Luckily, the density and temperature in a stellar core where these reactions take place is high enough for the third helium nucleus to encounter the ^{8}Be particle before it decays again, and fuse to ^{12}C. The final reaction,

$$^8\text{Be} + {}^4\text{He} \longrightarrow {}^{12}\text{C}^* \longrightarrow {}^{12}\text{C} + \gamma, \qquad (1.16)$$

however passes through an excited resonance state of carbon, ^{12}C*. Predicted by Fred Hoyle in the early fifties (Hoyle 1954), it is now called the Hoyle state. The state was experimentally found not much later at Caltech (Dunbar et al. 1953; Cook et al. 1957), but it took until 2011 to theoretically calculate its existence (Epelbaum et al. 2011). The fact that stellar nuclear burning has to go through ^{12}C* and that this state exists almost exactly at the sum of energies of ^{8}Be and ^{4}He is of considerable interest to philosophers in context of the anthropic principle and the fine-tuning of the universe (Kragh 2010). If this state did not exist, or had its properties altered by slightly different fundamental physical constants, the production of elements heavier than helium could have been radically different, or stalled altogether, and life as we know it might not have emerged.

In massive stars heavier than about 8 $M_\odot$, thermonuclear burning proceeds from ^{12}C to heavier elements like ^{16}O, ^{20}Ne, ^{28}Si, and finally ^{56}Fe. All these steps use multiple intermediary nuclei and interaction channels, too many to list in this thesis. Instead we refer to works on stellar nucleosynthesis, such as Clayton (1983). We do however note the reaction

$$^{12}\mathrm{C} + {}^{4}\mathrm{He} \longrightarrow {}^{16}\mathrm{O} + \gamma, \tag{1.17}$$

which has a particularly hard to determine reaction rate (deBoer et al. 2017), i.e., the speed with which this reaction occurs depending on the density and temperature. Because this rate establishes the precise carbon to oxygen ratio in the core, it has important implications for the final fate of the massive star (e.g., the mass of the black hole it will leave behind, Farag et al. 2022).

Figure 1.5 shows the binding energy per nucleon of the most stable isotope of each mass number A. It shows that the first step in the fusion chain, hydrogen burning to helium, is the most exothermic reaction since the jump in binding energy is by far the largest. This translates to the fact that a star will spend the vast majority of its life burning hydrogen in its core (around 90%). Each progressive burning stage releases less energy, which means these stages take less time to complete. As soon as ^{56}Fe is formed in the core of a massive star, nuclear fusion stops as it is

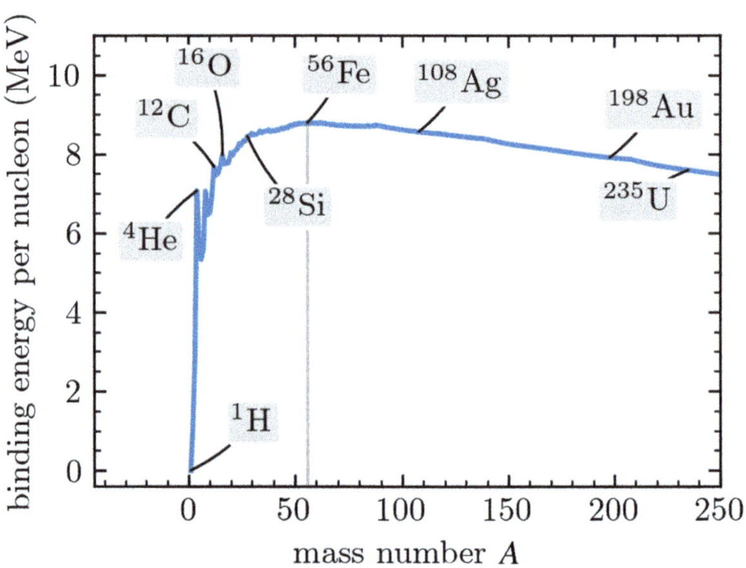

Fig. 1.5 Binding energy per nucleon of the most stable isotope per mass number A. Indicated are several important isotopes in stellar nucleosynthesis, as well as silver-108, gold-198 and uranium-235, the latter being the prime fissile material in modern nuclear power plants. The vertical line indicates iron-56, which is the most efficiently bound nucleus. *Nuclear data is from Huang et al. (2021) and Wang et al. (2021)*

no longer energetically favorable for the star. The whole machinery of producing energy to maintain a pressure gradient which counteracts gravity falls apart, much like a building that tumbles down when its foundation gives way. This announces the dramatic end of the life of the massive star with a core-collapse supernova (see Sect. 1.1.4).

In many of the nuclear reactions of the *pp* chain and the CNO cycle referred to above, neutrinos are produced as a result of interaction with the weak nuclear force. Furthermore, in burning stages of carbon and later, the conditions in the star are such that photons are energetic enough to create an electron-positron pair, which occasionally turn into a neutrino-antineutrino pair. However, once produced, neutrinos are notoriously non-reactive with protons and electrons, and therefore do not deposit their energy into the gas of the star. Stars thus lose energy to neutrino production, which can be substantial in the later phases of massive-star evolution. Additionally, gas could be heating, cooling, contracting or expanding, all of which affects the energy balance in the stellar layers. Rather than hiding all possible energy terms in the general function ε, the energy equation is more explicitly written as:

$$\frac{\partial L}{\partial r} = 4\pi r^2 \rho \left(\varepsilon_{\text{nuc}} - \varepsilon_{\text{neu}} - c_P \frac{\partial T}{\partial t} + \frac{\delta}{\rho} \frac{\partial P}{\partial t} \right), \tag{1.18}$$

where the four terms on the right hand side are, in order, the nuclear energy released from fusion, energy losses by neutrinos, thermodynamic heating or cooling with specific heat capacity c_P, and volumetric expansion or compression with coefficient $\delta = -\partial \ln \rho / \partial \ln T$.

1.1.1.3 Equation of Energy Transport

The transport of energy within the star, specified by the fourth equation, Eq. (1.5d), was first thought to be mediated solely by convective blobs of gas. Convection happens all around us, take for example boiling water on a stove. The intense heat near the bottom of the pot creates bubbles that rise to the surface. The same effect also powers ocean currents, weather patterns in Earth's atmosphere, and causes tectonic activity of Earth's crust as the mantle below convects heat from the core to the surface. However, it was put forward by Karl Schwarzschild that radiation can also play a significant role in the transportation of energy from the core to the outer layers of a star (Schwarzschild 1906). This idea was put on solid theoretical grounds some time later by Eddington (1916), where he also discussed the sources of opacity inside stellar material.

The type of energy transport, radiative or convective, is an important factor that determines the structure of a star. Considering first radiative transport, which is a diffusive process, Fick's law applies (Fick 1855), which, when applied to radiation, gives the radiative flux:

$$\mathcal{F}_{\mathbf{rad}} = -\frac{4ac}{3}\frac{T^3}{\kappa\rho}\boldsymbol{\nabla T}. \tag{1.19}$$

Here the opacity, κ, of the material is used, and $a = 8\pi^5 k_B^4/(15c^3h^3)$ is the radiation density constant, defined using the fundamental constants of Boltzmann, k_B, and Planck, h. Using then that the luminosity is $L(r) = 4\pi r^2 \mathcal{F}_{\rm rad}$, from Eq. (1.19) we have for a spherically symmetric configuration:

$$\left.\frac{\partial T}{\partial r}\right|_{\rm rad} = -\frac{3}{16\pi ac}\frac{\kappa\rho L}{r^2T^3}, \tag{1.20}$$

which, dividing by the equation of hydrostatic equilibrium, defines a radiative temperature gradient $\nabla_{\rm rad}$:

$$\nabla_{\rm rad} \equiv \left.\frac{\mathrm{d}\ln T}{\mathrm{d}\ln P}\right|_{\rm rad} = \frac{3}{16\pi acG}\frac{\kappa LP}{MT^4}. \tag{1.21}$$

If energy is transported fully by radiation, the temperature within the star will thus vary with pressure as given by Eq. (1.21).

Convective energy transport will occur if gas motion is unstable against adiabatic displacement. Put simply, if a blob of gas rises (or sinks) in the star and finds itself in an environment that makes it rise (or sink) even more, large-scale convective bubbles will be created. Defining the adiabatic temperature gradient at any point in the star:

$$\nabla_{\rm ad} \equiv \left(\frac{\mathrm{d}\ln T}{\mathrm{d}\ln P}\right)_s, \tag{1.22}$$

the subscript meaning to take the derivative with constant specific entropy, s, a stellar layer will be unstable to convection if:

$$\nabla_{\rm ad} < \nabla_{\rm rad}. \tag{1.23}$$

This is called the Schwarzschild criterion of convection (Schwarzschild 1906). The Belgian astrophysicist from Liège, Paul Ledoux, modified this criterion by also considering stabilizing molecular gradients (Ledoux 1947). He realized that when a convective blob moves to a region where the chemical composition is different, an extra term needs to be added to the stability criterion. The Ledoux criterion for convection is:

$$\nabla_{\rm ad} < \nabla_{\rm rad} - \frac{\varphi}{\delta}\nabla_\mu, \tag{1.24}$$

with the definitions $\varphi \equiv \partial\ln\rho/\partial\ln\mu$, $\delta \equiv -\partial\ln\rho/\partial\ln T$ and $\nabla_\mu \equiv \partial\ln\mu/\partial\ln P$ to write the stability condition concisely. The second term on the right-hand side is sometimes called the *Ledoux term*. The mean molecular weight,[7] μ, is defined as the amount of mass the average free particle has in units of m_u:

$$\mu = \frac{\rho}{n m_u}, \tag{1.25}$$

with n the number of free particles per unit volume in the mixture. A higher μ thus means that particles have more nucleons on average. Conversely, a lower μ means that particles are lighter, or that the mixture is more ionized, which increases the number of free electrons. Looking at Eqs. (1.21) and (1.24), we see that convection will set in if either the luminosity to mass ratio, L/M, or the opacity, κ, is particularly high, while steep composition gradients, ∇_μ (e.g., left behind by nuclear burning), have stabilizing effects.

Computing the actual temperature gradient, ∇, of the star in a convective zone requires a theory of turbulent motion. To emphasize, ∇ for a convective region *is not* a result of the stellar-structure equations, but is rather an input *required* to describe the temperature gradient of a convective region. Describing convection is hard, as it is a turbulent flow that is chaotic. In principle, the full Navier-Stokes equations need to be solved in three dimensions, after which parameterizations to one dimension can be attempted. We can however make the approximation that convection will occur on a typical length scale, and consider that all convective bubbles travel that length and then promptly deposit their excess heat into the environment. This set of assumptions defines the mixing length theory (MLT) of convection, and was pioneered by Erika Böhm-Vitense as she first described convection in the Sun (Vitense 1953; Böhm-Vitense 1958). MLT allows to compute a one-dimensional efficiency of convection, ζ, by comparing the convective heat flux with the radiative one. Following the description of Cox and Giuli (1968) of MLT, the convective temperature gradient is then:

$$\nabla = (1 - \zeta)\nabla_{\rm rad} + \zeta \nabla_{\rm ad}, \tag{1.26}$$

where all quantities, $\nabla_{\rm ad}$, $\nabla_{\rm rad}$ and ζ, can be evaluated from local quantities like density, luminosity, temperature and opacity. In convectively efficient regions (such as a core of a massive star), $\zeta \approx 1$, and thus the temperature gradient is nearly adiabatic, $\nabla \approx \nabla_{\rm ad}$. In convectively inefficient regions, such as sub-surface convection zones, $\zeta < 1$, and radiation still plays an important part in transporting energy outwards, and the region is said to be super-adiabatic (since we have still $\nabla_{\rm rad} > \nabla_{\rm ad}$).

Today, we have the general picture that stars of the lowest masses are fully convective, see Fig. 1.6, because of their (relatively) low temperatures, high densities and high opacities. Stars with masses around that of our Sun develop a split

[7] This is, in my opinion, another confusing misnomer, as, in the context of stellar structure, this has nothing to do with weight nor with molecules. I prefer *mean particle mass*.

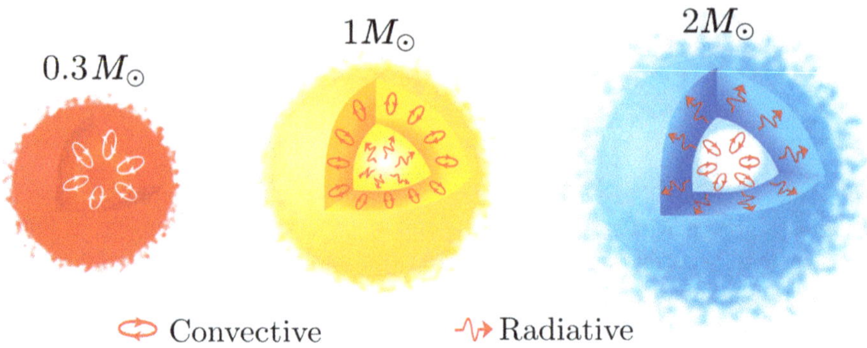

Fig. 1.6 Primary methods of heat transport from the core to the surface of a star. Stars with mass lower than about 0.3 $M_\odot$ are fully convective, while stars with mass around that of the Sun develop a radiative core within a convective envelope. For stars with mass above around 2 $M_\odot$, the situation is reversed. *Adapted from Sun.org, licensed under CC-BY-SA 3.0*

core-envelope structure, the core being radiative while the envelope still being convective.[8] As we keep climbing in mass, these roles reverse, as the core starts producing so much energy (very locally, because the CNO cycle is especially temperature sensitive) that radiation alone cannot carry it away, and the envelope loses opacity because its temperature is higher. It should be noted that, in convective regions, stellar material mixes very efficiently as the blobs travel large physical distances within the star. We can thus expect homogeneous abundance profiles in these regions. Conversely, in regions dominated by radiative energy transport, gas motion is limited to small scales, and any elements that are transmuted into others (e.g., due to fusion) remain in place, or move only very slowly by diffusion.

1.1.2 Star Formation

Stars do not spontaneously come into existence; they need to be formed first. Star formation is a whole branch of astrophysics on itself, distinct from stellar evolution. We therefore give the only a short introduction here and refer to detailed works like Stahler and Palla (2004) and Bodenheimer (2011) for the interested reader.

Stars are formed in molecular clouds, which are giant structures of gas and dust floating in galaxies, and have temperatures of around 100 K, densities of the order of 10^{-19} kg m^{-3} and physical sizes of up to several tens of light years. When such clouds are gravitationally disturbed, e.g., from a nearby supernova shock passing

[8] When stating "the envelope being convective" or "in the convective core," we mean that the dominant form of energy transport in that region is convection. "Radiative core" or "radiative envelope" are regions where radiative energy transport dominates.

through (McKee and Cowie 1975), the cloud can become unstable and start to collapse.[9] As the gas contracts, the temperature in the center gradually increases, and molecules are dissociated into their component atoms. As soon as the central region is dynamically stable, i.e., when it has produced a pressure gradient to oppose its own weight, as in Eq. (1.5b), the object is called a protostar. From that point, the core further accretes matter from its surroundings and gains in mass. At the same time, a protoplanetary disk of gas and dust is formed where planet formation takes place. When the star has accreted nearly all the material from its birth cloud, it becomes visible to far away observers, and we call the object a pre-main-sequence star. The energy source to sustain its luminosity is further gravitational contraction, as the central temperature is still not high enough to start nuclear fusion. Finally, as the central temperature reaches about 10^7 K, nuclear fusion of hydrogen starts and the star is said to have reached the zero-age main sequence, or ZAMS. From this point on, the object is a proper star, and is said to "be alive."

It should be noted that, for massive stars, appreciable accretion can still be going on while the star is already burning hydrogen in its core. As a result, the star is still embedded in its cloud, and is therefore difficult to observe young massive objects. Only later, which may be a significant fraction of its main-sequence lifetime (Bernasconi and Maeder 1996), does the intense radiation of the massive star fully disrupt its birth cloud, making the star visible to observers. This scenario is supported by the absence of very young massive stars in observational surveys (Holgado et al. 2020).

1.1.3 Main-Sequence Evolution

The main-sequence evolution of a star is defined by that part of its life where hydrogen is burned in its core. As this reaction requires the least extreme conditions in order to occur, this is the first stage of the life of the star. Since, compared to other fusion reactions, burning hydrogen to helium liberates the most energy per nucleon, this phase proceeds at a relatively slow pace and will take up around 90% of the total lifetime of the star. It should be mentioned here that objects with masses below about 0.08 $M_\odot$ never manage to create the conditions necessary to fuse hydrogen in their cores. These objects are not stars, and are called substellar objects or brown dwarfs instead (or, colloquially, failed stars).[10]

On the main sequence, low-mass stars like our Sun possess a radiative core, as pictured in Fig. 1.6. Since the rate of hydrogen fusion is heavily dependent on the local temperature, the very center burns it quicker than layers further out, creating

[9] It has been suggested that our own solar system formed this way (Cameron and Truran 1977).

[10] These objects do perform deuterium burning: $^2\text{H} + {}^1\text{H} \longrightarrow {}^3\text{He}$, with the little amount of deuterium that is naturally present. They do not do so in thermal equilibrium, however, and are thus not considered stars.

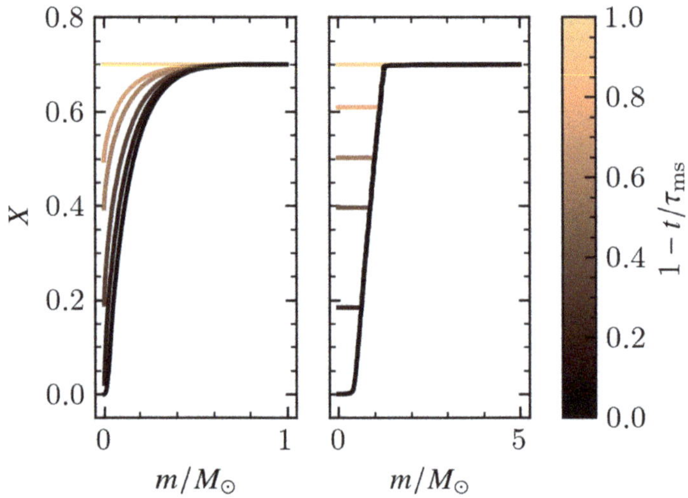

Fig. 1.7 Hydrogen content by relative mass fraction, X, as function of mass coordinate m at various points during the main-sequence evolution of a 1 $M_\odot$ (left) or 5 $M_\odot$ (right) star. The color indicates the fraction of the main-sequence lifetime $t/\tau_{\rm ms}$ that still remains

a sloped hydrogen profile, see the left panel of Fig. 1.7. This happens also in more massive stars, but, with convection continuously mixing the core, a flat bottomed hydrogen profile is established as fresh fuel from the edge of the core is transported inward, see the right panel Fig. 1.7.

As hydrogen burning progresses, more and more hydrogen nuclei are turned into helium, and so the structure of the star has to adapt accordingly. Because the mean molecular weight, μ, increases, the pressure, which obeys the ideal gas law, decreases:

$$P = \frac{\rho k_{\rm B} T}{\mu m_{\rm u}}. \tag{1.27}$$

The core thus has to contract, increasing the central temperature and density to again increase the pressure gradient and maintain hydrostatic equilibrium. This can be seen in the right panel of Fig. 1.7, where the flat-bottomed, convective core of the massive star becomes progressively smaller as time goes on. Because the rates of nuclear reactions are very sensitive to the temperature, the increase in core temperature causes an increase in luminosity of the star. As a result, the outer layers slowly expand to larger total radius. The increase in luminosity can be seen in Fig. 1.8 as tracks climb upward as the central hydrogen fraction, X_c, decreases.

No energy source is infinite however, and eventually all the hydrogen fuel in the center region is exhausted. For high-mass stars with a convective core, this is announced on the HRD by the characteristic 'hook' at the end of their main-sequence tracks (see Fig. 1.8). At this moment, the star falls out of thermal

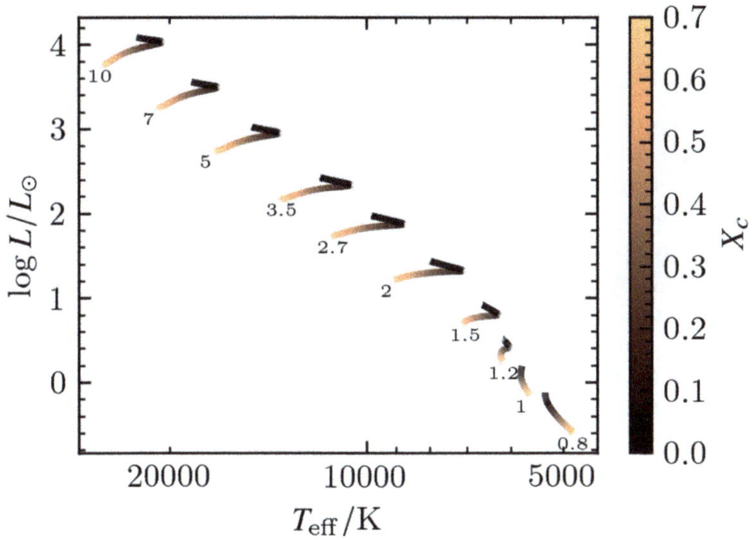

Fig. 1.8 HRD of the main-sequence evolution of stars of total mass (in $M_{\odot}$) annotated next to their tracks. The color indicates the central hydrogen mass fraction X_c, and can be used a proxy for age

equilibrium, and the terms with time derivatives in Eq. (1.18) become important to maintain the energy balance. This produces rapid changes in the structure of the star, and any further evolution is part of the post-main-sequence life of the star.

Finally, it is important to realize that the total main-sequence lifetime of a star is strongly dependent on its mass. As discussed above, the Sun will spend about ten billion years fusing hydrogen into helium in its core. Combining the estimation of the nuclear timescale (Eq. 1.4) with the mass-luminosity relation of main-sequence stars,

$$\frac{L}{L_{\odot}} \propto \left(\frac{M}{M_{\odot}}\right)^{3.5}, \tag{1.28}$$

Kippenhahn et al. (2012) derive:

$$\tau_{\mathrm{MS}} \approx 10^{10}\left(\frac{M}{M_{\odot}}\right)^{-2.5}. \tag{1.29}$$

From this equation we deduce that, even though a more massive star has more hydrogen available to burn, it has to do it much quicker to maintain equilibrium. A massive star of 10 $M_{\odot}$ will thus spend only around 30 Myr on the main sequence, while a very low-mass star of 0.1 $M_{\odot}$ can burn hydrogen for more than 3 Tyr.

1.1.4 Post-main-sequence Evolution

While all stars, at some point in their life, burn hydrogen to helium in their cores, the specific post-main-sequence evolution of a star is heavily dependent on its initial mass. For stars with mass below around $0.3\,M_\odot$, the end of the main sequence immediately means the end of their life. Since these stars are fully convective, almost all of the hydrogen has been turned into helium (maybe a small hydrogen atmosphere remains), and the star slowly cools and fades out as a helium white dwarf. Slightly heavier stars ($0.3 \lesssim M/M_\odot \lesssim 0.45$) have a core-envelope structure, and in an attempt to regain thermal equilibrium, the core contracts to create the conditions necessary to burn hydrogen in a shell around the core. Shell burning then moves through the layers around the core and turns the hydrogen of the inner part of the envelope into helium. The outer part of the envelope remains rich in hydrogen. At this point, the stars in this mass range have reached their final stop, a slowly cooling helium white dwarf, and it is supported by electron degeneracy pressure. It should be noted though that the low-mass stars of this caliber ($M \lesssim 0.45\,M_\odot$) burn their hydrogen so slowly on the main sequence that the universe is not old enough yet to have formed a helium white dwarf from a single star (Laughlin et al. 1997; Adams and Laughlin 1997). From the estimation of Eq. (1.29), we will have to wait some trillions of years! For fully convective stars this estimate is likely still an underestimation, as they have nearly their whole reservoir of hydrogen available for fusion, compared to the stars with radiative cores where only on the order of 10% is available. To close the discussion on very-low-mass stars, we note that helium white dwarfs can be made today in binary systems. A higher-mass star can lose its outer layers to a companion (see Sect. 1.2.2), such that what remains evolves toward a helium white dwarf.

For stars of higher mass, post-main-sequence evolution is more complicated. For masses between around $0.5\,M_\odot$ and $2.3\,M_\odot$, stars will burn hydrogen in a shell for long enough that the central temperatures is raised above the threshold to ignite helium inside the core. However, because the core is degenerate, the pressure does not increase with increasing temperature, and, for a short time, the nuclear reactions create a thermal runaway process, called the helium flash, until the degeneracy is lifted.[11] For intermediate-mass stars ($2.3 \lesssim M/M_\odot \lesssim 8$), no helium flash is necessary as the central pressure and temperature rises high enough from gravitational contraction alone to ignite helium in the core (and no degeneracy occurs). All stars in these mass regimes climb up the red giant branch (RGB, during hydrogen shell burning) and later the asymptotic giant branch (AGB, during core helium burning) on the right hand side of the HRD. One aspect of giant stars is that they suffer strong mass loss through their stellar wind, especially on the AGB. The

[11] It should be noted that, unlike what the name might suggest, a helium flash is not visible as a "flash" to an outside observer, as the energy released by the runaway fusion is used to restructure the star rather than to emit a lot of light. In fact, the star dims gradually after the helium flash (see, e.g., Kippenhahn et al. 2012).

Fig. 1.9 Planetary nebula NGC 3132, also known as the Southern Ring Nebula, as imaged by the James Webb Space Telescope. The orange-colored nebula is created by the shedding of the central star's outer layers on the RGB and AGB. *Image credit: NASA, ESA, CSA, STScI*

outer atmosphere of such stars is cool enough for dust grains to condense out of the stellar gas. Radiation from the star below encounters these dust particles and pushes them out, along with the gas in which they are mixed. In this way an AGB star can lose up to $10^{-4}\,M_{\odot}\,\mathrm{yr}^{-1}$ of material, a huge amount compared to our Sun's $10^{-14}\,M_{\odot}\,\mathrm{yr}^{-1}$.

When helium in the center is exhausted and fused into carbon and oxygen, a similar situation to when hydrogen was depleted arises. The core contracts in order to facilitate shell burning of helium, or core burning of the next available nuclear fuel. This time, degeneracy of the carbon-oxygen core happens for stars with initial mass up to around $8\,M_{\odot}$ and nuclear burning in the core halts. In the mass range of $0.5\,M_{\odot}$ and $8\,M_{\odot}$, the star will shed its outer layers, creating a beautiful planetary nebula,[12] an example of which is the Southern Ring Nebula (see Fig. 1.9). This will be the ultimate fate of our Sun too, and a carbon-oxygen white dwarf is left behind in the center of the nebula.

In stars with initial mass larger than around $8\,M_{\odot}$, the cycle of core contraction, heating and igniting a new fuel keeps going until iron is formed in the center. Electron degeneracy pressure is not high enough to prevent core contraction after any burning stage. This is because the core mass is higher than the Chandrasekhar limit of $M_{\mathrm{ch}} \approx 1.4\,M_{\odot}$. Since iron is the most stable nucleus, further fusion reactions toward higher mass nuclei do not release energy, but rather absorb energy (see Fig. 1.5). The core of the massive star therefore cannot find an energy source with which to stabilize itself against its gravity. It becomes unstable and collapses

[12] Another misnomer, planetary nebulae have nothing to do with planets. The name possibly comes from Herschel or Darquier and their early observations, who found these objects were "planet-sized nebulae" (Darquier de Pellepoix et al. 1782).

Fig. 1.10 Crab Nebula as imaged by the James Webb Space Telescope. It is the remnant of a supernova that erupted in 1054. *Image credit: NASA, ESA, CSA, STScI, T. Temim., licensed under CC-BY 2.0*

under its own weight, causing a violent supernova accompanied by a neutrino burst (however, in some cases a supernova can be avoided altogether, see below). In a supernova, a fraction of the material of the star (including its nuclear processed material, as well as heavy elements produced in the supernova itself) is ejected into the interstellar medium, making massive-star explosions one of the primary engines of chemical enrichment in galaxies (Burbidge et al. 1957). Some of the best studied supernovae are the ones that occurred in our galactic neighborhood. In 1054, Chinese and Japanese astronomers detected a "guest star" (Stephenson and Green 2003), which evolved into the Crab Nebula we see today (Fig. 1.10). More recently, SN 1987A erupted in February 1987 in the Large Magellanic Cloud (LMC), one of the satellite galaxies of our Milky Way. It was the first nearby supernova studied with modern telescopes and instruments, including neutrino detectors, and thus it provided excellent observational data about the physics of core-collapse supernovae.

The remnant of a core-collapse supernova is either a neutron star or black hole (BH), the former being a dense ball of nuclear material held up by neutron degeneracy pressure, the latter being a singularity that can currently only be described by Einstein's theory of general relativity (Einstein 1915; Schwarzschild 1916). The main parameters that determine what remnant (if any) will be left behind after a supernova is the mass of the stellar core, and the metallicity of the material, i.e., the fraction of mass in metals. Metallicity is important as it drives mass loss throughout the life of the star (see Sect. 1.1.5) and so it influences the mass of the carbon-oxygen core by the time helium is depleted. The explodability of the core also depends on the complex interaction between neutrinos and the core structure itself that is beyond the scope of this thesis (see e.g., Heger et al. 2003, 2023, for summaries). The combined result of these effects is that it is harder to form BHs at solar metallicity than at low metallicity, where carbon-oxygen cores are heavier for the same initial stellar mass. Those that do form are due to supernova fallback, where material that is launched in the supernova shock drops back onto the proto-neutron star and tips it over the mass limit to become a BH (sometimes called a "weak supernova"). If the core is too hard to explode, a supernova may not occur at all, and the star collapses directly into a BH without launching much material into interstellar space. This scenario is favored at low metallicity due to their strongly bound cores. However, the orbital characteristics of the binary systems Cygnus X-1 and VFTS 243, which are both located in the high-metallicity environments of the LMC and the Milky Way, indicate that BH formation through direct collapse is likely what occurred for one of the stars in these systems (Mirabel and Rodrigues 2003; Shenar et al. 2022). We draw this conclusion because the eccentricity of the orbit is very low, which suggests that there was very little mass loss during the formation of the BH, thus favoring the direct collapse scenario over a supernova.

1.1.5 Open Questions in Massive-Star Structure and Evolution

The physical processes outlined above give, generally, a good picture of the structure and evolution of stars in general, including those with initial masses greater than around $8\,M_{\odot}$, called massive stars.[13] We can explain the general trends in the life of the massive star, and understand the chemical enrichment of galaxies resulting from their winds and supernovae. We have also detected the remnants left behind by massive stars, neutron stars and black holes. The first neutron star was discovered by Jocelyn Bell using radio observations (Hewish et al. 1968), while the first convincing black hole candidate was Cygnus X-1, identified independently by Bolton (1972) and Webster and Murdin (1972) using a combination of X-ray observations from the UHURU space telescope and measurements from radio antennas.

Despite these successes, there are still multiple unsolved discrepancies between theory and observation regarding stellar evolution, particularly so in the massive-star regime, some of which we outline below.

1.1.5.1 Mass Discrepancy

A major problem present in the literature is the so-called mass discrepancy problem. For single stars, there are two main ways of calculating its mass. One involves measuring the luminosity and effective temperature of the star, and matching them to the position of evolutionary models in the HRD; the best fit giving the *evolutionary mass*. The other requires a measurement of the surface gravity, $g = GM/R^2$, through spectroscopic data, which, if the radius can be estimated from the luminosity and effective temperature, gives the *spectroscopic mass* of the star. Many

[13] Another definition of massive stars is that they are those stars that explode as core-collapse supernovae.

studies however find a systematic mismatch of both mass calculations (Herrero et al. 1992; Mokiem et al. 2007; Martins et al. 2012; Tkachenko et al. 2020; Mahy et al. 2020b). At this moment it is unclear whether the problem lies in the accuracy of determining surface gravities, which can be challenging for O-type stars, or if the evolutionary tracks suffer from uncertain physical modeling.

1.1.5.2 Mass Loss

Massive stars have a strong stellar wind on the main sequence (of order $10^{-6}\,M_{\odot}\,\mathrm{yr}^{-1}$ or more). This is in contrast with low-mass stars, which lose negligible mass on the main sequence (recall that the Sun has a mass-loss rate of just $10^{-14}\,M_{\odot}\,\mathrm{yr}^{-1}$).

Since stellar-structure and evolution theory does not (a priori) predict the mass-loss rates of stars, other theories or experiments need to be used. The first evidence of stellar winds in massive stars came from Morton (1967), who used measurements in the ultraviolet to infer outflows from three stars.[14] Later, it was proposed by Lucy and Solomon (1970) and Castor et al. (1975) that such outflows around hot stars are driven by the absorption and scattering of radiation by elements (in particular metals) in their outer layers. The radiation coming from the interior then deposits its momentum in these atoms (in the gas phase), driving them out. This is a very different mechanism than winds originating in AGB stars and cool supergiants, where it is dust that is accelerated to drive the wind.

Actually quantifying how strong a radiation field can accelerate atoms this way is a difficult problem. Vink et al. (2001) calculated the interactions of a large number of photons with over 10^5 different atomic lines of elements ranging from hydrogen to zinc. These computations then deliver a prediction for the mass-loss rate of the star, which depends on the composition of its atmosphere (in particular the metal fraction). More recent calculations however predict lower mass-loss rates (Sundqvist et al. 2019; Björklund et al. 2021, 2023). For example, a $15\,M_{\odot}$ star with luminosity of $\log(L/L_{\odot}) = 4.5$ and a surface temperature of 30,000 K has a predicted mass-loss rate by Vink et al. (2001) of $\dot{M} \approx 1 \times 10^{-8}\,M_{\odot}\mathrm{yr}^{-1}$, while Björklund et al. (2023) predict $\dot{M} \approx 2 \times 10^{-9}\,M_{\odot}\mathrm{yr}^{-1}$, almost an order of magnitude lower.

Another uncertainty that relates to wind-mass loss in massive stars is how homogeneous the wind is expected to be. Theoretical models of radiation transfer predict the outflows suffer from an instability that creates clumps (Owocki and Rybicki 1984). A simplified way to understand wind clumping is as follows. The turbulent wind will naturally cause small patches where the velocity is slightly higher. Due to the Doppler effect, the spectral lines with which the gas is accelerated then shift accordingly, and are thus de-shadowed! This is because radiation that

[14] Since ultraviolet light is largely absorbed by the atmosphere, he needed a rocket-mounted spectrograph to get to altitudes of 200 km to get sufficient signal!

once would not interact with the atoms is now absorbed by the Doppler-shifted gas. This causes that particular patch of gas to experience runaway acceleration because it is continuously de-shadowed. However, eventually, the gas parcel will run into slower gas, causing shocks and an increase in density. In 2D and 3D simulations, however, this so-called line-deshadowing instability is too expensive to model as it depends on non-local properties, and thus the degree of clumping (which may also arise from other instabilities) that is predicted from such simulations is lower. Furthermore, the choice of dimensionality, i.e., 2D vs. 3D, gives different answers for the clumping factor for otherwise similar stellar parameters (Moens et al. 2022). Because the amount of mass lost by the star is directly proportional to the mean density of its wind, the degree of clumping has a direct impact, and its value can range from 1 up to 40 as computed by Driessen et al. (2019).

It has become apparent that mass loss, both on the main sequence and beyond, can significantly change the evolution of a massive star. For example, if mass loss during the red supergiant phase is particularly strong and blows away the whole hydrogen-rich envelope, the hot helium core can become exposed to outside observers. If it does, such objects are called Wolf-Rayet (WR) stars, after Charles Wolf and Georges Rayet discovered they had strong and broad emission lines in their spectra, which are direct consequences of an optically thick wind. It is unclear if WR stars can be produced from a supergiant wind alone (e.g., Langer et al. 1994; Meynet and Maeder 2005), or if binary interaction is needed to take away the hydrogen layers (e.g., Pauli et al. 2022). Another area where mass-loss rates impact results is in predicting whether supernovae occur, and the masses of black holes and neutron stars left behind. In particular, the mass-loss prescriptions during the various phases of a star's life will determine a certain initial-to-final mass relation, with the final mass deciding whether a supernova occurs (e.g., Heger et al. 2023). Therefore, understanding the observable properties, as well as the evolutionary pathway and ultimate fate of a star, depends on knowing the precise mass-loss rates that massive stars have during their lives.

1.1.5.3 Rotation and Internal Mixing

In stating the equations of stellar structure in Eqs. (1.5), we assumed no forces other than self-gravity and internal pressure gradients act on the stellar material. However, as soon as a ball of gas is spinning about an axis, another term should be considered in the balance of forces of Eq. (1.5b). To sustain rotational motion, a centripetal acceleration is needed, which will be supplied by the gravitational attraction. Conversely, when viewed in the frame of reference of the rotating star, a centrifugal force appears that points away from the rotation axis. In the equation of hydrostatic equilibrium, we write it as:

$$\frac{\partial P}{\partial r} = -\rho \frac{\partial \Phi}{\partial r} - \rho r \Omega^2, \tag{1.30}$$

where Ω is the angular velocity about the axis of rotation. A major issue with Eq. (1.30) is that it is only valid in the equatorial plane of a star. By introducing rotation, we necessarily added an independent spatial coordinate, typically chosen to the the polar angle, θ. In 2D then, the equation of hydrostatic equilibrium, including rotation, reads:

$$\nabla \boldsymbol{P} = -\rho \nabla \boldsymbol{\Phi} - \frac{1}{2}\rho \Omega^2 \nabla (r \sin\theta)^2. \tag{1.31}$$

Note that the gravitational potential Φ is also no longer spherically symmetric, since the mass distribution is deformed. In principle this means all equations of stellar structure need to be solved in two dimensions, r and θ. While possible, this usually means sacrificing solving the time dependent problem, so no evolution is simulated, as is done by Roxburgh (2004) and Espinosa Lara and Rieutord (2013). On the other hand, Deupree (2001), and recently, Mombarg et al. (2023), succeeded in computing two-dimensional stellar-structure models while also following the time evolution along the main sequence. However, an alternative treatment of rotation exists, as Kippenhahn and Thomas (1970) provided a method to include the effects of rotation into 1D stellar-structure calculations, which we will introduce in Sect. 3.2.

Another uncertain physical process associated to rotation is rotationally induced mixing. Very early on, it was identified by Edvard Hugo von Zeipel that a rotating star cannot be in hydrostatic and radiative thermal equilibrium at the same time (von Zeipel 1924). This arises since, in hydrostatic equilibrium, the amount of energy that radiation can transport is proportional to the magnitude of the local effective gravity,

$$\boldsymbol{\mathcal{F}_{\mathrm{rad}}} \propto \boldsymbol{g_{\mathrm{eff}}} \equiv -\nabla \boldsymbol{\Phi} - \tfrac{1}{2}\Omega^2 \nabla (r \sin\theta)^2. \tag{1.32}$$

This relation is called the von Zeipel theorem, and explains temperature variations on the surface of a rotating star, which we call gravity darkening.[15] In radiative thermal equilibrium on the other hand, the equation of conservation of energy can be written as:

$$\nabla \cdot \boldsymbol{\mathcal{F}_{\mathrm{rad}}} = \varepsilon_{\mathrm{nuc}}, \tag{1.33}$$

which is a statement that any addition to the flux must be due to nuclear energy generation (and ignoring neutrino losses). If a star is in hydrostatic equilibrium, layers of constant density, temperature, pressure and composition align with surfaces of constant effective potential $\Psi = \Phi + \frac{1}{2}r^2\Omega^2$ (von Zeipel 1924). Von Zeipel then showed that, on an equipotential surface, the left-hand side of Eq. (1.33) is a function of the polar angle, θ, while the right-hand side is constant along that same equipotential, since $\varepsilon_{\mathrm{nuc}} = \varepsilon_{\mathrm{nuc}}(\rho, T, X_i)$ is a function of microphysical properties

[15] Some authors, e.g. Anderson and Shu (1977), call the term gravity darkening a misnomer, preferring *gravity brightening*, because increased gravity causes increased flux. While semantically more consistent, the term gravity darkening still dominates in the literature.

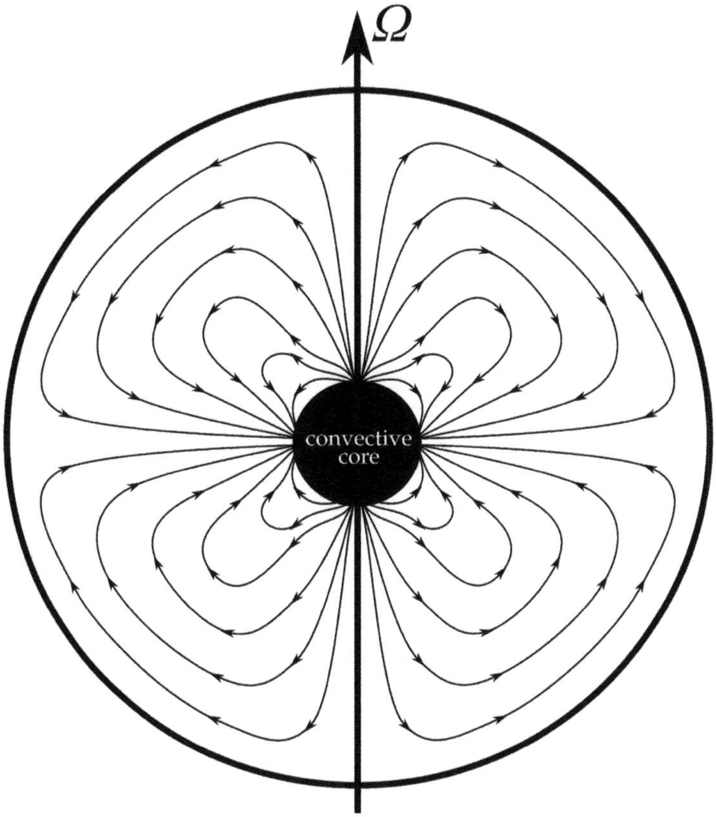

Fig. 1.11 Circulation currents in a slowly, uniformly rotating star. These currents balance out the non-uniform radiative energy transport as a result of von Zeipel's theorem

only. Therefore Eq. (1.33) cannot, generally, be satisfied on an equipotential surface of a rotating star.

Unless nuclear-fusion processes conspire with stellar rotation to keep isobars and surfaces of constant temperature aligned with the effective potential, in hydrostatic equilibrium then, under general conditions, radiative equilibrium cannot be satisfied. Instead, Eddington (1925) and Vogt (1925) independently proposed a solution to this problem by considering meridional circulation currents. These currents operate to carry away the excess energy generated in one region and deposit them in another so as to maintain radiative thermal equilibrium and hydrostatic equilibrium. Later, Sweet (1950) calculated the circulation velocities in the case of a slowly, uniformly rotating star, which corresponds to the currents shown in Fig. 1.11. Material wells up in the polar regions, and then flows along meridians toward the equator where it sinks again to the core. Such flows are now called Eddington-Sweet circulations.

While the origin of meridional circulation is reasonably well understood, the mixing of stellar material associated with these currents is not. From Fig. 1.11, it is clear that material near the core, which is altered by nuclear processes, can be transported up to the surface. However, this picture is a rather simplified view of the detailed physics that happens. Zahn (1992) shows that turbulent viscosity is at the root of the mixing process, rather than the meridional circulation, and derives it can only be efficient for very fast rotators. Furthermore, since fused elements near the core (like helium, carbon, etc.) are heavier than untouched hydrogen in the envelope of the star, there is a molecular weight gradient that inhibits these circulation currents. Different groups implement this interaction in different ways (compare, e.g., Heger et al. 2000; Meynet and Maeder 2000), and produce different results. While both models find elements like nitrogen are enhanced near the surface, only the models of Meynet and Maeder (2000) produce significant helium enrichment for fast-rotating, massive main-sequence stars (unless the stars rotate so fast they evolve chemically homogeneously, see Sect. 1.5.1). Using the 2D treatment, Mombarg et al. (2023) implemented the theory of Zahn (1992) and showed in their models that meridional flows are ineffective at transporting chemical elements from the core to the surface.

1.1.5.4 Convective-Core Overshooting

In massive stars, the core is unstable against convective motion of material, while in the envelope, radiative transport of energy occurs (see Fig. 1.6). At the edge of the core is thus a convective boundary, beyond which convective blobs should not exist anymore. It is however not unreasonable to think that the inertia of such blobs carry them past the boundary before they dissipate. This process is called convective-core overshooting. The question is how far convective elements can penetrate into the radiative envelope (if at all), and in so doing mix the material that would otherwise not participate in the nuclear burning of the core. The net effect of overshooting is the lengthening of the life of the star, since the more nuclear fuel a star has available, the longer it can maintain equilibrium against gravity. A byproduct of stronger overshooting is that the mass of the helium core increases, which has important ramifications for the further evolution (and fate) of the massive star (see Sect. 1.1.4).

Just like describing convection itself (see Sect. 1.1.1.3), theoretically deriving the properties of core overshooting is very hard. We note here the work of Canuto (1992), who derived a theory of convection that naturally includes a description of overshooting, but it has not found wide-spread use in stellar-structure calculations. Instead, overshooting is commonly parameterized by a dimensionless number, $\alpha_{\rm ov}$, which signifies how many pressure scale heights, H_P, convection extends beyond the convective boundary set by the Schwarzschild or Ledoux criterion, with

$$H_P = \left(-\frac{\mathrm{d}\ln P}{\mathrm{d}r}\right)^{-1}. \tag{1.34}$$

Using observations of clusters of stars, Brott et al. (2011) calibrated the overshoot parameter to $\alpha_{\rm ov} = 0.335$ for massive stars in the LMC, while Schroder et al. (1997)

constrained it to $\alpha_{ov} \in [0.24, 0.32]$ for intermediate-mass stars. More recently, Tkachenko et al. (2020) found values consistent with $\alpha_{ov} = 0.4$ for most stars in a sample of galactic eclipsing binaries.

Another way to constrain overshooting is by employing the theory of asteroseismology, the study of stellar oscillations (Aerts et al. 2010). Just as with Earth's seismology, the surface oscillations of stars carry information about the conditions inside them, which allows us to gauge the mass of the convective core (see, e.g., Johnston 2021 for a list of such convective core measurements), and thus constrain the strength of the overshoot (e.g., Aerts et al. 2003; Moravveji et al. 2016; Burssens et al. 2023).

Finally, with increasing efficiency of hardware and innovation in computational algorithms, in particular the use of CPU parallelization and GPU computing, more and more three-dimensional simulations of convection in stellar cores are performed (e.g., Browning et al. 2004; Cristini et al. 2017; Edelmann et al. 2019; Vanon et al. 2023; Andrassy et al. 2024). Such simulations could provide a direct mass and luminosity dependent measure of the overshoot parameter, as well as the temperature gradient in the core. Producing a single model like this still takes an immense amount of computing power, however, and so overshooting values that are applicable for the wide parameter space that 1D calculations can cover have not yet been obtained.

1.2 Binary Stars

The historical introduction in this section is based mainly on Aitken (1935).

While our own Sun is a lonely star, it is possible for stars to form and evolve near each other. Imagine a sunset, but instead of just the Sun descending over the horizon, there are now two balls of light sinking one after the other, like the stars in Fig. 1.12 Our vocabulary concerning the Sun, light, and maybe even day and night, would likely be different if two disks of light illuminated us rather than just one, not to mention the various deities associated to the Sun, which include Helios (who was later identified with Apollo, Larson 2007) in Greek mythology, Ra in Egyptian religion, or the Inca's Inti.

The study of binary stars, like regular single stars, is not a recent affair. As early as the Greek and Roman period, Ptolemy used the term *double star* to describe ν Sagittarii, which are two stars separated by only about half of the Moon's apparent diameter (which is around thirty arcminutes or half a degree). The first modern observations of double stars are usually attributed to the Italian Jean Baptiste Riccioli in 1650, who found the star Mizar in Ursa Major (better known as the big dipper) had two bright components. Some historians challenge this credit, and letters from Galileo to Benedetto Castelli indicate it was Castelli who first resolved Mizar into two components as early as 1617 (Ondra 1999).

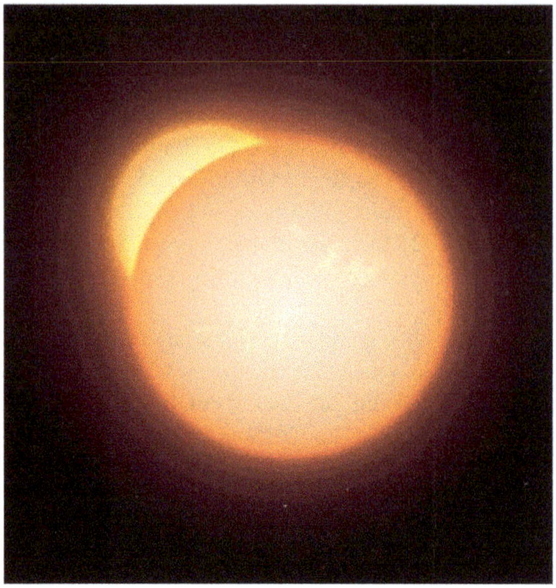

Fig. 1.12 Artist impression of a binary star, where one star passes in front of the other as viewed by the observer. *Image credit: ESO/L. Calçada, licensed under CC-BY 4.0*

Christiaan Huygens and Robert Hooke, among many others, made similar such observations in the seventeenth century. Initially, there was no specific interest in these objects, as it was thought it was mere coincidence that two stars happen to align when viewed from Earth.

It was in the mid eighteenth century that this perspective changed. Based upon general probability arguments, John Michell (an English priest) argued in 1767 that stars that appear close to each other are indeed very likely to be physically near each other too (as opposed to scattered randomly throughout the galaxy and aligning by chance), and thus compose a stellar system (Michell 1767). This realization coaxed William Herschel and Christian Mayer to systematically look for binary stars, and by 1782, hundreds of double stars were cataloged (Herschel and Watson 1782). Thanks to Newton's synthesis of motion and gravitation (Newton 1687), it was known that the forces that keep such stars revolving around each other are the same ones that make the Earth move around the Sun, or that make apples fall from trees. As a result, at the start of the nineteenth century, it was Herschel who made the formal distinction between double stars and proper binary systems, by requiring for the latter that the two stars are *"united by the bond of their own mutual gravitation toward each other"* (Herschel 1802). One year later, he published the first set of measurements that irrevocably proved that Newton's laws determine the motion of some of the observed double stars (Herschel 1803).

Up to this point, only *visual binaries* were discovered. These are binary systems that are resolvable by direct imaging with telescopes, where the size of the main mirror (and the quality of the atmosphere) determines how close stars can be distinguished. However, with the advent of spectroscopy, in 1887, Antonia Maury at Harvard noticed that the dark lines in the spectra from Mizar A appeared single

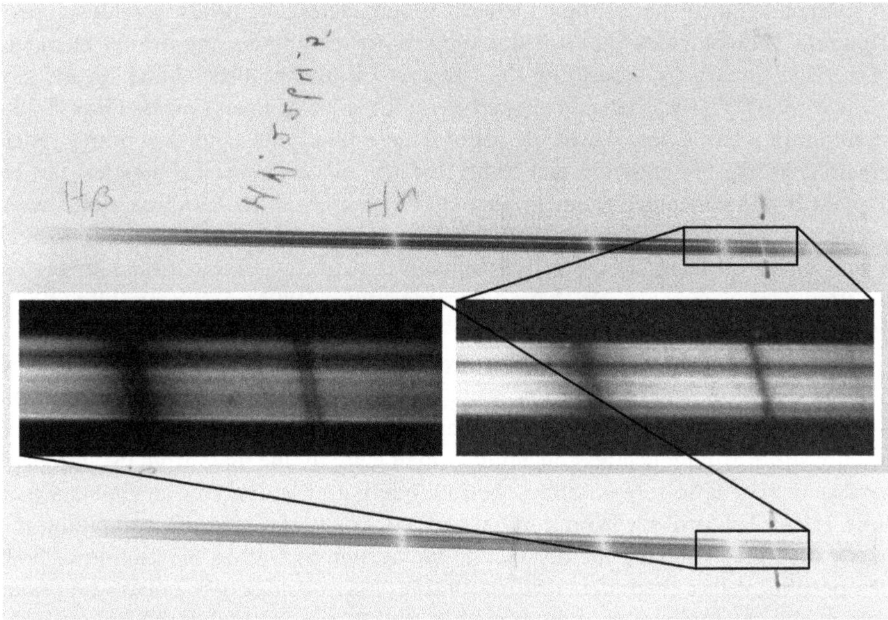

Fig. 1.13 Photographic plates showing the line splitting of Mizar A's spectra. On the lower plate, a the calcium K line can be distinguished as double, while on the lower right, the line appears single. The full spectrum on the top plate (C00441) is annotated with the hydrogen Hβ and Hγ spectral lines, as well as the (mirrored) catalog entry HP 2264-5. The plate was taken on Apr 5, 1887, at Harvard with the 11 inch Draper refractor. The bottom plate (C00427) was taken a week earlier, on March 29, 1887. *Images courtesy of Harvard Plate Stacks, Center for Astrophysics | Harvard & Smithsonian*

sometimes, and double other times. Figure 1.13 shows the original plates where they made this discovery. She and Pickering thus concluded that one component of the visual binary is in itself binary system as well (Pickering 1890). Based upon the Doppler-Fizeau principle, in a *spectroscopic binary* (SB) then, as a star alternately moves toward and away from the observer, a spectral line gets blueshifted to shorter wavelengths as the star is approaching, or redshifted to longer wavelengths when the star is receding. The position of the line correlates with the radial velocity (RV) of the star, i.e., the speed with which it approaches or recedes from the observer. The Doppler effect of sound waves we experience in daily life when hearing an ambulance, car, or airplane pass by is analogous. When the spectral line doubled, the Harvard team thus saw one star receding and the other star approaching the Earth, while when the line was single, both stars were moving tangentially to the plane of the sky (relative to the motion of the binary system as a whole). Only months later, Hermann Carl Vogel saw the spectra of Algol move in correlation with the periodic dimming of the star (Vogel 1890). The theory by John Goodricke that a dark(er) object passed in front of the star (Goodricke 1783), blocking some of its light, thus turned out to be correct, and the class of *eclipsing binaries* had its first discovery.

In the late nineteenth century, methods for characterizing orbits were developed. Thorvald Nicolai Thiele published a method for computing the orbital elements of a visual binary by describing the apparent ellipse the orbit draws on the sky (Thiele 1883). This was later developed on by Robert T.A. Innes (van den Bos 1926), resulting in what is now called the set of Thiele-Innes constants describing visual binaries. While the equation describing the RV of spectroscopic binaries can be derived from elementary geometry, actually fitting observed RVs was a challenge in the past. Today, we can use numerical methods and computing power to find a well fitting orbit, but in the early twentieth century several algorithms involving iteration were composed, e.g., from Lehmann-Filhés (1894), Schwarzschild (1900) and Russell (1902), among others.

The development of orbital fitting proved to be very important, as one of the main benefits of studying binary systems is that they provide a direct measurement of the masses of the stars through Kepler's laws. Such *dynamical mass* estimates usually suffer from much less uncertainty than gauging the mass from evolutionary models or having to rely on atmospheric modeling as is done for single stars (see Sect. 1.1.5). It is still far from a trivial task, as, for visual binaries, computing the masses requires knowing the distance to the system as well as the motion of both components, rather than their relative motion. For spectroscopic binaries, a major unknown is the inclination of the orbit (i.e., the angle of the orbital plane to that of the sky), and allows only to estimate a lower limit on the mass of the stars. Given these problems, eclipsing binaries are particularly useful, as, in order to be eclipsing, the inclination must be very close to 90°. Furthermore, the duration of the transits relates to the relative sizes of the stars to the orbit and so mass and radius can be measured simultaneously. We will introduce some of these methods briefly in Sect. 1.3, but we also refer here to Southworth (2020), who made a concise summary of the different parameters that can be estimated from studying particular types of binary systems.

From the mid twentieth century, observations with X-ray telescopes uncovered another way to detect binaries. When a star is donating material to a nearby companion, an accretion disc can form around the accretor where the in-falling matter heats up tremendously. This hot gas then emits high-energy radiation in the form of X-rays, and we call such systems X-ray binaries. Elevated X-ray emission from an object is thus a telltale sign of a binary system that is exchanging mass, and it also provided the first indirect evidence of black holes existing in nature.

Fast-forwarding to today, we have catalogs of hundreds of thousands of binary systems, coming from large, all-sky surveys such as OGLE[16] (Graczyk et al. 2011), ASAS[17] (Paczynski et al. 2006) and space missions like *Kepler* (Borucki et al. 2010), *TESS*[18] (Ricker et al. 2015) and Gaia (Gaia Collaboration et al. 2023). With

[16] Optical Gravitational Lensing Experiment.

[17] All-Sky Automated Survey.

[18] Transiting Exoplanet Survey Satellite.

such catalogs, statistical studies on the period, eccentricity, and mass distributions can be undertaken to gain a better understanding of stellar and binary evolution.

It would be naive to assume that a star that is part of a binary would evolve in precisely the same way as if it was single. In close binary systems, tidal interactions (which are absent in single stars) can directly influence the rotation rate of either star. There is a parallel with the Earth-Moon system, as tidal interactions between the Earth and Moon have made the rotation period of the Moon equal its orbital period around the Earth. It always shows the same face to us, and we call this process tidal locking. Binary stars that interact tidally thus could have their rotation, and its associated mixing, see Sect. 1.1.5, altered with respect to single stars. A more important aspect, however, is the proximity of the companion itself. As stellar evolution drives stars to become bigger, at some point the more massive star of the two, which evolves faster, grows to the point where the gravitational pull of the companion on its outer atmospheric layers supersedes its own. At that moment, a phase of mass transfer will be initiated, where material of the donor star falls toward the accretor star. Constructing detailed models of this form of binary interaction, and the impact on the evolution of the stars, is the subject of many current efforts in binary-star astrophysics. Of course, it is still possible that the orbital period is so long, that, even as the stars evolve through their respective giant phases, no significant tidal interaction or mass transfer occurs. In that case the stars evolve as "effectively single."

1.2.1 Stellar Multiplicity

Before embarking on the quest to accurately model binary evolution, it is worthwhile to consider whether binarity actually occurs frequently in the universe. While we have detected hundreds of thousands of binaries, we estimate there are hundreds of billions of stars in our galaxy. Naively, one might thus think it is unlikely for a star to be in a binary system. However, by carefully examining various samples of stars, in particular young ones in clusters, the inverse turns out to be true. One clever quote (of which the original author is unknown at this point) capturing this fact is that

Three out of two stars are in a binary system.

This paradoxical word play states that most stars in the galaxy (and in the universe) are actually part of binary or higher-order multiple systems.

In the last fifteen years or so, a clear picture has emerged on the multiplicity fraction, i.e., the probability of a star having at least one nearby companion: it is a strong function of the mass of the star. Moe and Di Stefano (2017) and Moe (2019) performed a thorough meta-analysis of all multiplicity studies (including, but certainly not limited to, Sana et al. 2012; Evans et al. 2013; De Rosa et al. 2014; Sana et al. 2014; Kobulnicky et al. 2014), and drew the following conclusions. At the low end of the stellar mass spectrum ($M \approx 0.1\,M_\odot$), around 80% of stars are

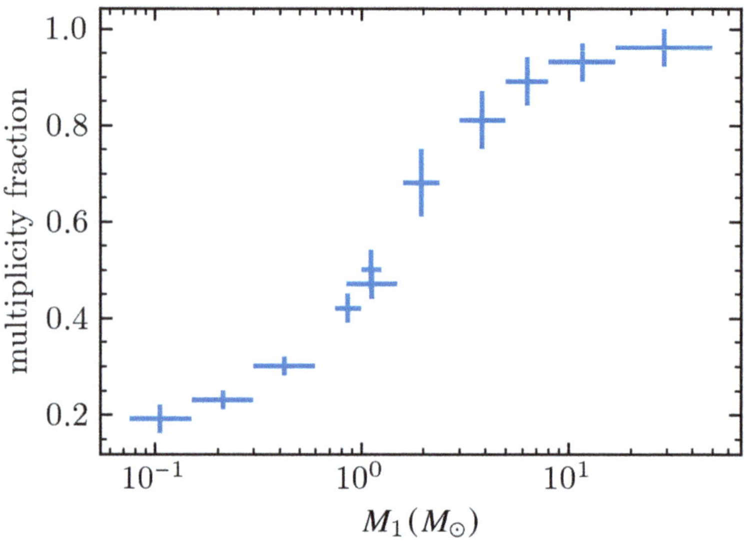

Fig. 1.14 Multiplicity fraction of stars as function of mass, using the data in Offner et al. (2023). Stars of mass lower than the Sun are primarily single, while stars much more massive than the Sun are overwhelmingly in binary and higher-order multiple systems

single, leaving only 20% that have a companion, see Fig. 1.14. For solar-type stars, it is split about fifty-fifty. Half of the stars are alone, like our own Sun, and the other half have at least one companion. At high mass however, $M \gtrsim 10\,M_\odot$, almost every star has a nearby companion, and it is very rare indeed to find a single massive star. In fact, the companion fraction of O-type massive stars is larger than two! This means the average O-type star is the primary of a triple star system.

The exact processes that makes the multiplicity fraction of massive stars so high are still uncertain. There is however consensus that it must be intimately linked with the star- and binary-formation processes. In one scenario, we think high-mass stars produce more companions because their protostellar clouds and disks are more massive and are thus susceptible to become unstable and fragment, forming a new core that can become itself a star (Kratter and Matzner 2006). However, it is rather difficult to produce companions at the orbital periods with which they are observed. When a star is forming, models indicate the large molecular cloud fragments on scales of thousands to tens of thousands of au,[19] and in a disk, the companion forms at a separation of hundreds to thousands of au (Krumholz and Bonnell 2007). While these findings are supported by observations of the young star-forming region M17 (Bordier et al. 2022), these distances are too large generally, since we observe many (unevolved) binaries that have separations less than one au (e.g., Sana et al. 2012).

[19] An au is a unit of length, defined as 1 au = 149,597,870,700 m, the average distance between the Earth and the Sun.

There must thus be very efficient mechanisms that bring the forming companions to closer separations in less than around one million years. Drag forces within the disk are one possible mechanism that tightens binary orbits, but this is challenged by numerical simulations (Muñoz et al. 2019).

Three-body interactions with an outer companion are also a possible scenario to form close binaries (Tokovinin and Moe 2020). Several conditions need to be fulfilled for this effect to be significant however (see, e.g., Naoz 2016). In particular, the outer star cannot be too far away with respect to the inner binary, but it cannot be too close either for the system to remain stable ($a_{\rm out}/a_{\rm in} \gtrsim 3$ for stability, Mardling and Aarseth 2001). The effect of a third body on an inner binary is to cause periodic variations in the eccentricity of the inner binary as well as the mutual inclination between the inner and outer orbits. This process we now call von Zeipel-Kozai-Lidov, or ZKL cycles (von Zeipel 1910; Lidov 1962; Kozai 1962). An increased eccentricity in a binary orbit, in turn, increases the strength of tidal interaction, which dissipates energy and permanently shrinks the separation of the components.

In another model of binary formation, two relatively low-mass proto-stars can be formed far from each other, but further accretion (to become massive stars in the end) actually drives the stars toward each other to conserve angular momentum (Bonnell and Bate 2005). These authors find it is possible to form two massive stars at a separation of the order of 1 au using this mechanism.

For massive stars, multiplicity plays an extra special role in their evolution. Sana et al. (2012) studied O stars in Galactic open clusters, and found that around 70% of these stars form in systems that will interact by mass transfer during their lives, or even merge with their companion. This is because the period distribution of massive binaries is skewed toward shorter periods. In subsequent studies of other samples, Sana et al. (2013a), Dunstall et al. (2015), Villaseñor et al. (2021) and Banyard et al. (2022) all found similar period distributions for O and B stars as Sana et al. (2012). Recently, this trend was found to persist in the lower metallicity environment of the SMC (Villaseñor et al. 2025; Sana et al. 2025).

The properties of interacting stars change dramatically as compared to single stars, and such processes thus need to be taken into account in evolutionary models. This, unfortunately, severely complicates the picture of the life of a massive star. If the majority of massive stars undergo strong interaction with a companion, our single-star evolution theory can only describe a small fraction of stars, and only up to when a strong interaction starts. From that point, the structure and evolution of the massive star is tied to the properties of the orbit, as well as (possibly) to the properties of the companion. The parameter space that determines the evolution of such a star has thus gained several new dimensions over that of a single star, making predicting the life and final fate of the star much harder (but also much more fascinating).

1.2.2 The Roche Potential and Mass Transfer

Describing the motion of two gaseous bodies is quite difficult without making approximations. In principle, each gas particle of both stars can have its own equation of motion, which, in its most basic form, is Newton's second law:

$$m\frac{\mathrm{d}^2\boldsymbol{x}}{\mathrm{d}t^2} = \boldsymbol{F}. \tag{1.35}$$

The main forces, $\boldsymbol{F}$, that change the position, $\boldsymbol{x}$, acting on the particle with mass m, are gravity, from both the parent star and the companion, as well as the gas and radiation pressure force from its surroundings. There are also other possible forces, like those that are magnetic or viscous in origin. Viscosity in particular adds a lot of complexity, as it is hard to compute from simple properties of the gas, like its temperature and density. Solving the equations of motion for the full stellar fluid thus seems like an intractable task, but it can be done with numerical codes when modeling timescales shorter than the thermal and nuclear timescales of stellar evolution.

However, even when we consider calculating just the motion of the center of mass of the stars under the influence of gravity alone, immediately some difficulties arise. After all, stars are not point masses, nor are they nicely spherically symmetric when in a binary due to tides and/or rotation. As a result, the total gravitational force on the center of mass of one star is not of the familiar GM^2/r^2 form. Instead we would need to use the full gravitational potential of the binary:

$$\mathcal{V}(\boldsymbol{x}) = -\int \frac{G}{|\boldsymbol{x}-\boldsymbol{r}_1|}dM_1 - \int \frac{G}{|\boldsymbol{x}-\boldsymbol{r}_2|}dM_2. \tag{1.36}$$

Then, the force is obtained by taking the gradient, $\boldsymbol{F} = -\nabla\mathcal{V}$, but the integrals making up $\mathcal{V}$ are not analytically solvable for general mass distributions dM_1 and dM_2. Once again, only numerically one can proceed.

It is thus common to approximate stars as point masses, in which the problem of the motion of one star around the other reduces to the classical two-body problem, which is solved since the times of Newton. Luckily, this approximation is not that terrible. Single-star models tell us that the vast majority of the mass of a star is concentrated in a small volume as compared to its total volume. For example, in the Sun, about 95% of the mass is located within 60% of the radius, or about 20% of the volume. For a massive $10\,M_\odot$ main-sequence star, 95% of its mass is within 10% of its volume, and this ratio only increases as the star evolves to become a giant. We say that stars are *centrally condensed* objects (Hilditch 2001). If we make the approximation that the dense cores of the stars are point masses, and that the total gravitational potential is unaffected by the tenuous outer layers of the stars, the potential simplifies to:

$$V(\boldsymbol{x}) = -\frac{GM_1}{|\boldsymbol{x}|} - \frac{GM_2}{|\boldsymbol{x} - \boldsymbol{a}|}, \tag{1.37}$$

where $\boldsymbol{a}$ is the vector separating the centers of mass of the stars. From this potential it follows that the centers of mass of the stars move around one another in ellipses, just as Johannes Kepler found for the planets around the Sun. If we make one final simplification, namely that the orbits of the stars are circular, and then move our reference frame into the rotating binary, we arrive at the Roche potential of the system:

$$V(\boldsymbol{x}) = -\frac{GM_1}{|\boldsymbol{x}|} - \frac{GM_2}{|\boldsymbol{x} - \boldsymbol{a}|} - \frac{1}{2}\left(\boldsymbol{\Omega} \times \boldsymbol{x}\right)^2, \tag{1.38}$$

where $\boldsymbol{\Omega}$ is the angular-velocity vector pointing along the axis of rotation. Its magnitude,

$$\Omega^2 = \frac{G(M_1 + M_2)}{a^3}, \tag{1.39}$$

follows from Kepler's third law. Equation (1.38) is named after nineteenth-century French mathematician Édouard Roche, who studied the restricted three-body problem using this potential. Figure 1.15 shows an impression of the Roche potential in the equatorial plane. We can see the two wells of the point masses that represent the heavy cores of the stars, as well as the outer regions that are dominated by the centrifugal force which appeared when changing into an accelerating reference frame. Without any other force, the slopes in Fig. 1.15 determine how a marble would start to roll if placed in this potential.

Fig. 1.15 3D representation of the Roche potential in the equatorial plane for $q = M_2/M_1 = 0.3$. The deep gravitational wells are created by the massive cores of the stars, while the outer regions are dominated by the centrifugal force

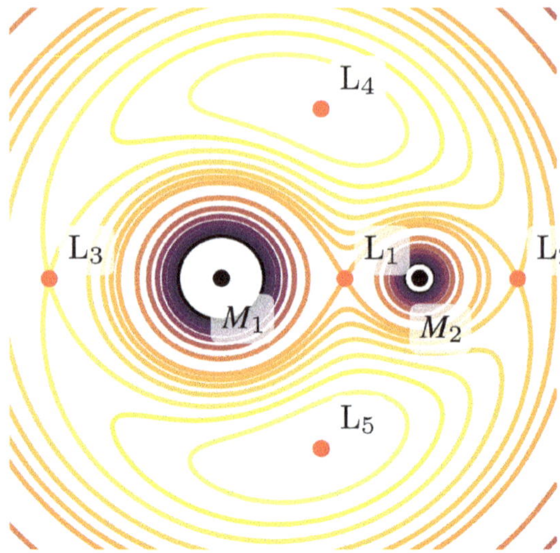

Fig. 1.16 Roche equipotential contours in the equatorial plane (for $q = M_2/M_1 = 0.3$), along with the five Lagrangian points marked as red dots

If we now consider surfaces of constant potential, i.e., equipotentials, in the equatorial plane, we arrive at the contours pictured in Fig. 1.16. Marked there also are the Lagrangian points, L_1 through L_5. These are special locations where the gradient (i.e., the slope) of the potential is zero. A particle that is exactly at any of the Lagrangian points would thus stay there indefinitely if not disturbed by an outside force. However, points L_1, L_2, and L_3 are unstable, meaning that the tiniest perturbation from them will amplify, and the particle will leave that point. In contrast, L_4 and L_5 can be stable thanks to the Coriolis force acting on a moving particle (which cannot be represented in a potential graph).[20]

The existence of the Lagrangian points has important ramifications for the life of stars that have a nearby companion. Generally, a star will grow in radius throughout its life. This is most apparent when a star exhausts its hydrogen in the core and becomes a giant, where the stellar size changes typically by two orders of magnitude. But, even during the main sequence itself, as the chemical mixture in the core is slowly changing from hydrogen to helium, the star needs to adapt and grow in size to accommodate for the increasing luminosity generated (see Fig. 1.8). The first Lagrangian point, L_1, is a showstopper however, as, when a star grows so its surface crosses L_1, mass will start funneling toward the companion. Figure 1.17 shows another view of the Roche potential, this time along the axis connecting the centers of the stars. In the left panel, both stars sit nicely inside their own wells,

[20] This is not a trivial calculation. Only for mass ratios bigger than $M_1/M_2 = \frac{25+3\sqrt{69}}{2} \approx 25$ are L_4 and L_5 stable. Furthermore, they are never asymptotically stable, meaning a particle will never exactly return to L_4 or L_5, but only stay in their vicinity. This nice result can be obtained from the restricted three-body problem by solving the linearized equations of motion around the Lagrangian points (Greenspan 2014).

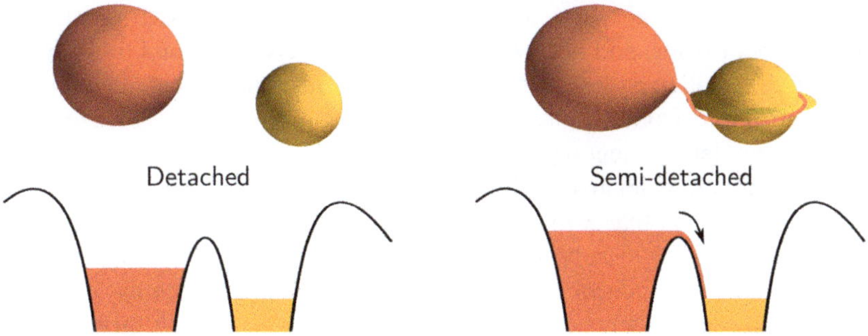

Fig. 1.17 Configurations in the Roche potential. On the left, both stars reside within their own Roche lobes, while on the right, mass transfer through Roche-lobe overflow is occurring

the so-called Roche lobes (RLs), while in the right panel, the more massive star has grown to fill its RL. We give these binary configurations the names 'detached' and 'semi-detached,' respectively. The size of the RL depends on the separation, a, and the mass ratio, q, of the binary, and can be computed by performing a numerical integration of the Roche potential of Eq. (1.38). Peter Eggleton determined a now widely used analytical fit for the volume-equivalent radius, i.e., the radius of a sphere having the same volume as the teardrop-shaped RL (Eggleton 1983):

$$R_{\mathrm{RL},1} = \frac{0.49q^{-2/3}}{0.6q^{-2/3} + \ln\left(1 + q^{-1/3}\right)} a, \tag{1.40}$$

where we defined $q = M_2/M_1$. The main feature of this equation is that the RL size increases as the separation (or orbital period) increases, and vice versa.

The process of mass spilling over the L_1 point is called *mass transfer (MT)*, or *Roche-lobe overflow (RLOF)*, and material that was once part of the donor star can end up on the accretor star. For main-sequence donor stars, we expect its radius to never significantly increase beyond the RL, because the mass-transfer rate, i.e., the amount of mass transferred per second, increases with the degree of overflow (Kolb and Ritter 1990). For giant donors however, whose outer layers are rarefied and tenuous, the overflow can become quite large even for moderate mass-transfer rates.

Mass-transfer phases are categorized depending on the evolutionary status of the donor star, coined by Kippenhahn et al. (1967). For main-sequence donors, we use the *case A* designation. For post-main-sequence donors, we use *case B* and *case C* for stars that have exhausted hydrogen and helium in their cores, respectively.[21]

[21] Although Lauterborn (1970) states Kippenhahn et al. (1967) categorized all three cases, to our knowledge, Kippenhahn et al. (1967) made no mention of case C. Instead, we find Lauterborn (1970) makes the first mention of case C mass transfer.

It is important to consider also the effects of mass transfer on the orbital parameters themselves, and here we focus on the case where mass transfer is fully conservative, i.e., all mass lost from the donor is accreted onto the companion. When mass is transferred from a more massive star to a less massive star, conservation of (orbital) angular momentum says the separation between the stars must shrink. Conversely, mass transfer from a lower-mass star to a higher-mass star will widen the orbit. This means that a mass-transfer phase of a main-sequence donor (case A) can be split into two sub-phases. First, as the more massive star reaches its RL, it will start mass transfer to a less massive companion, meaning the orbit will shrink, resulting in a decreasing RL size. The donor star cannot thermally adjust quick enough to this new RL size, and will overflow it even more, increasing the mass-transfer rate, which exacerbates the problem. This scenario is called fast or thermal-timescale mass transfer, as it takes (roughly) one thermal timescale for the donor to adjust its radius to its new mass (e.g., Pols 1994). In most binary systems, the accretor star will be gaining so much mass during fast case A it becomes the more massive star in the system. From that moment on, the binary separation widens as a result of further mass transfer, which increases the RL size of both components (Eq. 1.40). Once the donor has regained thermal equilibrium, and the star can immediately adjust its radius to its new mass, the mass-transfer rates remain rather benign as it is the nuclear expansion of the donor that drives mass transfer. This scenario is called slow case A mass transfer, and occurs on the nuclear timescale of the donor. In very close binary evolution, both phases occur one after the other (see, e.g., Sect. 4.4.1).

The transfer of mass between components in a binary is the main interaction process by which binary-star evolution differs significantly from single-star evolution. Binary interaction has been very successful in explaining various astrophysical phenomena that could not be understood in terms of single-star structure and evolution. One famous such case is the so-called Algol paradox.

Algol is both a spectroscopic and eclipsing binary,[22] and from its analysis the masses and radii of the two stars could be determined with good precision (see Sect. 1.3 below). The primary star, Algol A, has $M_A = 3.17 \pm 0.21\, M_\odot$ and $R_A = 2.73 \pm 0.20\, R_\odot$, while the secondary, Algol B, has $M_B = 0.70 \pm 0.08\, M_\odot$ and $R_B = 3.48 \pm 0.28\, R_\odot$ (Baron et al. 2012). If we make the assumption that these two stars were formed at the same time, then conventional, single-star evolution arguments should make the more massive star evolve faster and become larger before the less massive star. In Algol, this is clearly not the case, as $M_A > M_B$, but $R_B > R_A$, so it seems the less massive star has grown faster than the more massive one! In principle it could be that both stars were not born simultaneously, and that they later paired up (by multi-body interactions in a cluster) to become the binary star we see today. That scenario is impossible, however, as a $0.7\, M_\odot$ star has not had enough time yet to expand to $3.5\, R_\odot$; the universe is still too young. Instead, mass transfer can fairly easily resolve the paradox: In the past, Algol B was more massive than Algol A, and

[22] It is actually a triple star system, but we focus on the inner binary here.

evolved more rapidly, until mass was transferred from Algol B to Algol A, which has flipped the mass ratio. Algol A and B are thus a binary-interaction products. Algol A appears younger than its age suggests because, for much of its life, it was much less massive. Algol B, on the other hand, appears much older for its mass for the opposite reason.

1.3 Orbital Characterization and Dynamical Masses

In the historical introduction, we alluded to different methods of discovering and characterizing orbital motion of binary stars. This is invaluable as it allows to compare binary models to observed systems. After all, a model is only useful if it can reproduce certain aspects of reality. In the case of binary evolution, we would like to check whether the models of binary stars, potentially including interaction phases, are found in nature. This requires to know the orbital parameters of observed binary stars, which are the set of six numbers specifying the orbit (see Fig. 1.18). The shape parameters are the period, p, and the eccentricity, e. Three Euler angles determine the orientation of the orbit with respect to the observer: the inclination, i,

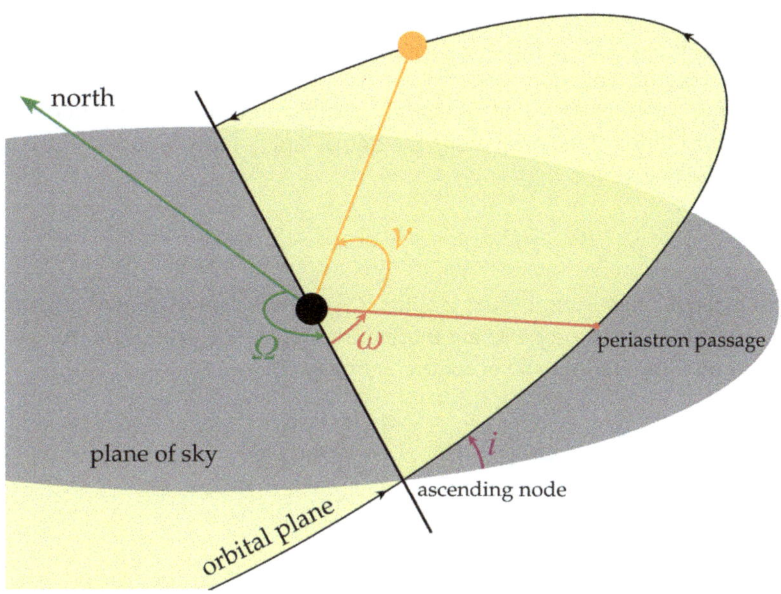

Fig. 1.18 Diagram of some of the orbital parameters of a binary. The longitude of the ascending node, Ω, is the angle between some reference direction (usually celestial north) and the line where the orbit ascends through the plane of the sky. The argument of periastron, ω, is the angle on the orbital plane between the ascending node and the point of closest approach of the bodies. The inclination, i, is the angle between the planes or the orbit and the sky. While not an orbital parameter, the true anomaly, $\nu = \nu(t)$, is the time-varying angle traced out by the orbiting body

the longitude of the ascending node, Ω, and the argument of periastron, ω. Finally, the epoch of periastron passage, T_0, defines a reference time at which the secondary is at its closest to the primary.

In this subsection, we briefly introduce the three main methods to characterize binary orbits.

1.3.1 Spectroscopic Binaries and Radial Velocities

In spectroscopic binaries (SBs), orbital motion is revealed in the Doppler shifts of spectral lines as the stars move around the common center of mass of the binary. When only one star is contributing to the observed spectrum, it is called single-lined, SB1, while when two components contribute, it is a double-lined spectroscopic binary, an SB2.

For an radiating body with a velocity v with respect to an observer, the Doppler effect will shift the emitted wavelength, λ_0, to an observed wavelength, λ, according to

$$\frac{\lambda - \lambda_0}{\lambda_0} = \frac{v}{c}. \tag{1.41}$$

This states that if a star is moving away from us (positive v), lines are shifted to longer wavelengths, and vice versa.[23]

The radial velocity (RV) of a star in a binary orbit can be determined by projecting the ellipse of the orbit along the line of sight of the observer. The equation for the RV then becomes

$$\mathrm{RV} = K\,(e\cos\omega + \cos(\nu + \omega)) + \gamma. \tag{1.43}$$

Here, K is the RV semi-amplitude, γ is the RV of the binary system as a whole, and $\nu = \nu(t)$ is the time-varying true anomaly of the star in the orbit. The true anomaly over time can be solved via the eccentric anomaly, E,

$$\tan\left(\frac{E}{2}\right) = \sqrt{\frac{1-e}{1+e}}\tan\left(\frac{\nu}{2}\right), \tag{1.44}$$

[23] The above is the non-relativistic Doppler effect. In cases where speeds are approaching the speed of light, time dilation needs to be taken into account, and the formula for the Doppler shift reads:

$$\frac{\lambda}{\lambda_0} = \sqrt{\frac{1 + v/c}{1 - v/c}}. \tag{1.42}$$

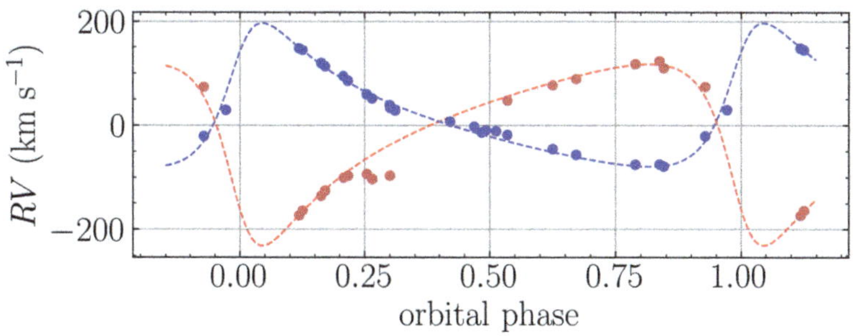

Fig. 1.19 Radial velocity curves of both components in HD100099. By measuring the Doppler shift in the spectral lines of the stars (the filled dots), Eq. (1.43) could be fitted to provide the orbital parameters of this system (the dashed lines). The horizontal axis is phase-folded with the orbital period, $P = 21.56$ d. *Reproduced with data from H. Sana, priv. comm*

and Kepler's equation (not to be confused with Kepler's laws):

$$E - e \sin E = \frac{2\pi(t - T_0)}{P}. \tag{1.45}$$

If you measure the RV of a star over a certain period of time, by measuring the shifts of spectral lines, and apply Eq. (1.41), you can fit the parameters K, e, ω, P, and T_0 of the orbit through Eq. (1.43). An example of this procedure is given in Fig. 1.19. There, Doppler shifts of the He I line at 5876 Å have been fitted, resulting in an estimated period of $P = 21.5585 \pm 0.0330$ d, an eccentricity of $e = 0.517 \pm 0.019$, and RV semi-amplitudes of $K_1 = 138.4 \pm 5.7\,\mathrm{km\,s^{-1}}$ and $K_2 = 174.8 \pm 7.3\,\mathrm{km\,s^{-1}}$. The full orbital solution is given in Sana et al. (2011a).

These five parameters unfortunately do not give the full information of the binary, since the semi-amplitudes

$$K_{1,2} = \frac{2\pi a_{1,2}}{P\sqrt{1-e}} \sin i, \tag{1.46}$$

depend on the semi-major axes, $a_{1,2}$, of their respective orbit, and the inclination, i, of the binary. As a result, only a lower limit on the masses of the stars can be determined. If both semi-amplitudes, K_1 and K_2, can be measured, by substituting Kepler's third law into Eq. (1.46), we can solve for, say, M_2, and arrive at:

$$M_2 \sin^3 i = \frac{1}{2\pi G}\left(1 - e^2\right)^{3/2} (K_1 + K_2)^2 K_1 P. \tag{1.47}$$

Only if the orbit is viewed edge on ($i = 90°$) can the true dynamical mass M_2 be computed. In other cases, $M_2 \sin^3 i < M_2$ gives the lower limit of the mass of the star. Either the inclination needs to be determined using other techniques (see below), or we could estimate the true mass by comparing the spectrum of the star to a stellar model (but this is sometimes risky, see Chap. 2). Still, independent of the inclination, in an SB2, the mass ratio of the components can be computed from the ratio of the semi-amplitudes,

$$\frac{M_2}{M_1} = \frac{K_1}{K_2}. \tag{1.48}$$

The success of determining the RV curve crucially depends on accurately measuring the the position of spectral lines as a function of time. For SB1s, this can done with methods like cross correlation or the fitting of Gaussian template profiles to the absorption lines. For SB2s, however, the situation is more complicated as lines of both stars can be blended when measuring them on the spectrograph. Traditional methods of RV-fitting can be biased in such cases and produce inaccurate, lower semi-amplitude estimates (Bodensteiner et al. 2021). This effect is called the SB2 bias, which can be mitigated by using spectral disentangling.

1.3.2 *Spectral Disentangling*

A more advanced technique that can determine the orbit of spectroscopic binaries is spectral disentangling. In this procedure, a series of spectra can be decomposed into the contributions of the two components of the binary, while simultaneously solving for the RV parameters. It is more advanced than 'standard RV fitting' in the sense that it does not measure RVs directly. Rather, the RV curve is a consequence of finding the best disentangled spectra, and this method thus suffers from fewer biases since it does not rely on fitting the center or the specific shape of a spectral line.

The problem of spectral disentangling can be tackled in two different ways. One involves solving a linear-least squares problem, while the other uses an iterative algorithm called shift-and-add. The core assumption of either method is that any variations in the time series of spectra comes from Doppler motion of the spectral lines. As such, line-profile variability due to stellar winds or pulsations cannot be taken into account.[24] In the following sections we introduce both disentangling methods.

1.3.2.1 Algebraic Method

The first approach to spectral disentangling was developed by Simon and Sturm (1994) so as to make the analysis of massive-star atmospheres possible. Atmosphere modeling was well developed, but it was poorly calibrated as function of stellar mass, in particular at high mass. Binaries provide dynamical masses, enabling

[24] One could however modify the algorithm to disentangle the signal from, say, pulsations, if one knows the variations these pulsations induce on the spectra as a function of time.

calibration, but because massive stars have intrinsically wide spectral lines and high rotational velocities, a composite spectrum of two such stars is impossible to fit to single-star atmospheric models. A method of separating the spectra was thus needed. Before stating the full problem, we start with a simple example to understand the disentangling process.

Consider a hypothetical situation where, at a certain time in the orbit of a binary system, the spectrum of component A is shifted by one wavelength bin (of our spectrograph) to the left, and the spectrum of B is shifted rightward one bin. If we assume the (unknown) spectra of the stars are $\boldsymbol{x}^A$ and $\boldsymbol{x}^B$, and those are the only stars contributing to the light that falls on our spectrograph, then the equation governing this observation, $\boldsymbol{c}_1$, can be written as:

$$\begin{pmatrix} 0 & 1 & 0 & \cdots & 0 \\ \vdots & \ddots & \ddots & \ddots & \vdots \\ \vdots & & \ddots & \ddots & 0 \\ \vdots & & & \ddots & 1 \\ 0 & \cdots & \cdots & \cdots & 0 \end{pmatrix} \boldsymbol{x}^A + \begin{pmatrix} 0 & \cdots & \cdots & \cdots & 0 \\ 1 & \ddots & & & \vdots \\ 0 & \ddots & \ddots & & \vdots \\ \vdots & \ddots & \ddots & \ddots & \vdots \\ 0 & \cdots & 0 & 1 & 0 \end{pmatrix} \boldsymbol{x}^B = \boldsymbol{c}_1, \tag{1.49}$$

from which it is clear that the first bin of the composite spectrum is actually just the second bin of $\boldsymbol{x}^A$. The secondary, $\boldsymbol{x}^B$, does not contribute because the common wavelength-base that we chose simply does not cover the wavelength with which B emitted in its rest frame. In the middle of the composite spectrum, both stars will contribute oppositely shifted parts of their individual spectra, as the two 1's on the off-diagonals select specific bins of $\boldsymbol{x}^A$ and $\boldsymbol{x}^B$, and combine to arrive at our spectrograph in the the bins of $\boldsymbol{c}_1$.

In more general cases, the location of the non-zero diagonal in equations like Eq. (1.49) depends on the orbital velocity of the components with respect to the wavelength bins used. Furthermore, when the orbital velocity is such that the Doppler shift falls between the grid spacing, two diagonals will be needed, of which the sum in each row is unity (this implements simple linear interpolation of the flux *w.r.t.* the wavelength bins).

When multiple observations are involved, we can concatenate them into one large column vector on the right-hand side of the equation, and build further diagonal matrices that correspond to the RVs at those observation times. Finally then, the full linear-algebra problem will, schematically, look like this:

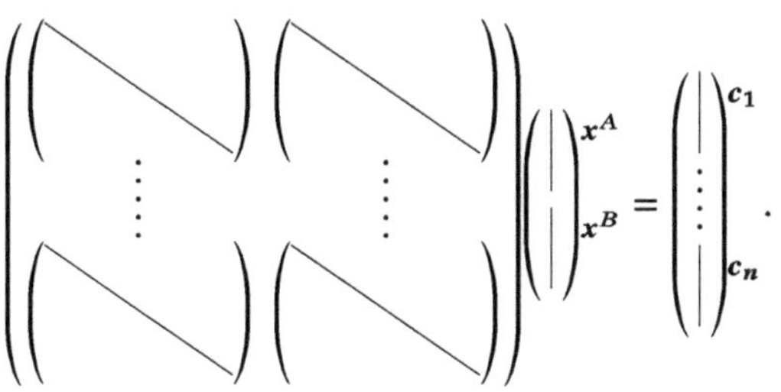

(1.50)

The slanted lines represent the 1's that are located on the (shifted) diagonals in each of the sub-parts of the matrix. The matrix on the left will have size $(nk \times 2k)$, with k the number of wavelength bins, n the number of observations, and 2 appears because we disentangle two components. In principle we can add an arbitrary number of unknown components below $\boldsymbol{x}^B$, as long as we also add the blocks corresponding to the movement of that component as extra columns in the big matrix. Equation (1.50) is a verbose way to write a system of equations of the form,

$$\boldsymbol{M}\boldsymbol{x} = \boldsymbol{c}. \tag{1.51}$$

The matrix $\boldsymbol{M}$ is very sparse and contains many rows that are zero. Such systems are most easily solved using the singular value decomposition,

$$\boldsymbol{M} = \boldsymbol{U}\boldsymbol{\Sigma}\boldsymbol{V}^T. \tag{1.52}$$

We then find the disentangled spectra, $\boldsymbol{x}$, by inversion,

$$\boldsymbol{x} = \boldsymbol{M}^{-1}\boldsymbol{c} = \boldsymbol{V}\boldsymbol{\Sigma}^{-1}\boldsymbol{U}^T\boldsymbol{c}. \tag{1.53}$$

Since $\boldsymbol{M}$ generally has more rows than columns, the system is overdetermined and the least-squares solution, $\boldsymbol{x}_1$, is returned. It minimizes the value

$$r^2 = \left(\frac{\boldsymbol{M}}{\sigma}\boldsymbol{x}_1 - \frac{\boldsymbol{c}}{\sigma}\right)^2, \tag{1.54}$$

with σ the standard error on the fluxes of the observed spectra.

In a subsequent paper, Hadrava (1995) generalized this disentangling method to any number of components that might have their spectra convolved. Additionally, he reformulated the problem in Fourier space rather than in wavelength space, which results in being able to solve much smaller matrix problems independently for each Fourier component. This is because Doppler shifts in wavelength space only add a phase factor (which can be divided out) in Fourier space. We still have an equation of the form,

$$\boldsymbol{A}(y)\tilde{\boldsymbol{x}}(y) = \tilde{\boldsymbol{c}}(y), \tag{1.55}$$

but the matrix $\boldsymbol{A}$ now has the size of just $(n \times m)$, with m the number of components one wants to disentangle. Because solving linear-algebra problems typically scales with the size of the matrix squared, it is cheaper to solve k problems of size $(n \times m)$ in Fourier space, than it is to solve one problem of size $(nk \times mk)$ in wavelength space.

To finish, we only need to undo the Fourier transform, $\mathcal{F}$, to get the disentangled spectra,[25]

$$\boldsymbol{x}(\lambda) = \mathcal{F}^{-1}\{\tilde{\boldsymbol{x}}(y)\}\,. \tag{1.56}$$

In this section, we ignored the relative normalization of $\boldsymbol{c}$ vs. $\boldsymbol{x}$, and the fact that the relative contribution of the individual spectra is a free parameter. In a binary system, f^A and $f^B = 1 - f^A$, called the light factors that multiply $\boldsymbol{x}^A$ and $\boldsymbol{x}^B$, respectively, must be supplied, as the equations do not constrain them. Usually, this can be done by scaling the spectra so that the depth of certain characteristic lines match spectral models (see, e.g., Shenar et al. 2020).

1.3.2.2 Shift-and-Add

Another method to disentangle composite spectra is the so-called shift-and-add algorithm. It was introduced by Marchenko et al. (1998), and has recently gained popularity in the literature (see, e.g. González and Levato 2006; Mahy et al. 2012; Shenar et al. 2018, 2020; Bodensteiner et al. 2020; Janssens et al. 2023; El-Badry et al. 2025).

Shift-and-add is an iterative approach where the set of observed spectra are, as the name implies, RV-shifted into the rest frame of one component (say the primary) after which they are co-added. This boosts the signal of the primary, while diluting the signal of the companion, since, when in the rest frame of the primary, its contributions are positioned at the same wavelength, while the secondary star will appear at different locations for each different observation. If we write $\boldsymbol{x}_i^A$ as the obtained spectrum for the primary star A in the ith iteration of the algorithm, we first compute the next trial solution of the secondary by subtracting the guess from the composites (taking into account the light factors), shifting the remainder into the secondary's rest frame, and finally adding the results:

$$f^B \boldsymbol{x}'^B_{i+1,j} = \boldsymbol{c}_j - f^A \boldsymbol{x}'^A_i, \tag{1.57a}$$

$$\boldsymbol{x}^B_{i+1,j}(\lambda) = \boldsymbol{x}'^B_{i+1,j}\left(\lambda \cdot \left[1 - \frac{v_j^B}{c}\right]\right), \tag{1.57b}$$

$$\boldsymbol{x}^B_{i+1} = \sum_j \boldsymbol{x}^B_{i+1,j}. \tag{1.57c}$$

[25] The extra Fourier transforms that need to be taken do not slow down the algorithm much, as computers implement the Fast Fourier Transform (FFT, Cooley and Tukey 1965), which has complexity $O(k \log k)$, which is still smaller than $O(k^2)$ for the wavelength algorithm.

We write primed quantities in the observer's frame, and un-primed in the rest frame of the component in question. With the new guess of the secondary, we can compute a better spectrum for the primary using the same procedure, but with the roles reversed: Subtract the secondary from the composites, shift into the rest frame of the primary (with its RVs, v_j^A), and add the results:

$$f^A \boldsymbol{x}'^A_{i+1,j} = \boldsymbol{c}_j - f^B \boldsymbol{x}'^B_{i+1}, \tag{1.58a}$$

$$\boldsymbol{x}^A_{i+1,j}(\lambda) = \boldsymbol{x}'^A_{i+1,j}\left(\lambda \cdot \left[1 - \frac{v_j^A}{c}\right]\right), \tag{1.58b}$$

$$\boldsymbol{x}^A_{i+1} = \sum_j \boldsymbol{x}^A_{i+1,j}. \tag{1.58c}$$

To get the algorithm started, the initial guess for the secondary can be made by assuming $\boldsymbol{x}_0^B = 1$ in Eq. (1.58), i.e., a featureless spectrum. $\boldsymbol{x}_0^A$ is then the sum of the observed spectra shifted into the rest frame of the primary.

After sufficient iterations, the signal of the other component is so diluted that each further iteration barely changes the resultant disentangled spectra. A criterion can thus be designed that stops the algorithm when the changes in the spectra drop below a certain (small) value.

We show an example of shift-and-add disentangling in Fig. 1.20. We "entangled" two model spectral lines, one box-like, the other Gaussian (left panel), and mocked observing this system eight times on random phases of a circular orbit. We show four such "observed" spectra in the middle panel. In the right panel, the dashed lines give the initial guesses $\boldsymbol{x}_0^A$ and $\boldsymbol{x}_0^B = 1$, and the full lines are the spectra obtained at each iteration, colored progressively heavier for larger i. Notice how the initial "blocky" wings for the red spectrum steadily approach the true shape of the Gaussian as the signal of the box-like line is diluted. This disentangling run took 18 iterations to reach a maximal relative change in the spectra of only 1×10^{-3}.

1.3.2.3 Applications

The technique of spectral disentangling can be exploited to uncover previously hidden companions. Consider the following. We have a spectroscopic binary in which we see one component moving with a well constrained orbit. Unfortunately, we do not see doubly-split lines as in Fig. 1.13, meaning there is no obvious companion around the visible star (this is thus an SB1 system). However, if we hypothesize that there is a faint, secondary source, the only unconstrained parameter is its RV semi-amplitude, K_2. Because spectral disentangling sums up the spectra taken at various moments in time, increasing the signal-to-noise ratio (S/N), even faint sources can become apparent when choosing the correct value for K_2. Applying this strategy to the system called LB-1, Shenar et al. (2020) found that, instead of the companion being a truly dark, 70 $M_\odot$ black hole, it was a rapidly

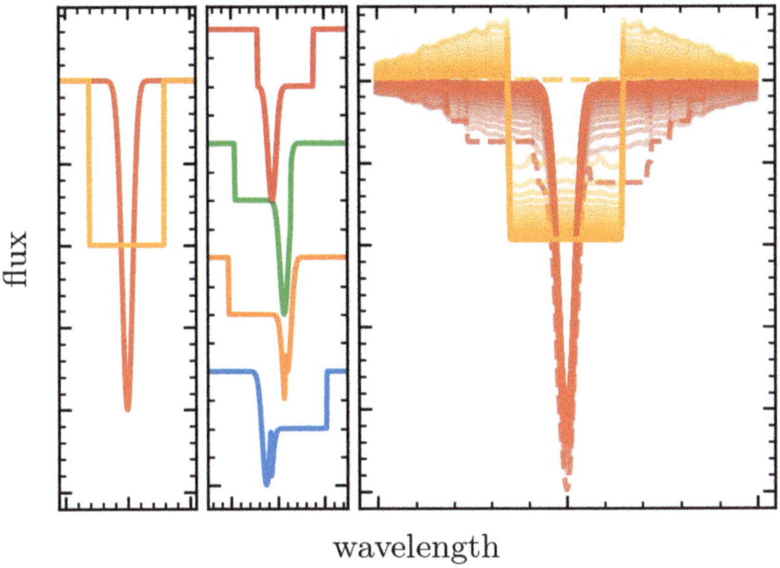

Fig. 1.20 Example of shift-and-add disentangling on two toy spectral lines. The synthetic spectra are given in the left panel, with four mock observations shown in the middle panel. In the right panel, the disentangling iterations are shown. The initial guesses for are given by he dashed lines, and the subsequent iterations are indicated with progressively heavier, full lines. The units on the axes are arbitrary in this toy model

rotating B-type star of only around 8 $M_{\odot}$. A similar strategy has been used by Bodensteiner et al. (2020) to disprove the existence of a black hole in the system HR 6819.

The reverse strategy can be used as well. If spectral disentangling reveals no obvious spectrum for a hypothesized companion at any value for K_2, then it is likely that indeed a dark object resides in the system. All other scenarios in which a dimmer star could elude detection must be eliminated though, such as a potential inner binary,[26] a binary-interaction product, or simply a low-mass main sequence star (see, e.g., Mahy et al. 2022).

1.3.3 Astrometric Binaries

With interferometric techniques, we can obtain relative astrometric measurements of binary stars, where we observe the position of one component with respect to the

[26] Two lower mass stars are dimmer than one star of the combined mass, because the exponent of the mass-luminosity relation is larger than one, see Eq. (1.28).

other. This is useful, because it allows to compute an astrometric orbit, as was done for visual binaries by Thiele and Innes (Thiele 1883; van den Bos 1926).

If we project the 3D, relative orbital ellipse of a binary star on the plane of the sky (instead of the line of sight, as for SBs), we can derive expressions specifying the relative eastward and northward displacement of one component to the other:

$$\Delta \mathrm{N} = AX + FY, \tag{1.59a}$$

$$\Delta \mathrm{E} = BX + GY. \tag{1.59b}$$

Here, X and Y are the so-called rectangular elliptical coordinates,

$$X = \cos E - e, \tag{1.60a}$$

$$Y = \sqrt{1 - e^2} \sin E, \tag{1.60b}$$

and A, B, F, and G are the Thiele-Innes constants, which depend on the orbital parameters e, i, ω, and Ω:

$$A = a_{\mathrm{app}}(\cos\Omega\cos\omega - \sin\Omega\sin\omega\cos i), \tag{1.61a}$$

$$B = a_{\mathrm{app}}(\sin\Omega\cos\omega + \cos\Omega\sin\omega\cos i), \tag{1.61b}$$

$$F = a_{\mathrm{app}}(-\cos\Omega\sin\omega - \sin\Omega\cos\omega\cos i), \tag{1.61c}$$

$$G = a_{\mathrm{app}}(-\sin\Omega\sin\omega + \cos\Omega\cos\omega\cos i), \tag{1.61d}$$

The apparent semi-major axis, a_{app}, is an observed quantity (it is the angle subtended by the two objects), and, if the distance, d, to the system is known, allows for the computation of the true semi-major axis, $a = a_{\mathrm{app}}d$.

We can now do a similar exercise as for SBs, by fitting Eqs. (1.59) to measured east- and north-ward displacements on the sky. In Chap. 2, we show Fig. 2.1, which is the result of the astrometric fitting of the binary star 9 Sagittarii. This fit provides a firm constraint on the inclination of the orbit, $i = 86.5 \pm 0.5°$ (almost edge on!), which is instrumental in finding accurate dynamical masses by combining it with a spectroscopic analysis that gives only RV semi-amplitudes. The only quantities that a combined astrometric-SB2 system cannot provide are the radii of the stars. To obtain these, the system needs to be eclipsing.

1.3.4 Eclipsing Binaries

In an eclipsing binary, orbital motion is revealed through the periodic dimming and brightening of the system. To be eclipsing, such systems must have an inclination

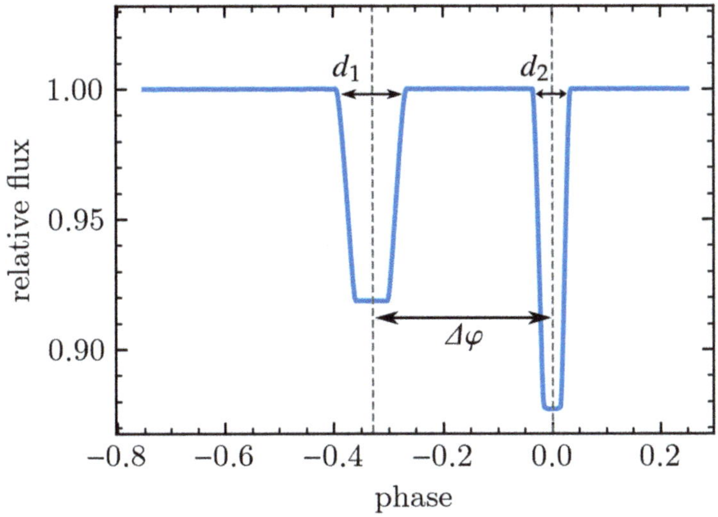

Fig. 1.21 Synthetic light curve of a binary system with $e = 0.4$ and $\omega = 50°$. Note that the durations of the eclipses are unequal, and that the secondary eclipse does not occur at $\varphi = -0.5$

angle close to 90°, i.e., viewed edge on. To see at least one (partial) eclipse, mathematically we require:

$$\sin(90° - i) \leqslant \frac{R_1 + R_2}{a}, \tag{1.62}$$

with $R_{1,2}$ the radii of the two stars of the system. Stars that orbit very close to one another, or those that are physically large, thus have a higher probability to be eclipsing.

A light curve model of an eclipsing binary system is shown in Fig. 1.21. From such a light curve (obtained for real binary stars through time-series photometry), we can determine various properties of its orbit. For example, from the duration and separation of the eclipses, the eccentricity and the argument of periastron can be calculated. Kopal (1959) computed that

$$e \sin\omega = \frac{d_1 - d_2}{d_1 + d_2}, \tag{1.63}$$

with d_1 and d_2 the duration of the eclipses. The phase difference $\Delta\varphi$ of the mid-eclipses are related to the tangential part of the eccentricity ($e\cos\omega$) by:

$$2\pi\,\Delta\varphi = X - \sin X, \tag{1.64}$$

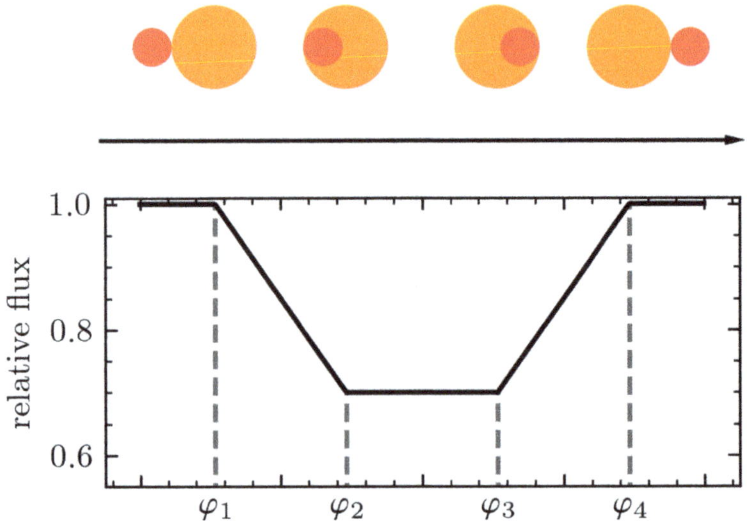

Fig. 1.22 Geometry of a simple eclipse. Time runs from left to right (with arbitrary units). As the secondary ingresses at phase φ_1, the flux received by an observer reduces, with totality occurring between φ_2 and φ_3

with X the auxiliary variable

$$X = \pi + 2\tan^{-1}\left(\frac{e\cos\omega}{\sqrt{1-e^2}}\right). \tag{1.65}$$

Another aspect unique to eclipsing binaries is that they allow for the estimation of some stellar parameters. First, the ratio of the stellar radii with the semi-major axis can be calculated. In the simple case of total eclipses in a circular orbit with $i \approx 90°$, and where the radii of the stars, $R_{1,2}$, are much smaller than the semimajor axis of the relative orbit, a, the timing of ingress, totality and egress connect the radii to the semi-major axis. Consider the secondary star starting to eclipse the primary, see Fig. 1.22. From the moment of ingress (with phase $\varphi_1 = t_1/P$) to the moment of first totality (at φ_2), the secondary has traveled $2\pi a(\varphi_2 - \varphi_1)$ in distance, which is also equal to twice its radius:

$$2\pi a(\varphi_2 - \varphi_1) = 2R_2. \tag{1.66}$$

Equivalently, the duration of totality determines the size of the eclipsed body, the primary:

$$2\pi a(\varphi_3 - \varphi_2) = 2R_1, \tag{1.67}$$

where φ_3 corresponds to the moment at which totality stops. Next, with the values of $R_{1,2}/a$, the temperature ratio can be computed from the ratio of the depths of the eclipses. Assuming both stars are black bodies (and we ignore any effects of limb- and gravity darkening), the flux each star emits is $\mathcal{F} = \sigma T_{\text{eff}}^4$. Without loss of generality, if we have that $R_1 > R_2$, the ratio of the fluxes at total eclipses is:

$$\frac{\mathcal{F}_1}{\mathcal{F}_2} = \left(\frac{R_2}{R_1}\right)^2 \left(\frac{T_{\text{eff},2}}{T_{\text{eff},1}}\right)^4 + 1 - \left(\frac{R_2}{R_1}\right)^2, \tag{1.68}$$

which can be inverted to find $T_{\text{eff},2}/T_{\text{eff},1}$.

Importantly, from a light curve alone, the masses of the stars are undetermined (not even the mass ratio[27]), and one must resort to evolutionary models. However, if an eclipsing binary is also a spectroscopic SB2 system (SB2-EBs), a wealth of information about the stars can be obtained (see, e.g., Southworth 2020). Because the eclipse modeling constrains the inclination, the minimum masses of Eq. (1.47) turn into true masses. Next, the semi-major axis can be computed from Eq. (1.46), which then determines the radii of the stars in units of length through Eqs. (1.66) and (1.67). SB2-EBs provide the only way to determine the masses and radii in a model-independent way.[28] If one then models the spectra to obtain absolute effective temperatures (not only their ratio), the luminosities are calculated via $L = 4\pi R^2 \sigma T_{\text{eff}}^4$, and the distance to the binary system can be computed by comparing the absolute and apparent magnitudes. Hence SB2-EBs can be used as distance indicators (so-called standard candles), for example to the LMC (Pietrzyński et al. 2019).

The equations we derived above are unfortunately only valid under specific conditions. For example, Eqs. (1.63) and (1.64), are only valid for $i \approx 90°$. In general then, more involved numerical techniques must be used to model the entire light curve numerically, rather than fitting only the timings of the eclipses. Wilson and Devinney (1971) first described procedurally how to compute light curves by modeling both stars as collections of small surface elements, and summing up their light contributions in the direction of the observer. They also provided methods to compute the effects of reflection, limb and gravity darkening, and they could also allow for the distorted shape of close binary stars. Today, PHOEBE (Prša and Zwitter 2005; Prša et al. 2016; Horvat et al. 2018; Jones et al. 2020; Conroy et al. 2020) is the most widely used light-curve modeling software tool. It is based on the Wilson-Devinney methods, but implements sophisticated atmospheric, geometrical, and even relativistic effects that greatly improve the fidelity of the

---

[27] There is one exception to this, namely the case of totally eclipsing contact binaries. In this scenario, the radius relation of the Roche geometry constrains the mass ratio.

[28] With model independent, here we mean independent of evolutionary or spectroscopic models. Light-curve fitting still involves models for physical effects, such as limb-darkening, rotational or tidal distortion, and irradiation, to name only a few. However, generally, light-curve models suffer from much smaller uncertainties than evolutionary or spectroscopic ones.

models. It includes different stellar atmosphere prescriptions, gravity- and limb-darkening laws, and photometric passbands, among others.

1.4 Contact Binaries

When a star grows to fill its RL, it will start a phase of mass transfer to its companion in a semi-detached configuration. However, it is possible that both stars *simultaneously* want to overfill their respective RLs. In this case, both stars will grow to have their surface layers on the same equipotential somewhere between L_1 and L_2, see Fig. 1.23, and a *contact binary* forms. Their shape merits the colloquialism "peanut stars."

In the literature, both the terms *overcontact* and *contact* are used for the same configuration. Wilson (2001) summarizes the history that gave rise to the two terms, and he argues that "contact binaries" are these systems where both stars are exactly filling their RL, i.e., $R = R_{\mathrm{RL}}$, while "overcontact" systems have $R > R_{\mathrm{RL}}$. He even argues that in an Algol system where the accretor is spun up to critical velocity, we should call those systems "double contact" (because both stars are in contact with their limiting surfaces). While these terms might make logical sense to describe the morphology of binary systems, they do so at the price of obfuscating the physical processes that happen in binaries. In this thesis, we therefore stick to the three-term set of binary configurations. Detached systems are those that do not engage in strong mass transfer (at most through wind accretion), and have $R < R_{\mathrm{RL}}$ for both stars. Semi-detached systems are those that undergo mass transfer through RLOF, where one star has $R = R_{\mathrm{RL}}$ and $R < R_{\mathrm{RL}}$ for the other (these include the mass-transfer systems that might potentially be "double contact"). Finally, contact systems are those systems where both stare are in physical and thermal contact with each other, and can thus exchange both mass and energy, and both stars have $R \geqslant R_{\mathrm{RL}}$.

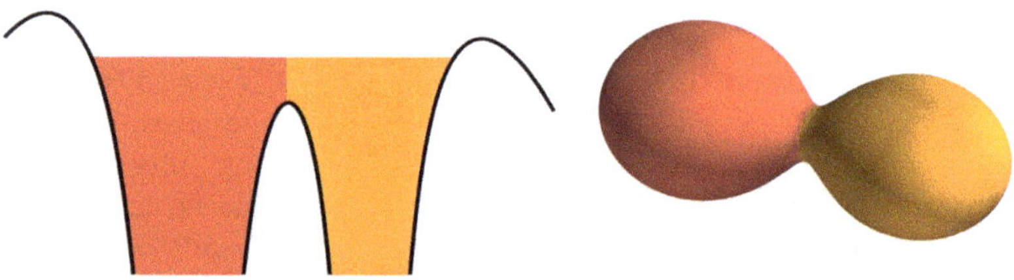

Fig. 1.23 Contact configuration in the Roche potential. Both stars are overflowing their respective RLs, and share common layers above L_1. The surface layer of the contact binary is a Roche equipotential

1.4.1 Observed Systems

The characterization of massive binaries is not an easy task. It requires using several observational techniques to accurately determine the orbital and stellar parameters. In Sect. 1.3, we introduced the three main techniques with which one can get orbital information. In particular, a combined spectroscopic and photometric analysis can fully determine a system, including the radii of the stars. For contact binaries, this is the most widely used approach (e.g., Lorenzo et al. 2014; Abdul-Masih et al. 2021; Kobulnicky et al. 2022).

One of the best studied massive contact binaries is VFTS 352, in the Tarantula star-forming region of the LMC. By simultaneously fitting RV and light curve data (Fig. 1.24), Almeida et al. (2015) found it is a $28.6 \pm 0.3\,M_{\odot} + 28.8 \pm 0.3\,M_{\odot}$ binary, i.e., a nearly equal-mass system. They also found both stars are overluminous and hotter than what single-star theory would predict. Abdul-Masih et al. (2021) characterized two more contact binaries, and similarly found overluminous components in both of them. A potential source of this discrepancy is that 1D, spherical atmosphere models were fitted to (clearly) non-spherical, highly distorted stars (cf., Fig. 1.23). Abdul-Masih et al. (2020) developed a spectroscopic patch model that more accurately fits the spectra than the 1D models. This method assigns a "local" spectrum to patches across the surface, thus taking into account temperature

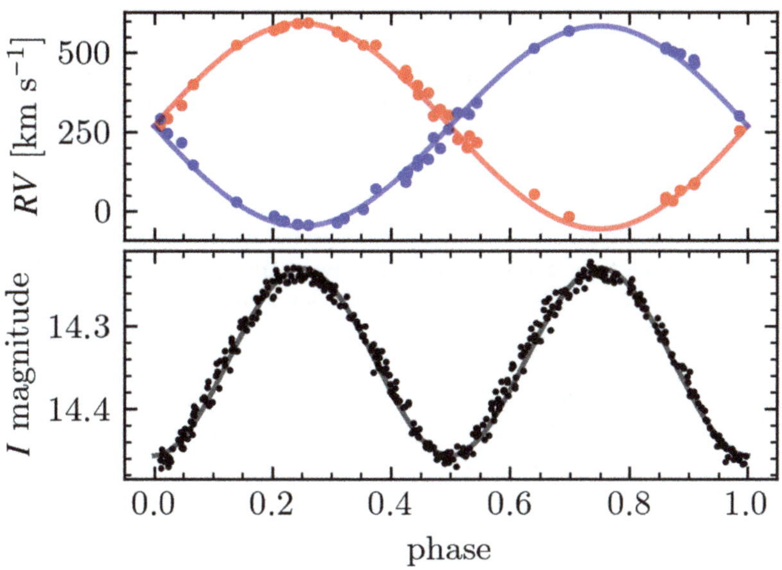

Fig. 1.24 RV and light curve of VFTS 352, a massive contact binary in the LMC. The primary RV curve is in red, while that of the secondary in blue, and the light curve of the whole system is in black. The dots are measured values while the solid lines are models to the respective data. *RV values are taken from Almeida et al. (2017), while the photometry is from the OGLE IV catalog (Wrona, M., priv. comm.)*

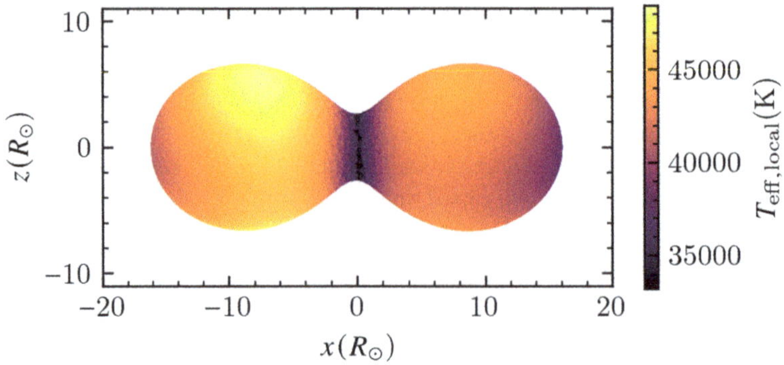

Fig. 1.25 Surface model of VFTS 352. The color indicates the local effective temperature, which varies across the surface because of the loss of a spherically symmetric gravitational field. *After Abdul-Masih et al. (2021). © ESO, adapted with permission*

variations due to tidal distortion (see Fig. 1.25), and integrates the contributions of all patches to come to a composite spectrum.

Usually, contact-binary candidates are first identified through their photometric light curves obtained from stellar variability surveys like OGLE, ASAS, MACHO, *Kepler*, *Gaia*, and *TESS* (Szymanski et al. 2001; Paczynski et al. 2006; Rucinski et al. 2007; Prša et al. 2011a; Ricker et al. 2015; Gaia Collaboration et al. 2016, 2023). Such surveys provide hundreds of thousands of candidates. In photometry, contact binaries are characterized by smoothly varying light curves, which is due the system being in a constant (or near-constant) state of partial eclipse. Because low-mass stars are more numerous, usually low-mass contact binaries are found, but recently, Menon et al. (2024) searched through the whole OGLE database, and found 61 contact-binary candidates with orbital periods that are commensurate with containing massive-star components. This is a significant increase in the total number of observed systems (increasing from around two dozen).

However, from a light-curve alone, it is quite challenging to confidently say an observed contact binary star is indeed experiencing a contact phase, or whether it is nearly so in a (semi-)detached configuration. Furthermore, the light curve by itself does not constrain the mass ratio (except in the special case where the contact binary is totally eclipsing, see, e.g., Poro et al. 2025). To confirm the status of the binary, spectroscopic measurements need to be taken to determine the RV curve, which has its own challenges alluded to in Sect. 1.3.1. Even with a large, homogeneous dataset of spectra, Mahy et al. (2020a) could only confirm the contact configuration of VFTS 352. Three other candidates (out of a sample of thirteen targets) have measured parameters whose uncertainty ranges make them consistent with a detached configuration.

Despite the observational challenges, over the past decade, the following picture of observed, massive contact binaries emerged. From spectroscopic analysis, it is found that a good fraction of contact binaries are observed with unequal-mass components, meaning their mass ratio, $q = M_2/M_1$, is significantly away

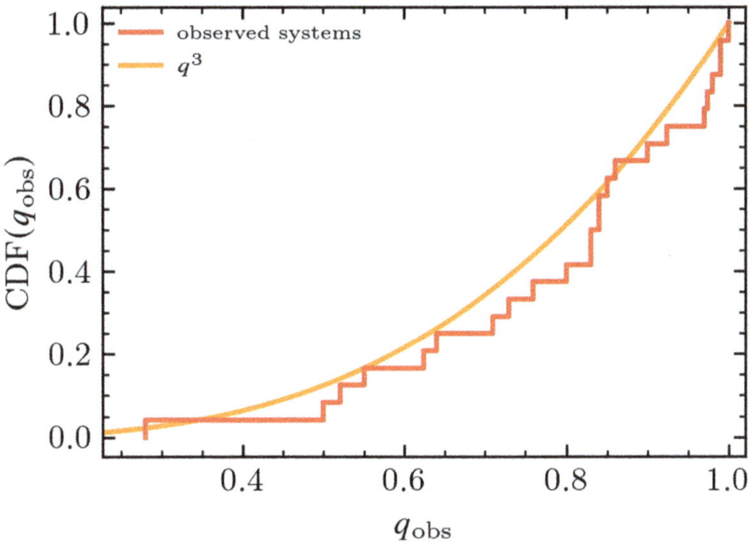

Fig. 1.26 Cumulative probability distribution function of the observed sample of massive (near-)contact binaries (see Tables D.1 and D.2). It follows nearly a q^3 distribution. About a half of the systems are observed at mass ratios more extreme than $q = 0.85$

from unity. In Fig. 1.26, we plot the cumulative distribution of observed, massive (near-)contact binaries (Tables D.1 and D.2), and deduce, under the assumption this is a representative sample, that the probability of finding contact binaries with mass ratios less than, say, $q = 0.85$, is around 50%. Furthermore, from studying the evolution of the orbital period over the past thirty years, Abdul-Masih et al. (2022) and Vrancken et al. (2024) determined that the period derivative in most of these contact binaries is rather low, on the order of the nuclear timescale, meaning their mass-ratio evolution also occurs on the nuclear timescale. It thus seems that contact binaries with unequal-mass components are stable in that configuration. This is an important observational constraint that models of contact binaries would ideally want to reproduce.

1.4.2 Modeling Efforts

Modeling of contact binaries got well underway in the late 1960s, thanks to growing surveys of variable stars. Many of these variable stars turned out to be partially eclipsing, low-mass contact binaries, called W UMa stars (Eggen 1967; Binnendijk 1970). The surveys revealed many systems contained unequal-mass components. Theoretically, this was not expected, as Kuiper (1941) identified that the mass-radius relation of single stars in equilibrium, combined with the geometric constraint from the Roche geometry, did not allow for contact binaries with mass ratios different from unity. This discrepancy between theory and observation is called Kuiper's paradox. To resolve it, Lucy (1967b) first considered the process of energy transfer in the common layers of the contact binary. Biermann and Thomas (1972), Vilhu (1973), Flannery (1976), Shu et al. (1976) and Webbink (1977) also worked on advancing the models of W UMa stars. Unfortunately, all of these models contained inconsistencies, or were not applicable to the massive contact binaries. For example, Hazlehurst (1993) showed that the model of Shu et al. (1976) violated the second law of thermodynamics, while the energy-transfer model of Lucy (1967b) is not applicable to the radiative envelopes of massive stars.

Since the initial efforts to model W UMa stars, little progress has been made at obtaining a complete model of contact binary stars, especially the massive ones. As a result, much of the theoretical work from fifty years ago is ignored in recent modeling. Now that it has become clear that massive stars undergo contact phases frequently (see below), many studies focus on close binary evolution with specific considerations of the contact phase, e.g. Marchant et al. (2016), Menon et al. (2021), Sen et al. (2022), and Henneco et al. (2024b). However, they treat contact phases in a rudimentary way only. Mass transfer is computed consistently, by requiring that the mass-transfer rate is adjusted so that the surfaces of both stars lie on a common Roche equipotential. Unfortunately, other effects are ignored in all models and calculations. For example, the spherically symmetric stellar-structure equations, Eqs. (1.5), or their equivalents for single rotating stars, are solved, so that any effect of tidal deformation cannot be taken into account. Furthermore, Hastings et al. (2020) calculated that internal mixing currents in tidally deformed stars are stronger than the original prescription of Eddington (1925) and Sweet (1950) by about a factor of two. Tidally distorted stars could thus live longer and have larger cores as they mix more fresh fuel into their stellar cores. Finally, the process of energy transfer, which was heavily discussed for W UMa stars in the seventies, and taken up again by a few papers in the 2000s (e.g., Kähler 2004; Yakut and Eggleton 2005), has thus far never been considered in massive binary evolution.

1.4.3 Evolutionary Pathways—Mergers

Contact binaries represent a unique, but not-uncommon phase in binary evolution. Recent calculations estimate that at least 40% of all massive stars undergo at least one contact phase throughout their lives (Henneco et al. 2024b). Generally speaking, there are two ways a contact configuration might form two with main-sequence components (see Henneco et al. 2024b, for a complete view of contact phases in all stars). First, when the accretor in a semi-detached configuration swells as a consequence of it gaining mass, it can overflow its Roche lobe too and engage in contact with its donor (this was dubbed case AR in Nelson and Eggleton 2001). This happens in vigorous, thermal-timescale mass-transfer phases. Second, in cases of slow mass transfer, the accretor, which is becoming ever more massive, can "overtake" the evolution of the donor (cases AS/AE of Nelson and Eggleton 2001).

In this case, the accretor, growing on its nuclear timescale, enters contact with the donor, and the mass-transfer direction will soon reverse. Both these scenarios fall under the "accretor expansion" outcome of Henneco et al. (2024b).

Around half of all case A binary systems that enter a contact phase are expected to merge (Henneco et al. 2024b), and this number rises to essentially all systems if one considers nuclear-timescale contact phases only (Menon et al. 2021). The merger process of two stars is a violent and dynamical occurrence, resulting in matter being flung outward, and a temporary brightening of the system (e.g., Pejcha et al. 2016). However, because the probability of observing a phenomenon is proportional to its duration, stellar mergers are hard to catch. They happen on the dynamical timescale of the system, which is approximately

$$\tau_{\rm dyn} \approx \sqrt{\frac{R^3}{GM}}, \tag{1.69}$$

with R and M the typical radial extent and total mass of the system in question. Plugging in $R \approx 30\,R_\odot$ and $M \approx 50\,M_\odot$ as a rough estimate for the total mass and radial extent of VFTS 352, we get $\tau_{\rm merge} \approx 1 \times 10^{-3}$ yr ≈ 1 d. Stellar mergers thus occur on the timescale of days, way faster than the thermal or nuclear timescale of stellar evolution. It should be noted, however, that the aftermath of the merger event (i.e., the brightening and illumination of the ejecta) can stick around for much longer. Despite being short-lived events, thanks to long-term monitoring programs and transient surveys, which use dedicated telescopes that scan the whole sky every few days, we have caught several events called Luminous Red Novae (LRNe; see Kasliwal et al. 2017; Pastorello et al. 2019). Such events are dimmer than supernovae, but more luminous than classical novae occurring in binary systems with white dwarfs.[29]

The eruptions of V838 Monocerotis and V1309 Scorpii are the canonical examples of LRNe. Both were rather unremarkable stars that quite suddenly brightened by several magnitudes. V838 Mon erupted in January 2002, and was first noticed by Brown et al. (2002) as a "peculiar variable." In the following years, various models for the event were put forward, like an expanding red giant swallowing its close planets (Retter and Marom 2003), or the merger of a low-mass star with an intermediate-mass one (Soker and Tylenda 2003). Because the distance to this source was adjusted by analysis of the light echo (i.e., the delayed illumination of the surrounding gas and dust; Tylenda 2004), Tylenda et al. (2005) concluded that the progenitor had a massive, $\sim 8\,M_\odot$ star as the primary. By comparing the various proposed models, Tylenda and Soker (2006) conclude a merger of a $8\,M_\odot$ star with a $0.4\,M_\odot$ pre-main-sequence star is the most promising (although it is unsure whether

[29] Classical novae erupt as an accreting white dwarf accumulates hydrogen from a companion. The atmosphere heats as it gets denser, and suddenly it reaches the threshold for nuclear burning to occur. This is a runaway process and consumes all the hydrogen in the atmosphere in a short time span. After the nova dims, the cycle can repeat if mass transfer continues.

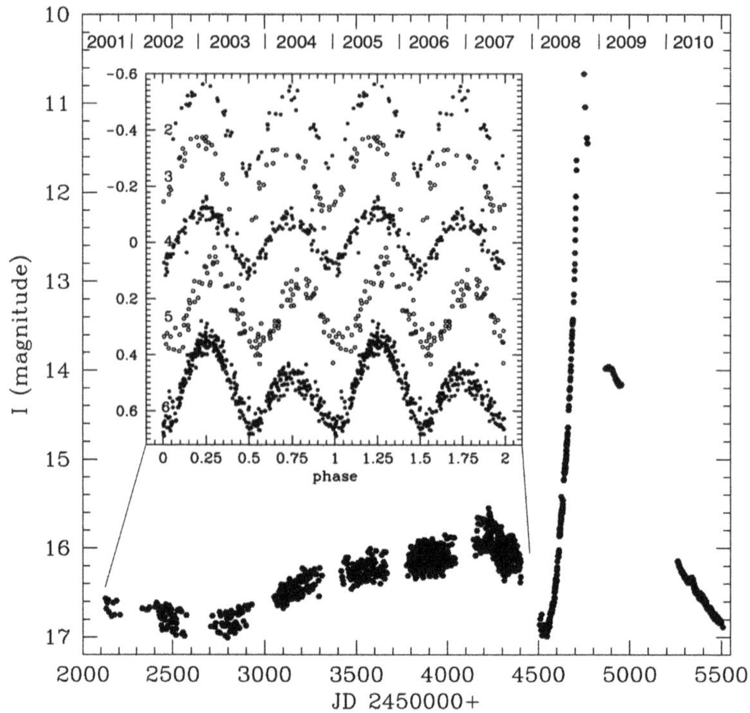

Fig. 1.27 Light curve of V1309 Sco. Both the outburst and the pre-merger phases were measured for this system over around ten years. The inset axes show the contact-binary light curve evolving top to bottom over the successive years. *Adapted from Tylenda et al. (2011). © ESO, adapted with permission*

a contact binary formed). Apart from its similarity to V1309 Sco, the precise nature of the eruption of V838 Mon has not been uncovered. More recently, mass loss from the outer Lagrangian point is also considered as the origin for the eruption (Pejcha et al. 2016).

V1309 Sco is perhaps even better studied than V838 Mon, because there is well-resolved OGLE data of the progenitor, see Fig. 1.27. Its outburst in 2008 was discovered by Nakano et al. (2008), after which Mason et al. (2010) established that it was not a classical nova. Tylenda et al. (2011) analyzed both the outburst and the progenitor data, and conclusively showed that this object was indeed a contact binary that experienced a dynamical merger event. The OGLE data revealed that, in the years preceding the merger, the orbital period was decreasing exponentially, which hints at a runaway, self-strengthening process. During the outburst, the system brightened by a total of about seven magnitudes in around six months, concluding with a sharp, ten-day rise of three magnitudes. In the two years following the outburst, the system gradually dimmed to its progenitor levels.

Soon after the merger scenario was confirmed, Stępień (2011) computed evolutionary models for the progenitor, and estimated initial masses of $M_1 \approx 1.2\,M_\odot$ and $M_2 \approx 0.6\,M_\odot$ for the components, respectively. These stars first reverse their mass ratio by case A mass transfer, after which the second phase of mass transfer becomes unstable, resulting in the observed inspiral and merger. Pejcha (2014) focused on the exponential rise of the outburst, and connected it to rapid mass loss from the outer Lagrangian point L_2. Several more LRNe have been detected and connected to mergers or common-envelope events (Smith et al. 2016; Pastorello et al. 2019; Blagorodnova et al. 2017, 2020, 2021). In common-envelope evolution, like contact binaries, both components share their outer layers, but a crucial difference is that there is loss of co-rotation (Ivanova et al. 2013; Röpke and De Marco 2023). While a contact binary can be described as a (quasi-)hydrostatic object, a common-envelope phase is always a hydrodynamical affair, where the core of a star, or a compact object,[30] is moving through the shared envelope of the system, and it experiences drag as it plows its path through the gas. Furthermore, in common-envelope events, the Roche potential is no longer a suitable approximation of the geometry of the system.

The result of a merger, once it has thermally relaxed, is often a rather peculiar star. If it is part of a stellar cluster, it will appear younger than its fellow cluster members. It will also appear hotter ("bluer") on the HRD, giving these stars the name *blue stragglers*, first identified in the globular cluster M3 by Sandage (1953). While there is agreement that stellar mergers contribute to the number of blue stragglers (Hills and Day 1976), the precise amount is unclear (Leonard 1989). Other possibilities are that they are stars with particularly well mixed material (see Sect. 1.1.5; Wheeler 1979), that they have accreted a significant amount of mass from their dense neighborhood, or that they are less extreme binary-interaction products (like Algol, see Sect. 1.2.2; McCrea 1964; Schneider et al. 2013; de Mink et al. 2014). Further work is required to distinguish between these channels.

Finally, mergers of contact binaries have been put forward as a potential source of magnetic fields in massive stars, found in around 8–10% of massive stars (Wade et al. 2016; Grunhut et al. 2017; Shultz et al. 2018). Because they have radiative envelopes, the magnetic dynamo that occurs in the turbulent convective envelopes of low-mass stars, such as our Sun (Parker 1955), cannot power large-scale magnetic fields of massive stars. Instead, suggested origins include the remnant magnetic field of the molecular cloud the star was born in, but it is rather unlikely that such a field survives the star-formation process (Moss 2003; Schleicher et al. 2023). However, it was suggested that mergers can generate strong differential rotation needed to generate large-scale magnetic fields in radiative stars (e.g., Ferrario et al. 2009; Wickramasinghe et al. 2014). To test this hypothesis, Schneider et al. (2019) simulated the merger of two massive stars with a 3D magneto-hydrodynamical code—starting from a contact-binary configuration—and found that an initially

[30] A compact object is an umbrella term that includes black holes, neutron stars, and white dwarfs. They are much denser than regular, "non-compact" stars that have no degenerate matter.

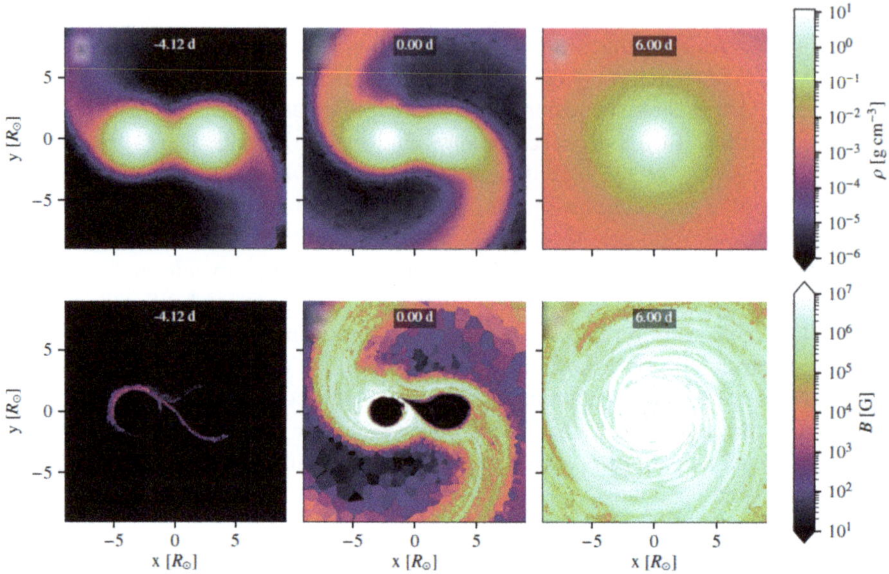

Fig. 1.28 Merger simulation of an 8 $M_\odot$ and a 9 $M_\odot$ star. The top row shows the density, while the bottom row the magnetic field strength, B. The bottom panels show the extreme amplification of the magnetic field due to turbulence as the right-hand star is torn apart during the merger. *Adapted from Schneider et al. (2019). © Schneider et al., adapted with permission*

tiny magnetic field of only 1 G is amplified to 10^7 G![31] Figure 1.28 shows the density and magnetic field strength at three stages of the merger. The bottom rows show that, as one of the stars gets torn apart by tidal forces, turbulence from the so-called magneto-rotational instability (Chandrasekhar 1960; Balbus and Hawley 1991) results in the extreme amplifications of the magnetic field. Recently, the analysis of the binary system HD148937 revealed that only one star of the two is magnetic (Frost et al. 2024). The magnetic star also appears younger than its companion, leading to the conclusion that a merger occurred in what was originally a triple system. This is the first observational evidence that mergers of binary stars can create strong magnetic fields.

1.5 Gravitational Wave Astronomy

Up until September 13, 2015, any and all information about the deep universe came to us observers on Earth in the form of electromagnetic waves (i.e., photons)

[31] In astrophysics, the unit "gauss" (G) is used for magnetic field strength instead of tesla (T). 1 G is equivalent to 10^{-4} T in SI units.

and neutrinos. Photons are the force carriers of electromagnetism, while neutrinos are light particles generated by the weak nuclear force. Throughout history, most observations have been made with photons, the eye being an excellent detector of visible photons (albeit poorly calibrated on an absolute scale). Today, we have several sensor technologies (e.g., charged-coupled devices, CCDs, or complementary metal-oxide semiconductors, CMOS), used in cellphone cameras as well as in scientific instrumentation, that efficiently convert incoming photons into digital information.

Neutrinos have escaped our view until 1956 when Clyde Cowan and his collaborators detected neutrinos coming from a nuclear fission power plant on the Savannah river in South Carolina, USA (Cowan et al. 1956). Neutrinos are elusive particles, interacting very weakly with ordinary matter, and therefore very large detectors with lots of water need to be constructed. Super-Kamiokande is located in an old Japanese mine and has a tank holding more than 50,000 tons of purified water (Fukuda et al. 2003), while IceCube (Abbasi et al. 2009) uses the Antarctic ice sheet itself as the detecting medium. With these massive detectors, we have been able to detect neutrinos coming from astrophysical sources, including from the nuclear reactions of the Sun (cf. Eq. (1.11); Fukuda et al. 2001, and see Davis et al. 1968 for the first solar neutrino detection) and from supernova 1987A (Hirata et al. 1987; Bionta et al. 1987). There are also tentative detections from active galactic nuclei, tidal disruption events and γ-ray bursts (see, e.g., Kurahashi et al. 2022, for a recent review). A detection coming from a blazar (i.e., an active galactic nucleus with its jet pointed at Earth) is confirmed as it coincides with a flare of high-energy γ-rays (IceCube Collaboration et al. 2018).

On September 14, 2015, however, GW150914 rolled by the Earth, a gravitational wave coming from the merger of two black holes. At that time, the Laser Interferometer Gravitational-wave Observatory (LIGO) was performing its inaugural observation run, and both detectors, one in Louisiana and the other in Washington state, USA, captured the gravitational wave passing by (Abbott et al. 2016). The four-kilometer-long arms of the detectors stretched and squeezed by just one thousandth the diameter of a proton, and did so in an oscillating pattern that increased in frequency, see Fig. 1.29. This monumental detection provided further experimental verification of Einstein's theory of general relativity, who himself predicted the existence of gravitational waves (Einstein 1918). Up to that point, gravitational waves were only indirectly observed by the shrinking of the orbit of binary systems with a pulsar, like the Hulse-Taylor pulsar (Hulse and Taylor 1975). However, an (arguably) more important consequence of the discovery of gravitational waves was that it also gave a direct view of astrophysical black holes. Black holes are, well, black, and can only been seen by light passing or being emitted near them, for example in accretion discs of, for example, X-ray binaries. Only very recently, the Event Horizon Telescope team were able to image the black hole shadow in the accretion disks of the supermassive black holes at the center of M87 and our own galaxy (Event Horizon Telescope Collaboration et al. 2019, 2022). A non-accreting black hole is not visible with photons, but now, gravitational

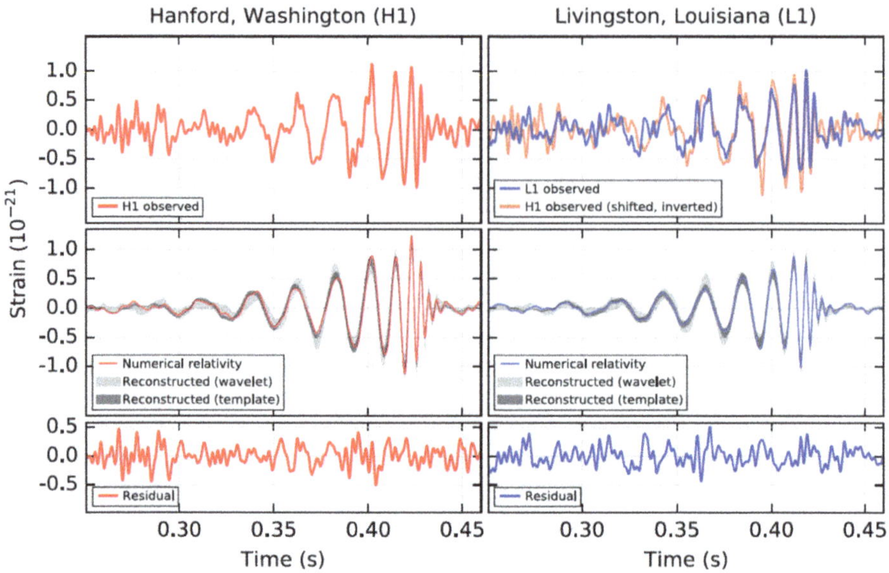

Fig. 1.29 Gravitational wave GW150914, observed on September 14, 2015 by LIGO. The horizontal axis gives time, while the vertical axis gives the strain in the detector arms, i.e. the fractional change of their 4 km long tubes. The top row gives the measured signal, the second row the prediction from general relativity, and the third row the difference between the two. *Adapted from Abbott et al. (2016), licensed under CC-BY 3.0*

waves give us another observational window to study them, heralding the era of gravitational-wave astronomy.

At the time of writing, there are about two hundred gravitational-wave events detected by the LIGO and VIRGO[32] detectors (LIGO Scientific Collaboration et al. 2023; LIGO Scientific Collaboration et al. 2025), an impressive number given we have observed for only a dozen years or so. KAGRA, the Japanese detector, has joined observational runs alongside LIGO and VIRGO, and they have completed their fourth run mid November 2025. As the detectors are constantly being improved, raising their sensitivity, the amount of detectable gravitational waves will increase dramatically. Current estimates place the amount of compact binary mergers observed with LIGO/VIRGO/KAGRA at their design sensitivity to be several hundreds per year (Abbott et al. 2018). With next-generation observatories however, like the American-led Cosmic Explorer (Reitze et al. 2019) and Europe's Einstein Telescope (Maggiore et al. 2020),[33] the full universe comes into view, raising the observed merger rate up to tens of thousands per year. In this scenario,

[32] Located near Pisa, Italy.

[33] There is a proposal being evaluated to build the Einstein Telescope partly in Belgium, underground in the region between Liège, Maastricht and Aachen. Ground layers there are particularly stable, which minimizes seismic noise.

astrophysicists will have large, statistical samples of compact-object properties that can be used to verify binary-evolution models, among many other scientific use-cases. This is possible as gravitational waves arriving on Earth carry information about the masses and spins of the merging objects. Furthermore, by carefully timing the arrival of the gravitational wave to different detectors around the world, we can triangulate from which direction the wave came from.

1.5.1 Gravitational-Wave Progenitors

Thanks to gravitational-wave astronomy, we get an unprecedented view of the end products of massive-star evolution: black holes and neutron stars. However, it is still uncertain how the gravitational-wave mergers that are observed actually form. There are several channels in which merging compact objects are proposed to form. First we have isolated binary evolution, where the stars interact in such a way so as to leave two compact objects in a close orbit. The second is cluster dynamics, where the black holes form separately, and later pair up through dynamical interactions in dense star clusters. The third option is that the black holes are cosmic in origin, being formed right after the big bang. For the latter two, and other channels we did not list, we refer to the review of Mandel and Broekgaarden (2022) for the current state of the art, while we will discuss the first channel in more detail here.

If two massive stars are born close enough together, and can stay close throughout their evolution and core-collapse supernovae, the compact objects left behind can spiral in on each other and merge. The time it takes two black holes to eventually merge under gravitational wave emission can then be calculated from (e.g., Landau and Lifshitz 1975):

$$\tau_{\rm merger}(r_{\rm init}) = \frac{5}{256}\frac{c^5}{G^3}\frac{r_{\rm init}^4}{M_1 M_2 (M_1 + M_2)}. \tag{1.70}$$

If we assume the maximal time allowed is the evolutionary timescale of the universe, also called a Hubble time, $\tau_{\rm H} = 14.5\,{\rm Gyr}$ (Planck Collaboration et al. 2020),[34] two black holes of $30\,M_\odot$ each can be no farther apart than $r_{\rm init} = 48\,R_\odot$, or, equivalently, have an orbital period larger than 4.7 d after the second black hole forms. Unless they evolve chemically homogeneously (see below), it is impossible for stars that start out in a binary and leave behind two black holes on a five-day orbit, not to interact during their lifetimes. In fact, many models predict that there have to be multiple phases of mass transfer throughout the evolution of the binary. Within the isolated-binary-evolution channel for gravitational-wave progenitors, four scenarios are being explored.

---

[34] This is close, but not equal, to the age of the universe, $\tau_{\rm uni} \approx 13.8\,{\rm Gyr}$, because the expansion of the universe is not constant.

1. **Chemically homogeneous evolution.** If a massive binary is born very close, with periods of around one day, the components rotate so quickly that it causes efficient rotational mixing (see however Sect. 1.1.5). This prevents the formation of a core-envelope boundary, so that, essentially, the whole star is uniformly converted to helium, which is called chemically homogeneous evolution (CHE, Maeder 1987). More importantly, it also prevents the stars from becoming giants, meaning there is enough room in a one-day orbit for these stars to evolve all the way to their core collapse. The second required ingredient in this channel is low metallicity, as this reduces the strength of the stellar winds of the stars. This is necessary to avoid significant mass loss, which would widen the orbit and may lead to the stars losing their homogeneous character. de Mink et al. (2009) first showed that in close binary evolution, the rotational velocities achieved by tidal spin-up are high enough to trigger CHE. Brott et al. (2011) modeled this type of evolution in low-metallicity, rapidly rotating single stars. Later, both Marchant et al. (2016) and Mandel and de Mink (2016) modeled CHE in binary stars, and found that the most massive binary-black-hole mergers could be dominated by progenitors from this channel. Figure 1.30 shows an example model of CHE, where two very massive stars both go supernova in a γ-ray burst (GRB), and merge within 1.7 Gyr. Additionally, during CHE, the formation of main-sequence contact binaries is the norm, because of their very close orbit. Therefore, modeling the physics of a contact configuration accurately plays an important role in predicting the properties of the merging black holes this channel produces.
2. **Common-envelope evolution.** If instead the initial period of the binary is moderate, or if the stars lose their chemically homogeneous character, the primary star grows and starts mass transfer to its companion, see Fig. 1.31. The donor star then collapses into a black hole as its nuclear fuel is expended. At this point in the evolution, the orbital period is still on the order of hundreds of days, meaning no gravitational wave merger is expected if the second star also became a black hole at this period. We thus need a mechanism that shrinks the orbit by around two orders of magnitude. As time moves on, the second star starts evolving significantly, causing it to fill its Roche lobe. If the mass ratio is low, $M_{\rm BH} \ll M_2$, the second phase of mass transfer is unstable, forming a common envelope (CE). In common-envelope evolution, the black hole spirals in through the envelope of the giant secondary, and experiences drag. It is then postulated that the gravitational energy liberated from the inspiral is converted to unbind the outer layers of the giant. In this process, the orbit of the black hole can shrink by orders of magnitude, enabling a gravitational-wave merger event to happen in a Hubble time (e.g., Tutukov and Yungelson 1993; Stevenson et al. 2017; Giacobbo and Mapelli 2018). While we understand the qualitative picture of a common-envelope inspiral, there are still many uncertainties in its quantitative evolution, in particular on the efficiency of energy conversion, and on the endpoint of the common-envelope phase (see, e.g., Ivanova et al. 2013, for a review). One of the possible outcomes is that the envelope is not ejected,

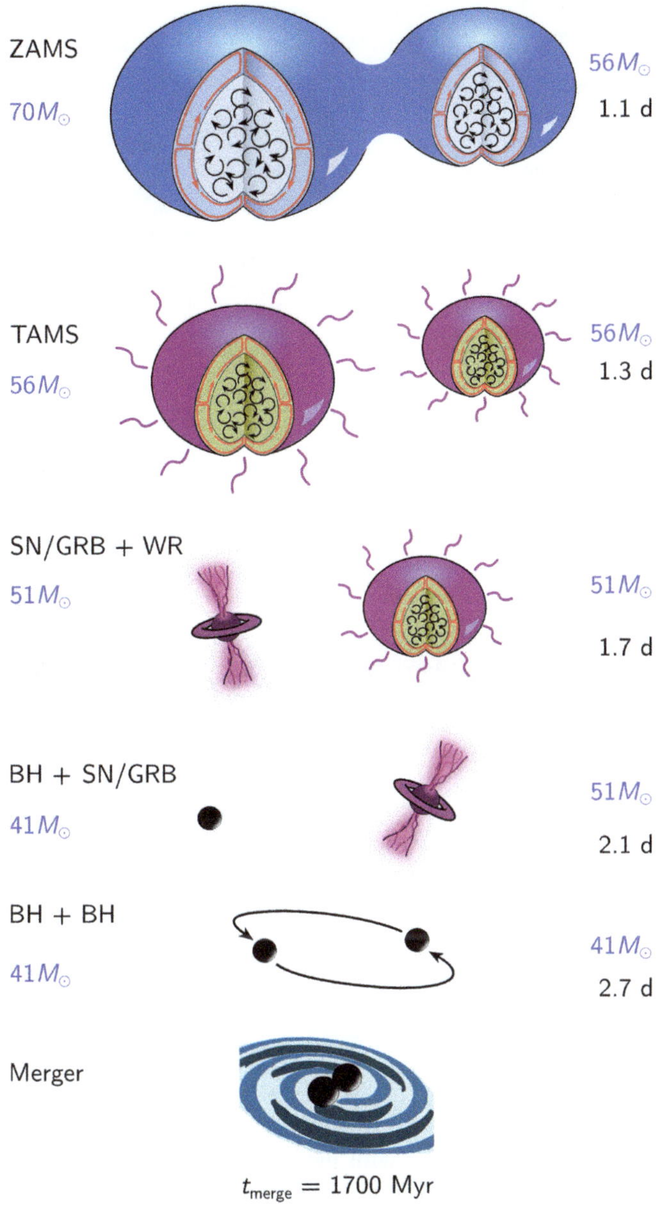

Fig. 1.30 Pathway of a binary-black-hole merger through the chemically-homogeneous-evolution channel. Masses are indicated in blue and the orbital period is in black on the right. Two very massive stars start in a close orbit, do not expand due to efficient mixing, and form black holes close enough to merge in a Hubble time. Acronyms are explained in the Preface. *After Marchant (2018), reproduced with permission*

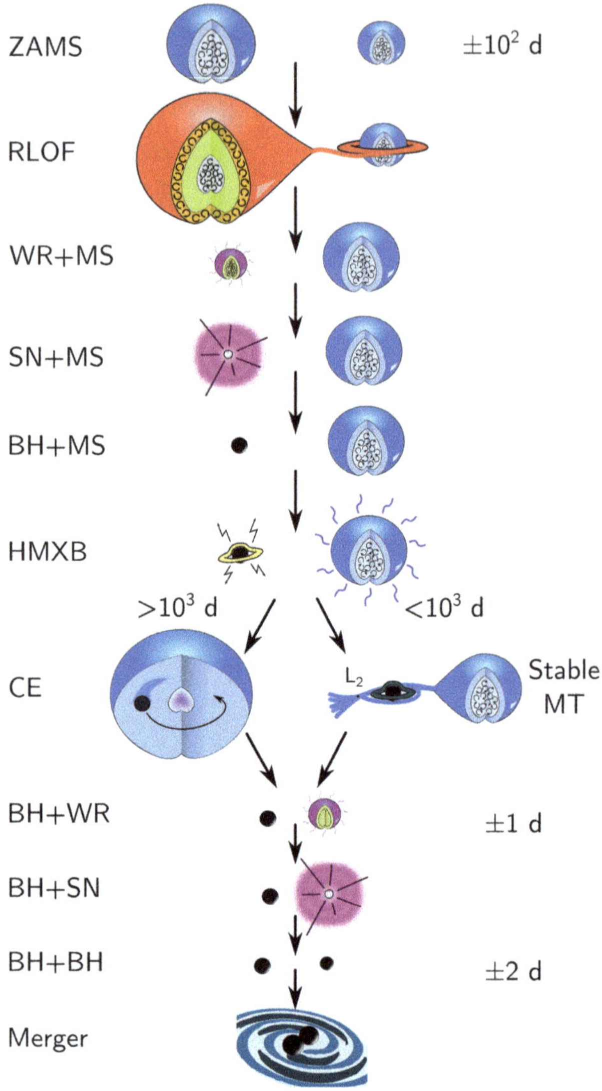

Fig. 1.31 Pathway of a binary-black-hole merger through the common-envelope and stable-mass-transfer channels. Two phases of interaction are required to produce merging black holes in this scenario. First a case A or B mass-transfer phase occurs, later followed by either a common-envelope phase or a stable mass-transfer phase. Representative periods are indicated at various phases in the evolution. Acronyms are explained in the Preface

and that the black hole merges with the core of the giant star, preempting any gravitational-wave merger happening from the system (e.g., Klencki et al. 2021).

3. **Stable mass transfer.** A sister of the common-envelope channel, this scenario requires that the second phase of mass transfer is stable. Under certain conditions, if mass lost from the donor is *not* accreted by the companion, the period of the binary can shrink without the mass-transfer rate increasing uncontrollably (van den Heuvel et al. 2017). In essence, the non-conservative mass transfer acts as an energy and an angular momentum sink, with which the binary can also shrink in period by several orders of magnitude. Characterizing exactly when

mass transfer is stable is crucial to understand what parameter space of binaries might go through this channel (Marchant et al. 2021; Klencki et al. 2021; Olejak et al. 2021; Gallegos-Garcia et al. 2021).

4. **Population III stars.** As early as 1926 by Jan Hendrik Oort, and later in 1944 by Walter Baade, stars were categorized in two classes depending on their measured velocity. Population I stars are the slowly moving, metal-rich stars of the galactic disk, while population II stars were faster moving, metal-poorer stars located in the galactic bulge and globular clusters in the halo. Since globular clusters are very old, population II stars are older than population I stars. Later, the third class, population III, was added for the stars that were formed very soon after the big bang (in the first billion years or so, Abel et al. 2002). Because they are the very first generation, and only hydrogen and helium was formed during the big bang (with trace amounts of lithium), population III stars are thought to be essentially metal-free (Bromm et al. 2002). Population III stars should be more compact at the same mass as population I stars, so they more easily avoid interaction during their lives. Kinugawa et al. (2014) modeled a large set of population III stars, and computed the compact-object merger rates from it. They find that, even though the merger rate from population III stars is lower than for population I stars, they are more easily observable because the average masses of the merging black holes are much higher for the population III stars. This conclusion however depends on cosmological parameters (i.e., the amount of dark matter) and the star-formation-rate history.

The community has not yet reached consensus on which of the channels produces the observed gravitational-wave mergers. Most probably, all of them will contribute to a certain degree, but uncertainties in the modeling prevent us from making precise predictions. While the increasing number of gravitational-wave detections can help constrain the physical processes of the progenitors (Mandel and Farmer 2022), efforts to improve the modeling itself, and understanding its associated physics, are the main way to progress in this growing field.

1.6 This Thesis

The main goal of this thesis is to advance on the theoretical modeling of massive contact binary stars. Given the frequency of contact phases, their pivotal role in stellar mergers and associated exotic phenomena, and occurrence in the CHE channel of gravitational-wave progenitors, it is timely to concentrate efforts on the accurate modeling of contact binaries.

However, we will first make a brief, but important, excursion into the observational characterization of binary stars. Most model-independent masses of (very) massive stars come from short-period eclipsing binaries. However, such measurements might be affected by additional physical effects, such as tidal deformation and unconstrained internal mixing. In this sense, young, long-period binaries, which are unaffected by close binary interaction, offer an excellent alternative to calibrate massive-star models with. In Chap. 2, we will analyze the long-period, massive binary star 9 Sagittarii (9 Sgr). The central question we want to solve is whether the spectroscopic analysis of Rauw et al. (2012) suffered from biases using line-profile fitting of the highly blended lines of 9 Sgr. We will use relative astrometry and spectral disentangling to determine all the orbital parameters, and in so doing calculate the dynamical masses of both stars. Afterwards, we will analyze the atmospheres of the stars, deriving their temperature and abundances of elements like carbon, nitrogen and oxygen, and compare the stellar parameters with evolutionary models.

In Chap. 3, we develop the methodology to accurately represent contact binaries in a one-dimensional stellar-evolution code. We will use the method of Kippenhahn and Thomas (1970) and Endal and Sofia (1976) to represent tidally distorted stars under the shellular approximation, and adapt the equations of stellar structure accordingly. We will then quantify how the surface properties of a stellar model change when going from a spherically symmetric geometry to a rotating or tidally deformed geometry.

Chapter 4 deals with the process of energy transfer in contact binaries. We will give a brief overview of the theory of energy transfer, after which we implement an energy-transfer scheme into our stellar-evolution models. Here we note the difference between the terms *energy transfer* and *energy transport*, the former occurring in contact binaries where energy moves from one component to the other, the latter being the process where stars move heat from their cores to their envelopes (Sect. 1.1.1). The effects on the shared layers of the stars is carefully studied, and differences in the evolutionary pathways of a binary model with and without energy transfer are identified.

In Chap. 5, we will compute a large grid of binary-evolution models, where, in one grid, we include the process of energy transfer, while in another, we exclude it. In this way, we will get a measure of the effects of energy transfer across the entire population of massive binary stars. In particular, we aim to answer whether including energy transfer in theoretical modeling can resolve the discrepancy on the mass-ratio distribution introduced in Sect. 1.4.2.

While the bulk of this thesis is focused on massive stars and binaries, Chap. 6 introduces the evolution of low-mass contact binaries. We discuss the effects of magnetic braking and present the first grid of contact-binary models in the low-mass regime.

Finally, Chap. 7 concludes this thesis with a summary and outlook.

Chapter 2
The Dynamical Mass of 9 Sagittarii

This chapter is mainly based on:

Resolving the dynamical mass tension of the massive binary 9 Sgr
M. Fabry, C. Hawcroft, A. J. Frost, L. Mahy, P. Marchant, J-B. Le Bouquin, H. Sana
Astronomy & Astrophysics, 651, A119 (2021)

2.1 Introduction

Massive stars drive the chemical enrichment of heavy elements in galaxies and inject large amounts of kinetic energy into their neighborhoods through their strong, line-driven winds and final explosions as supernovae and $gamma$-ray bursts. Obtaining accurate mass measurements of stars in the upper part of the HRD has been a challenge, however, as evolutionary models of high-mass stars are riddled with physical uncertainties. Effects of rotation, precise wind-mass-loss rates, and internal mixing strengths are but the most well known problems (cf. Sect. 1.1.5). Furthermore, spectroscopic masses, obtained through atmospheric model fitting, typically suffer from the mass discrepancy problem. Therefore, direct mass measurements that are independent of atmosphere or evolutionary models offer valuable constraints to gauge the quality of both type of models.

In Sect. 1.3, we introduced how model-independent masses can be found through Kepler's laws, which provide so-called dynamical masses. Unfortunately, no single observational technique can fully characterize the orbit and dynamical masses of the two components. Either a double-lined spectroscopic binary (SB2) has to be eclipsing, *or*, astrometric data, either absolute or relative, must be available. In both cases, a time-series of observations is required.

M. Fabry, *Detailed Structure and Evolution Modeling of the Tightest Massive Binary Stars*, Springer Theses, https://doi.org/10.1007/978-3-032-28788-5_2

Traditionally, eclipsing SB2s are considered to provide the best constraints on the orbital parameters. Yet, they are rare, and uncertainties about the effects of tidal deformation, mutual illumination and/or binary interaction may pollute the obtained results, making it challenging to confront these objects to single-star models. This is particularly the case in the realm of massive stars. In this context, astrometric binaries are a valuable alternative to eclipsing SB2s. Recent advances in optical long-baseline interferometry have identified a number of such systems (e.g., Sana et al. 2013b, 2014; Mayer et al. 2014; Maíz Apellániz et al. 2017; Mahy et al. 2018; Lanthermann et al. 2023), which offer new opportunities to obtain dynamical mass constraints of stars in the high-mass regime.

9 Sagittarii (9 Sgr, HD164794) is such a long-period, astrometric SB2 system in the Lagoon Nebula. It was first studied by Abbott et al. (1984) in the context of its variable synchrotron emission, which is interpreted to be a result of wind-wind collisions of binaries (see, e.g., Pittard and Dougherty 2006), hinting that the then-presumed single star 9 Sgr was, in fact, a binary. Subsequent studies by Rauw et al. (2002a,b) and Nazé et al. (2008) confirmed the presence of elevated X-ray emission, a typical indication of colliding winds in O + O binaries (Rauw et al. 2002c; Sana et al. 2004, 2006; Rauw and Nazé 2016). Rauw et al. (2012) first confirmed the long-period binary nature of 9 Sgr through RV measurements and classified its components as O3.5V((f$^+$)) and O5.5V((f)). Rauw et al. (2016) studied the periastron passage of 9 Sgr and reported a maximum in the X-ray emission coming from shocked gas in the interaction zone of the stellar winds, as expected from a wind-wind collision in a wide binary where the shocked material cools adiabatically (Stevens et al. 1992).

The components in the system were resolved for the first time in 2009 with optical interferometry from the Astronomical Multi-BEam combineR (AMBER). In 2013, a new interferometric instrument, the Precision Integrated-Optics Near-infrared Imaging ExpeRiment (PIONIER), continued observing 9 Sgr (Sana et al. 2014). Le Bouquin et al. (2017) constrained the astrometric orbit using all AMBER and PIONIER measurements, uncovering a discrepancy with the spectroscopic analyses of Rauw et al. (2012, 2016). While the interferometric measurements of Le Bouquin et al. (2017) firmly excluded inclinations below 80°, the RV curve semi-amplitudes of Rauw et al. (2012, 2016) resulted in an estimated inclination of about 50° if the stars were to have masses representative of their spectral types.

Another long standing issue is the distance to 9 Sgr, and whether or not it is a member of the young open cluster NGC 6530. Prisinzano et al. (2005) and Kharchenko et al. (2005) measured distances to the cluster of around 1.25 kpc, while earlier measurements indicated a distance close to 1.8 kpc (van den Ancker et al. 1997; Sung et al. 2000) as a result of differences in the adopted reddening laws. The combination of the astrometric and spectroscopic measurements provide a direct constraint on the distance, allowing us to confirm the current Gaia eDR3 measurement of 1.21 kpc. Gaia does not suffer from systematics owing to an assumed reddening law, and thus provides a powerful constraint, but its measurement can still be impacted by the multiplicity of the system.

In this chapter, we aim to resolve the existing discrepant results that cast doubt on the current mass estimates, evolutionary status, and cluster membership of 9 Sgr. To do so, we leverage the accuracy of relative astrometry and we perform a grid spectral disentangling analysis on spectroscopic data to fully constrain the orbit of 9 Sgr. We further use the atmospheric properties of the stars in the system to derive its evolutionary status and test stellar-evolution models at high masses. Section 2.2 covers the observational data that were used. We present and discuss the orbital analysis and spectral disentangling in Sect. 2.3, the atmosphere modeling of both components in Sect. 2.4, and the evolutionary status of 9 Sgr is discussed in Sect. 2.5. Section 2.6 presents our conclusions and final remarks.

2.2 Observations

We combine archival and new optical spectroscopy with near-infrared (NIR) interferometry of 9 Sgr. Most of these measurements were part of long-term monitoring programs of massive stars in the Milky Way.

2.2.1 Optical Spectra

Archival data consist of 57 spectra from the High Efficiency and Resolution Mercator Echelle Spectrograph (HERMES, Raskin et al. 2011, used in Rauw et al. 2016), 20 Fiber-fed Extended Range Optical Spectrograph (FEROS) spectra (Kaufer et al. 1999) and 49 Ultraviolet and Visual Echelle Spectrograph (UVES) spectra (Dekker et al. 2000) in the blue and red arms (used in Rauw et al. 2012). Observations that are not previously analyzed are listed in Table A.1 and consist of five additional HERMES spectra and one spectrum from the High Accuracy Radial velocity Planet Searcher (HARPS) spectrograph (Mayor et al. 2003). The FEROS spectra cover the spectral domain between 3700 and 9000 Å and have a resolving power of $R \approx 48{,}000$. The UVES spectra cover the wavelength range $\lambda\lambda = 3500$–5000 Å with its blue arm and $\lambda\lambda = 5000$–7000 Å with its red arm, and each have a resolving power of $R \approx 40{,}000$. The HERMES and HARPS spectra both have $R \approx 85{,}000$ with a coverage of $\lambda\lambda = 3800$–9000 Å and $\lambda\lambda = 3750$–6900 Å, respectively. The FEROS, UVES, and HARPS spectra were obtained through the ESO archive science portal and were pre-reduced with their respective pipelines. The HERMES spectra are reduced using the HERMES Data Reduction Software (DRS) pipeline. Finally, all spectra were normalized over their whole spectral domain by fitting a cubic spline function through selected continuum regions.

2.2.2 *Near-Infrared Interferometry*

We used the previously published AMBER and PIONIER dataset obtained from Jun 2009 to Aug 2016 (Le Bouquin et al. 2017), along with two new PIONIER observations obtained in May and August 2017. These new data show for the first time the system turning back on the apparent ellipse and provide an almost complete coverage of the nine-year orbit. They were obtained with PIONIER (Le Bouquin et al. 2011) at the Very Large Telescope Interferometer (VLTI) using the four 1.8 m Auxiliary Telescopes (ATs) in configurations B2-K0-D0-I3 and A0-G1-J2-J3, offering a maximal projected baseline of 120 and 130 m respectively. The PIONIER data were reduced and analyzed, as described in Le Bouquin et al. (2017), using the `pndrs` package.[1] Each observation produces six visibilities and four closure phases, delivering relative astrometry with sub-milliarcsecond precision and an H-band flux ratio of $f_{\mathrm{H}} = 0.62 \pm 0.02$. The full journal of interferometric observations is given in Table A.2 along with the measured astrometric properties of the system.

Additionally, three VLTI/GRAVITY (GRAVITY Collaboration et al. 2017) measurements were taken, of which two were obtained in June 2016 and the other in September 2016. These observations were part of the science verification (SV) program and used the ATs in configurations A0-G1-D0-C1 and A0-G1-J2-K0. As GRAVITY is a spectro-interferometer, it provides six visibilities and four closure phases for each wavelength bin in the 2.0–2.4 μm NIR K band with a spectral resolving power of $R \approx 4000$. These SV data were reduced with the standard GRAVITY pipeline (Lapeyrere et al. 2014, version 1.0.11) and fitted parametrically to a binary model with PMOIRED.[2] The uncertainties on the relative astrometry were estimated by adding a bootstrapped error and a systematic error of 0.1 mas in quadrature. In the bootstrapping procedure, data were drawn randomly to create new datasets and the final parameters and uncertainties were estimated as the average and standard deviation of all the fits that were performed. Unfortunately, in this data the spectral signal was not of sufficient quality to infer RVs.

2.3 Orbital Analysis

2.3.1 *Astrometric Orbit*

The interferometric observations allow us to constrain the orbital parameters of the relative orbit by solving for the Thiele-Innes constants; see Eqs. (1.61). These determine the period, p, eccentricity, e, longitude of the ascending node, Ω, argument of periastron, ω, and the inclination, i. Even without knowing the precise

[1] http://www.jmmc.fr/pndrs.

[2] https://github.com/amerand/PMOIRED.

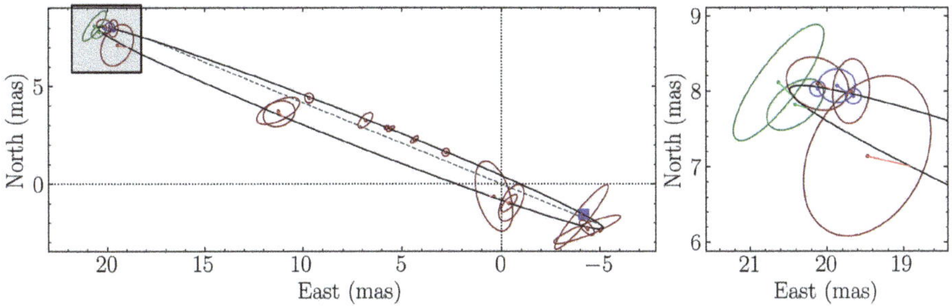

Fig. 2.1 Relative astrometric orbit of 9 Sgr. Archival observations are denoted in red, the new PIONIER data in green, and the GRAVITY observations in blue, all with their respective error ellipses. The blue square indicates the periastron passage and the dashed line represents the line of nodes. The right panel shows the zoom-in of the shaded region

parameter values, the plot of the astrometric data in Fig. 2.1 shows that the apparent ellipse is very eccentric. This can only be if the eccentricity is high, the inclination is high, or both.

In fact, a simple estimate of e can be obtained as follows (see also, Hilditch 2001). At periastron, the eccentric anomaly is zero, $E = 0$, and thus $X = 1 - e$ and $Y = 0$ per Eqs. (1.60). Using the Thiele-Innes equations (Eqs. 1.59), that makes $\Delta N_{peri} = A(1-e)$, and $\Delta E_{peri} = B(1-e)$. Similarly, for apastron, we have $E = \pi$, and thus $\Delta N_{ap} = A(-1-e)$. and $\Delta E_{ap} = B(-1-e)$. That means that the center of the apparent orbit is located at $(-Ae, -Be)$. Therefore, a displaced center implies a non-zero eccentricity. For 9 Sgr, Fig. 2.1 shows that, roughly, $\Delta N_{peri} \approx -2$ and $\Delta N_{center} \approx 3$, so that[3]

$$e = \frac{\Delta N_{center}}{\Delta N_{center} - \Delta N_{peri}} \approx \frac{3}{3+2} = 0.6. \tag{2.1}$$

Obtaining precise values for the parameters must be done using numerical fitting. We do so through a Levenberg-Marquardt least-squares minimization algorithm implemented in the newly developed Python package SPINOS that can simultaneously solve the radial-velocity equations and the Thiele-Innes equations (Eqs. 1.43 and 1.59).[4] For this chapter, we use the astrometric mode only. Errors on each parameter are determined by running a Markov chain Monte Carlo (MCMC) analysis of a hundred thousand samples. The resulting apparent orbit and orbital elements are shown in Fig. 2.1 and Table 2.1, respectively. The scatterplot matrix of the MCMC analysis is shown in Fig. 2.2.

[3] The shape of the apparent ellipse alone does not reveal where periastron occurs, but the timing of the observations does, as the secondary moves fastest around periastron.

[4] https://github.com/matthiasfabry/spinOS, https://www.ascl.net/2102.001.

Table 2.1 Relative astrometric solution of 9 Sgr, where the usual notation of the orbital elements have been adopted (see Sect. 1.3). The reported errors correspond to the 1σ confidence interval determined by the MCMC analysis

Element [Unit]	Value	Error
p [d]	3261	69
T_0 [MJD]	56,547	12
Ω [°]	67.3	0.4
ω [°][a]	210.7	2.3
i [°]	86.5	0.5
e	0.648	0.009

[a] By construction, the argument of periastron of the relative orbit ω is shifted by 180° compared to the argument of periastron ω_1 that is traditionally adopted by fitting the RV curve of the primary star

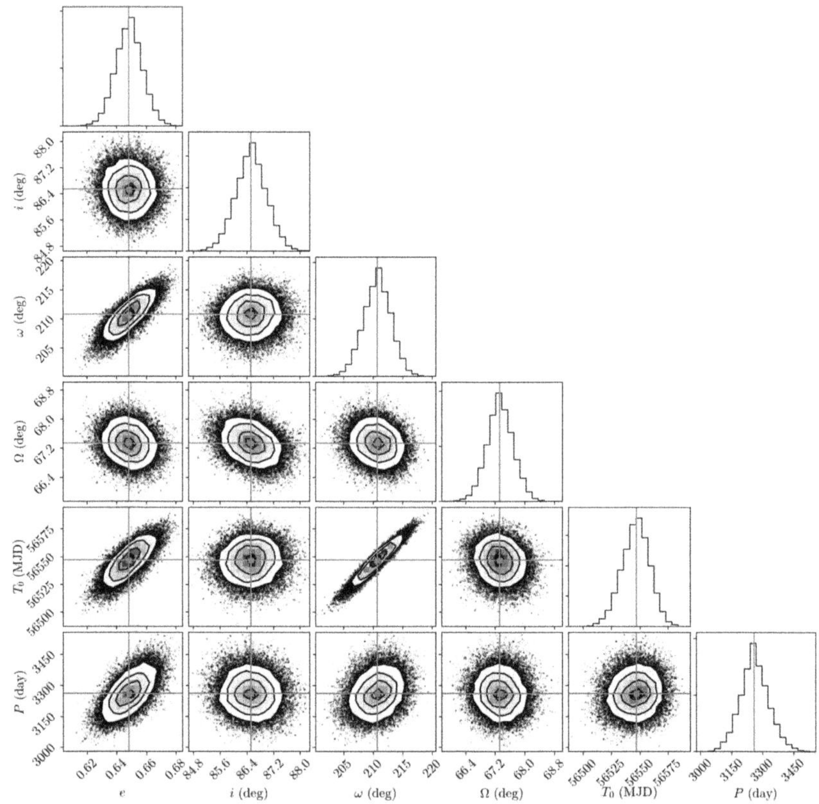

Fig. 2.2 Scatterplot matrix of the MCMC sampling of the orbital solution minimization in Sect. 2.3.1. While the expected heavy $T_0 - \omega$ correlation is apparent, no unexpected degeneracies in the parameter space are observed

As Le Bouquin et al. (2017) find, the orbit is very well constrained with a high eccentricity, $e = 0.648 \pm 0.009$, a near edge-on inclination of the orbital plane with the plane of the sky, $i = 86.5° \pm 0.5°$, and an orbital period of $P = 8.9 \pm 0.2$ yr. The scatterplot matrix of the MCMC confirms this solution is robust (Fig. 2.2). Importantly, we corroborate the findings of Le Bouquin et al. (2017) that inclinations of 85 ° or lower are ruled out. From Eq. (1.47), this also means that the minimum masses that would be obtained from spectroscopy are nearly the true masses of the components.

2.3.2 Spectral Disentangling

The astrometric solution of Sect. 2.3.1 only provides the relative size of the orbit. Determining the absolute size of the orbit independently of an assumed distance requires additional information about the system. In the case of 9 Sgr, this information is contained in the Doppler motion of its spectra, specifically, in the values of the semi-amplitudes of its RV curves, K_1 and K_2. In the standard approach, we would fit the profiles of selected spectral lines, or determine the line barycenter, and measure the red- or blue-shift as a function of the observation time to obtain an observational RV curve.

However, in long-period systems with broad spectral lines such as 9 Sgr, the lines of the two components may never fully deblend, and a small amount of contamination by the other component may significantly impact the measured RVs and therefore the orbital solution and derived masses. Spectral disentangling can help to improve the measured RVs in this respect because it is able to account for the cross-contamination of the spectral lines of the components. Additionally, spectral disentangling provides the full spectra of the components (not just the RVs), so that their atmosphere may be characterized.

Spectral disentangling however still faces challenges when the spectral lines never fully deblend or when the orbital solution is poorly defined. The most advanced spectral disentangling algorithms can optimize the orbital solution at the same time as separating the spectra, but the methods still suffer from degenerate solutions and local minima when applied to long-period systems as briefly discussed in, for example, Le Bouquin et al. (2017).

To disentangle the spectra of 9 Sgr, we use the Fourier disentangling code `fd3` (formerly `FDBinary`; Ilijić et al. 2004) by Ilijić (2017). Given the orbital elements $(e, P, T_0, \omega, K_1, K_2)$, `fd3` computes the separated spectra independently of any spectral templates. Internally, this is achieved by solving a linear least-squares problem introduced in Sect. 1.3.2. We solve the equations:

$$M\tilde{\boldsymbol{x}} = \tilde{\boldsymbol{c}} \tag{2.2}$$

in Fourier space, $\tilde{\boldsymbol{c}}$ containing the Fourier components of the observed composite spectra. M is the matrix that maps the unknown disentangled spectra $\tilde{\boldsymbol{x}}$ to $\tilde{\boldsymbol{c}}$, and depends on the orbital solution of the system and times of observation. The solver then uses the singular value decomposition of M (Eq. 1.52, but in Fourier space) and returns the solution, $\tilde{\boldsymbol{x}}_1$, that minimizes the two-norm:

$$r = \left\| \frac{M}{\boldsymbol{\sigma}} \tilde{\boldsymbol{x}}_1 - \frac{\tilde{\boldsymbol{c}}}{\boldsymbol{\sigma}} \right\|_2 . \tag{2.3}$$

The number r^2 follows a chi-squared statistic and measures how well the orbital elements map the disentangled spectra $\boldsymbol{x}_1$ to the observed spectra $\boldsymbol{c}$, weighted with its measurement errors $\boldsymbol{\sigma}$.

A disadvantage of this method is that the continuum level of the individual stars is undetermined with respect to the continuum of the composite spectra. Not only must a flux ratio be supplied to set the individual contributions to the total flux, but numerical instabilities can create large undulations of the disentangled spectra. Therefore, the output of the `fd3` algorithm must be normalized again by fitting a smooth function through the regions where

$$\boldsymbol{x}_1^A + \boldsymbol{x}_1^B = 1. \tag{2.4}$$

Our novel strategy is to fix the elements (e, P, T_0, ω) to those found from the astrometric orbit, and vary K_1 and K_2 in a grid of $K_1 \in [15, 45]\,\mathrm{km\,s^{-1}}$ and $K_2 \in [35, 65]\,\mathrm{km\,s^{-1}}$ with $1\,\mathrm{km\,s^{-1}}$ increments. This grid is chosen with the expected total mass from the astrometric orbit and the mass ratio from Rauw et al. (2012) in mind. We decided to use a flux ratio equal to that observed in the H band, namely $f \equiv f^B/f^A = f_\mathrm{H} = 0.62$. Thus $f^A = 0.6173$ and $f^B = 1 - f^A$. We recorded the reported reduced chi-squared distance $\chi^2 = r^2$ of Eq. (2.3) for each grid point, and these are represented in a contour map. We wrapped the C code of Ilijić (2017) and the grid setup in a Python package available though GitHub.[5] Other applications of this disentangling strategy were presented on other challenging systems (Shenar et al. 2020; Bodensteiner et al. 2020).

Instead of running the above procedure on the full wavelength domain of the spectra, which would be computationally expensive, we selected strong spectral lines and did the analysis on these lines separately. We then summed the obtained chi-squared distances, and reduced them by dividing with the summed degrees of freedom, effectively clipping unwanted continuum regions. The errors were estimated by a Monte Carlo (MC) sampling analysis, in which we drew 1500 samples of the full grid-disentangling procedure, where the spectra are perturbed with Gaussian additive noise representative of their signal-to-noise ratio (S/N). The orbital elements are perturbed with their 1σ error given in Table 2.1.

The considered lines and wavelength ranges are shown in Table 2.2. The criteria for selection are that they have high S/N, and that they are absorption lines. Therefore, contrary to the analysis of Rauw et al. (2012), weak metal lines like

[5] https://github.com/matthiasfabry/gridfd3.

Table 2.2 Spectral lines and their respective wavelength ranges that were disentangled to do the grid search described in Sect. 2.3.2

Spectral line	λ_{min} [Å]	λ_{max} [Å]
He I 4026	4020.0	4031.0
Hδ	4091.8	4112.5
He II 4200	4193.0	4206.0
Hγ	4327.8	4354.5
He I 4471	4465.0	4477.0
He II 4541	4532.5	4550.0
He II 4686	4677.0	4692.0
Hβ	4841.6	4878.0
He I 4922	4917.0	4927.0
He I 5016	5011.0	5026.0
He II 5411	5399.2	5424.1
He I 5875	5871.0	5879.5

C III λ 5696, Si IV λ 4116, O III λ 5592 were neglected owing to their poor S/N with respect to the deep H and He II lines.

Performing the fd3 grid disentangling on the lines of Table 2.2 gives the reduced chi-squared contour plot in Fig. 2.3. We find that the (K_1, K_2) pair that yields the minimal χ^2 value is $K_1 = 36\,\mathrm{km\,s^{-1}}$, $K_2 = 49\,\mathrm{km\,s^{-1}}$ and lies in a rather shallow minimum. We used the difference of the (marginalized) 0.84 and 0.16 percentiles of the MC sampling as estimators of the 1σ confidence interval, yielding $K_1 = 36^{+4}_{-1}$, $K_2 = 49 \pm 3\,\mathrm{km\,s^{-1}}$, resulting in dynamical masses of $53^{+7}_{-6}\,M_\odot$ and $39^{+6}_{-3}\,M_\odot$ for the primary and secondary respectively. This differs to the values of Rauw et al. (2012), who measured $K_1 = 25.7 \pm 0.6\,\mathrm{km\,s^{-1}}$ and $K_2 = 38.8 \pm 0.9\,\mathrm{km\,s^{-1}}$ and needed to invoke a much lower inclination to reconcile these with the inferred masses from their spectral types. While the mass ratio K_1/K_2 is similar to our results, the semi-amplitude values from Rauw et al. (2012, 2016) differ by 30%, that is, more than the statistical uncertainties of the fit. In this sense, in this work we present a different orbital solution. Rauw et al. (2012, 2016) use the mean of the RVs from the Si IV λ 4116 and N V $\lambda\lambda$ 4604, 4620 lines for the primary and similarly the mean from the He I $\lambda\lambda$ 4471, 5876, O III λ 5592, C III λ 5696, and C IV $\lambda\lambda$ 5801, 5812 lines for the secondary. The authors argue that, while most of these lines are in fact SB2 lines (as visible near maximum RV separation), the line blend is dominated by only one component such that they treated them as SB1 lines tracing only the motion of that component. The spectral disentangling results however show that both components contribute to these lines, sometimes significantly. For example, Fig. A.2 shows the O III λ 5592 line, to which the primary contributes about 25% as compared to the total depth of the line. As discussed in Sana et al. (2011b) and quantitatively investigated in Bodensteiner et al. (2021), the effect may be more subtle than this example because an even smaller contamination impacts the measured RVs. While for many systems this is of minor importance, in the present case, even a bias as small as $5\,\mathrm{km\,s^{-1}}$ on the RVs is large compared to the measured semi-amplitudes ($\sim$10%).

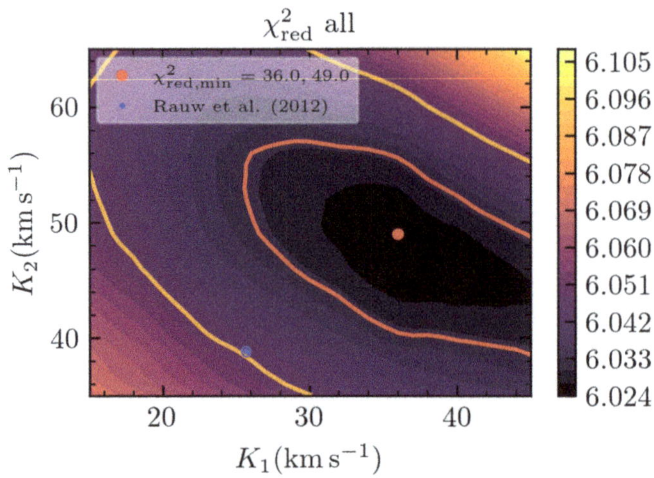

Fig. 2.3 Reduced chi-squared contour map of the grid disentangling combining all lines of Table 2.2. The minimal $\chi^2_{\rm red}$, at $K_1 = 36\,\rm km\,s^{-1}$, $K_2 = 49\,\rm km\,s^{-1}$, is denoted with a red dot. The solid red and orange lines represent the 1σ and 2σ contours respectively, after scaling the χ^2 map so that the minimum satisfies $\chi^2_{\rm red} = 1$

Table 2.3 Results of the fd3 grid search for the RV semi-amplitudes. Column two gives the grid point that has the minimal reduced chi-squared value when disentangling the original data. Columns 3, 4 and 5 give the MC median and 1σ confidence intervals

Semi-amplitudes	Data	MC median	σ_-	σ_+
$K_1\,(\rm km\,s^{-1})$	36	36	35	40
$K_2\,(\rm km\,s^{-1})$	49	50	47	53

The results of the fd3 grid search are summarized in Table 2.3. The error analysis shows K_1 has a slightly asymmetric 1σ uncertainty, as indicated in Table 2.3 and shown in the 2D (K_1, K_2) histogram in the top panel of Fig. 2.4. Yet, for both K_1 and K_2, the median of the retrieved MC values lies within 1-σ of the input value, giving us confidence in the accuracy of the method. For comparison, the marginalized histograms are shown in bottom panel of Fig. 2.4.

We show the broad-range disentangled spectra in Fig. 2.5. From visual inspection of the line depth ratio of He I λ 4471 to He II λ 4541, we immediately notice that the primary is the hottest star of the two. This fact is seen in the N V $\lambda\lambda$ 4604, 4620 lines as well. Comparing equivalent widths of lines in the disentangled spectra with classification tables of Conti and Alschuler (1971) and Sana et al. (in prep.), we update the classification of the primary by Rauw et al. (2012) from O3.5V((f$^+$)) to O3V((f$^+$)), and narrow down the subtype of the secondary from O5-5.5V((f)) to O5V((f)). This is in excellent agreement with the recent work of Quintero et al. (2020), who used an updated version of the shift-and-add disentangling technique in wavelength space and also obtained spectral types of O3V((f$^+$)) and O5V((f)).

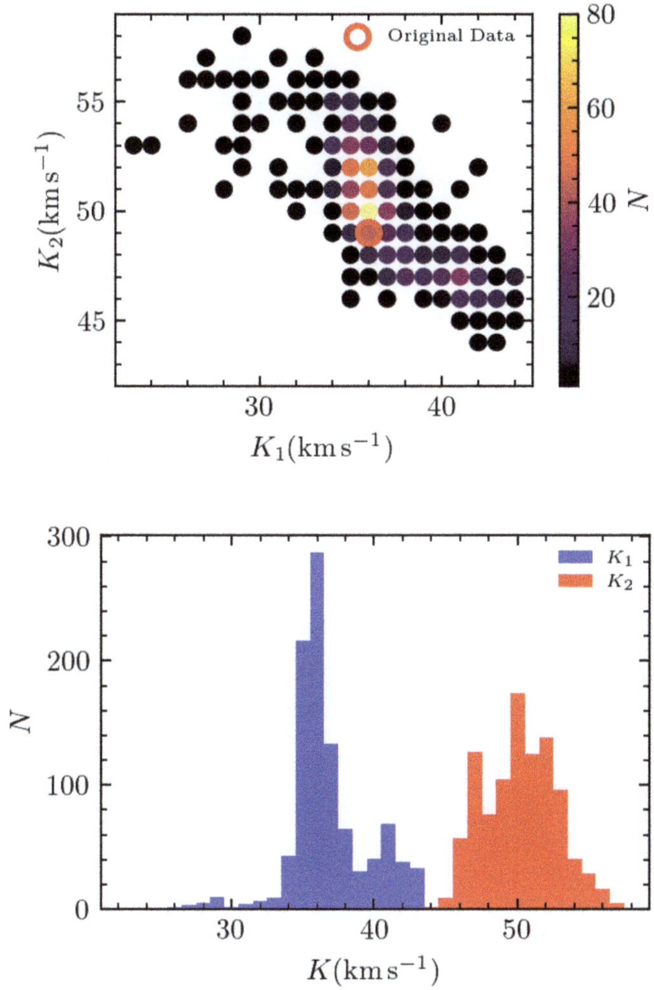

Fig. 2.4 *(Top)* Two-dimensional histogram of the results of the MC sampling in (K_1, K_2) space. The color indicates the number of samples N (out of a total of 1500) at the corresponding (K_1, K_2) pairs. The result from the original data is encircled in red. *(Bottom)* Marginalized histograms

To confirm the quality of the disentangling procedure, we recombine the disentangled spectra at various phases and compute residuals. To avoid mixing different instrumental wavelength calibrations, only the HERMES spectra are used. Reconstructed lines are shown in Figs. A.1 and A.2. The average residuals of the reconstructed lines to the original composite spectra are on the order 1%. While very good, this is larger than expected from pure S/N considerations and probably results from small uncertainties in the normalization of individual spectra that impacts the disentangling results of fd3.

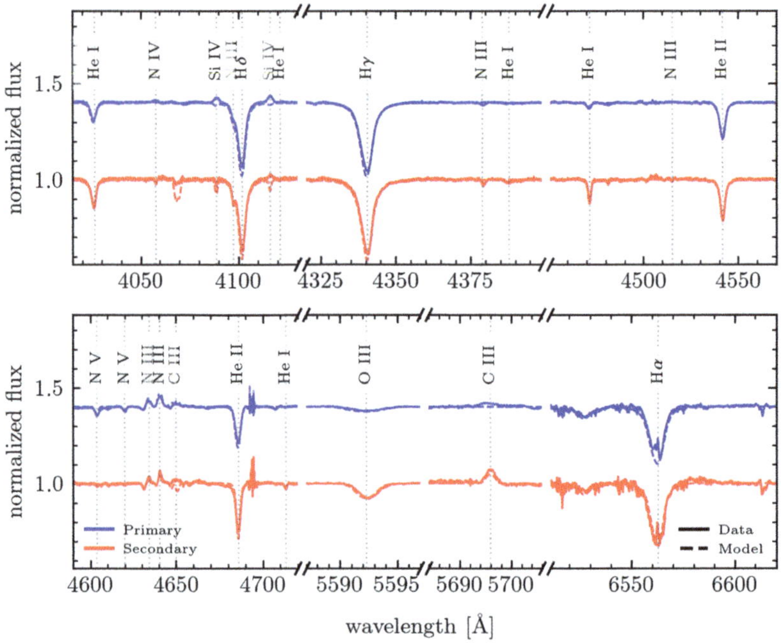

Fig. 2.5 Disentangled, renormalized spectra of both components using $K_1 = 36\,\mathrm{km\,s^{-1}}$ and $K_2 = 49\,\mathrm{km\,s^{-1}}$. The spectrum of the primary is shifted up by 0.4 units with respect to the secondary for clarity. In Hα, the core is contaminated by nebular emission present in the FEROS and HERMES spectra. The model spectra are those obtained with the best-fit atmospheric parameters from the FASTWIND analysis (Sect. 2.4)

2.3.3 Distance

Because the semi-amplitudes K_1 and K_2 set the absolute scale of the orbit, we can infer the distance to 9 Sgr by measuring against the apparent orbit found through the interferometry (Fig. 2.1), via the total mass of the system and Kepler's third law, as follows:

$$d = \frac{1}{a_{\mathrm{app}}} \sqrt[3]{\frac{G(M_1 + M_2)P^2}{4\pi^2}}, \tag{2.5}$$

where a_{app} is the semimajor axis of the apparent orbit in angular units. The values $(K_1, K_2) = (36, 49)\,\mathrm{km\,s^{-1}}$ correspond to a total dynamical mass of $92\,M_{\odot}$, which results in a distance of

$$d = 1310 \pm 60\,\mathrm{pc}, \tag{2.6}$$

where we propagated the MC errors.

This value lies within the Gaia eDR3 geometric distance reported by Bailer-Jones et al. (2021) of 1218^{+100}_{-108} pc. With these results, it seems likely that 9 Sgr is a member of NGC 6530, confirming the measurements of Prisinzano et al. (2005) and especially Kharchenko et al. (2005), who quoted a distance of $d = 1322$ pc. Conversely, we can firmly exclude distances to 9 Sgr of over 1500 pc because that would require a total mass of over 130 $M_\odot$, which our grid-disentangling results do not support. Similarly, the distance of 1780 ± 80 pc of Sung et al. (2000), adopted by Rauw et al. (2012), is incompatible with the interferometric and Gaia measurements of the apparent size of the orbit and the derived component masses.

2.4 Atmosphere Modeling

2.4.1 Setup

Using the disentangled spectra, we adjust theoretical line profiles computed with the FASTWIND non-local thermal equilibrium atmosphere code suitable for the expanding atmosphere of O-type stars (Santolaya-Rey et al. 1997; Puls et al. 2005; Carneiro et al. 2016; Puls 2017; Sundqvist and Puls 2018). To reduce the dimensionality of the parameter space, the rotational and macroturbulent velocities are estimated using the `iacob-broad` tool (Simón-Díaz and Herrero 2014). Analyzing (with the goodness-of-fit method) the O III λ 5592 and N V λ 4603 lines for the primary and O III λ 5592, He I λ 4713 and He I λ 5876 for the secondary, we find a projected rotational velocity $v \sin i = 102^{+8}_{-12}$ km s^{-1} and a macroturbulent velocity $v_{\rm mac} = 77^{+23}_{-20}$ km s^{-1} for the primary star, while we take $v \sin i = 67^{+6}_{-13}$ km s^{-1} and $v_{\rm mac} = 48^{+21}_{-14}$ km s^{-1} for the secondary. Similar values are obtained when using the Fourier transform method.

The stellar atmosphere models are then iterated using a genetic algorithm (Charbonneau 1995; Mokiem et al. 2005) within a predefined parameter space to optimize a χ^2 fitness metric until a convergence to the best fit with the spectrum is reached. The version of the genetic algorithm used is detailed in Abdul-Masih et al. (2019). We set the β exponent of the wind-acceleration law to 0.85, as appropriate for main-sequence stars (Muijres et al. 2012). The microturbulence velocity is fixed to $v_{\rm mic} = 10\,{\rm km\,s^{-1}}$ in the computation by FASTWIND; in the formal integral it is selected on the criteria $v_{\rm mic} = \max(10\,{\rm km\,s^{-1}}, 0.1 v_{\rm wind})$. We include optically thin wind clumping, with a near-constant clumping factor $f_{\rm cl}$ throughout the wind. Lastly, we opted to clip the core of the Hα line to avoid fitting the nebular emission remnant (visible in the bottom-rightmost panel of Fig. 2.5). The full list of fitted spectral lines is shown in Table A.3.

To provide an absolute magnitude anchor point for the atmospheric model, from which the component luminosities will be computed, we adopt the photometric data from the Two Micron All-Sky Survey (2MASS, Skrutskie et al. 2006), giving an apparent magnitude in the NIR K$_{\rm S}$ band of $m_{\rm K_S} = 5.731 \pm 0.024$ for the total system. Using the interstellar absorption coefficient $A_V = 1.338 \pm 0.021$ from Maíz Apellániz and Barbá (2018), the color correction $A_{K_S}/A_V = 0.116$ from Fitzpatrick (1999) and a distance of 1.31 ± 0.06 kpc calculated in Sect. 2.3.3, the absolute magnitude in the K_S band comes to $M_{K_S} = -5.01 \pm 0.10$. We correct for the measured PIONIER flux ratio $f = 0.62$ between the components, where we assume that it remains unchanged between the H and K_S band (as expected for such hot objects). This yields absolute component magnitudes of $M_{K_S} = -4.49 \pm 0.10$ and -3.96 ± 0.10 for the primary and secondary, respectively. We note that these values correspond to fainter stars than their derived spectral types suggest. Synthetic photometry of Martins and Plez (2006) give $M_K = -4.98$ and $M_K = -4.39$ for the O3V primary and O5V secondary, respectively, which suggests the stars are slightly more compact. Finally, using the bolometric correction in the K-band from Martins and Plez (2006), $BC_K = 28.80 - 7.24 \log(T_{\text{eff}})$, and the magnitude-luminosity relation

$$\log(L/L_\odot) = -0.4(M_K + BC_K - 4.75), \tag{2.7}$$

we compute the component luminosities from the K-band absolute magnitudes and the modeled effective temperatures.

2.4.2 Results and Discussion

We list the spectroscopic parameters of the resulting best-fit atmospheric models and the resulting inferred parameters in the leftmost column of Table 2.4. The corresponding theoretical spectra are plotted in Fig. 2.5 with dashed lines. We note the obvious nebular contamination of Hα, as well as the general trend that the disentangled spectra are slightly shallower than the model spectra in the deep and broad lines (like Hδ, Hγ and He II λ 4686), suggesting issues in the normalization of these broad lines.

The best-fit parameters depend on which line features were considered in the fit and with what weights. For example, giving more weight to He I lines would result in lower inferred effective temperatures and vice versa. Correspondingly, the inferred surface gravities would be lower for lower T_{eff} and vice versa. Therefore, conservative errors of 1 kK and 0.2 dex on T_{eff} and $\log g$, respectively, are adopted. Furthermore, the determination of the quality of the CNO abundance measurements is challenging. The carbon abundance for the secondary, for example, is fitted to $\varepsilon_C = \log(N_C/N_H) + 12 = 9.12$; this is an unusually high measurement that has (to our knowledge) never before been observed. We note that this measurement is driven by the C III λ 5696 line, which is in emission. At $T_{\text{eff}} = 42$ kK, FASTWIND can only reconcile this line in emission by boosting the carbon abundance. Keeping the issues presented by this line in mind, however (see Martins and Hillier 2012), we adopt the minimum of a formal 0.1 dex error and the statistical error of the grid of models. Since this formal error is somewhat arbitrary, even these uncertainties should be interpreted with great care. Therefore we can only argue for qualitative enrichment of nitrogen in the primary and enrichment of carbon in the secondary. For added justification, we show in Figs. A.3 and A.4 the comparison of the model spectra, their error ranges along with spectra using the Brott et al. (2011) CNO baseline abundances in various diagnostic lines of the CNO elements.

The best-fit $\log g$ values then provide the spectroscopic masses of the stars, which are found to be $32 \pm 16\,M_{\odot}$ and $19 \pm 10\,M_{\odot}$ for the primary and secondary, respectively. These masses are significantly lower than their dynamical counterparts, albeit with large error bars, and are not representative of dwarf stars of that luminosity. The main reason for this discrepancy is the low inferred $\log g$, which should be raised by about 0.25 dex for both stars, that is, slightly beyond the adopted uncertainty, to match the dynamical masses. The mass discrepancy problem (Herrero et al. 1992) is still an open issue in massive-star spectroscopy, and while more recent studies (e.g., Mahy et al. 2020a) show that for stars above $\sim 35\,M_{\odot}$, the discrepancy largely disappears, in this analysis, it is still present. The repeated normalization of the spectra before and after disentangling could be the root cause of this fact, as the reconstruction plots in Figs. A.1 and A.2 and the comparison to the model spectra (Fig. 2.5) hint towards.

2.5 Evolutionary Modeling

We compare our previous results with the Milky Way evolutionary tracks of Brott et al. (2011), using the Bayesian search tool BONNSAI (Schneider et al. 2014).[6] The BONNSAI tool allows us to search the rotating single-star evolution tracks of Brott et al. (2011) for the highest-likelihood stellar model that corresponds to measured quantities. We input the observed $\log L$, $T_{\rm eff}$, $Y_{\rm He}$, and $v \sin i$ from the left column of Table 2.4 and request the highest likelihood models of both stars in the grid. To avoid biasing the Bayesian search, we refrain from using the $\log g$ due to the uncertainties posed by the mass discrepancy. In a first search, we do not input the CNO abundances obtained from FASTWIND. The parameters from the highest likelihood models replicated from our spectroscopic and photometric observables of this search are given in the middle column of Table 2.4.

The comparison with evolutionary tracks point toward relatively compact and coeval stars with an age of about 1 Myr. We find that the evolutionary masses are within error of the dynamical masses, which provides a further indication that the spectroscopic mass likely suffers from systematic errors. Additionally, the evolutionary models favor lower CNO surface abundances than are spectroscopically inferred, especially nitrogen in the primary and carbon in the secondary. The rather modest rotational velocities and young ages do not allow for rotational mixing to modify the surface composition; the CNO abundances returned by BONNSAI

[6] The BONNSAI web-service is available at www.astro.uni-bonn.de/stars/bonnsai.

Table 2.4 Parameters of the best-fit, genetically evolved FASTWIND atmospheric model (described in Sect. 2.4), along with replicated observables from the Brott et al. (2011) models using BONNSAI (Sect. 2.5). The errors correspond to the 1σ confidence level. Empty entries are indeterminable for that parameter

	FASTWIND		BONNSAI		
	Primary	Secondary	Primary	Secondary	Primary, Scaled CNO
$T_{\rm eff}$/kK	46.0 ± 1.0	42.0 ± 1.0	$45.9^{+0.6}_{-0.9}$	41.9 ± 0.9	$46.0^{+0.6}_{-1.0}$
$\log(g/[\rm cgs])$	3.87 ± 0.20	3.87 ± 0.20	4.11 ± 0.05	$4.12^{+0.06}_{-0.07}$	$4.10^{+0.02}_{-0.06}$
$\log \frac{\dot{M}}{M_\odot\,{\rm yr}^{-1}}$	-6.6 ± 0.2	-6.6 ± 0.2	...	...	...
$f_{\rm cl}$	29 ± 5	22 ± 3	...	...	...
$v \sin i/{\rm km\,s}^{-1}$ [a]	102^{+8}_{-12}	67^{+6}_{-13}	...	...	...
$v_{\rm rot}/{\rm km\,s}^{-1}$	...	...	110^{+59}_{-26}	70^{+8}_{-15}	330^{+26}_{-30}
$Y_{\rm He}$	0.25 ± 0.04	0.24 ± 0.03	0.26 [b]	0.26 [b]	$0.28^{+0.08}_{-0.02}$
$\varepsilon_{\rm C}$	$8.17^{+0.60}_{-0.55}$	9.12 ± 0.10 [c]	$8.14^{+0.01}_{-0.03}$	8.13 [b]	$7.12^{+0.55}_{-0.05}$
$\varepsilon_{\rm N}$	$8.45^{+0.10}_{-0.29}$	7.42 ± 0.10	$7.63^{+0.09}_{-0.01}$	7.64 [b]	$8.72^{+0.10}_{-0.27}$
$\varepsilon_{\rm O}$	$8.63^{+0.10}_{-0.70}$	$8.64^{+0.10}_{-0.13}$	$8.55^{+0.01}_{-0.02}$	8.55 [b]	$8.55^{+0.01}_{-0.61}$
$\log(L/L_\odot)$	5.68 ± 0.08	5.35 ± 0.08	$5.64^{+0.07}_{-0.06}$	$5.33^{+0.08}_{-0.06}$	$5.67^{+0.06}_{-0.07}$
$R/R_\odot$	10.8 ± 1.0	8.9 ± 1.2	$10.45^{+0.88}_{-0.59}$	$8.73^{+0.75}_{-0.67}$	$10.73^{+0.79}_{-0.61}$
$M_{\rm spec}/M_\odot$	32.1 ± 16.0	18.9 ± 10.1	...	...	...
$M_{\rm evol}/M_\odot$	...	...	$53.4^{+3.2}_{-3.3}$	$37.0^{+2.0}_{-2.3}$	53.8 ± 4.7
Age (Myr)	...	...	$0.52^{+0.32}_{-0.33}$	$1.00^{+0.48}_{-0.58}$	$1.00^{+0.80}_{-0.41}$

[a] Determined using IACOB-BROAD, not FASTWIND
[b] Very small error, see Sect. 2.5
[c] Highly uncertain measurement, this formal error is likely not representative, see Sect. 2.4.2

correspond to the baseline value of the Brott et al. (2011) models with very small uncertainties.

From the atmosphere models in Table 2.4, it is clear that the CNO composition between the primary and the secondary is different. This is not reflected in the evolutionary tracks because both models prefer the baseline values of the Brott et al. (2011) tracks, namely $\varepsilon_{\rm C} = 8.13$, $\varepsilon_{\rm N} = 7.64$ and $\varepsilon_{\rm O} = 8.55$ (middle column of Table 2.4). Furthermore, if we assume the abundances of the secondary are baseline for 9 Sgr, this source has a different CNO baseline than the Brott et al. (2011) tracks. It is hard to justify observationally that the observed abundances of the secondary are baseline for 9 Sgr or the cluster NGC 6530. But we expect the least massive star with the lowest rotational velocity to be least contaminated by surface enrichment from a theoretical standpoint. We thus test if BONNSAI finds different models for the primary if we scale down the observed CNO abundances to the Brott et al. (2011) baseline. For the abundances of the primary to maintain the same fractional difference versus the secondary, this amounts to calculating $\varepsilon'_{\rm X} = \varepsilon_{\rm X,base,Brott} - \varepsilon_{\rm X,base,obs} + \varepsilon_{\rm X,prim,obs}$. Keeping the doubtful C abundance measurement of the secondary in mind (Sect. 2.4.2), we refrain from scaling C and compute $\varepsilon'_\mathrm{N} = 8.67^{+0.14}_{-0.31}$ and $\varepsilon'_\mathrm{O} = 8.54^{+0.14}_{-0.71}$. Using then again $\log L$, T_eff, Y_He, and $\upsilon \sin i$ from the FASTWIND models, along with these new N and O abundances as input for BONNSAI, we obtain other highest likelihood evolutionary parameters; these are listed in the rightmost column of Table 2.4. These results point to a different scenario. Here the rotational velocity is significantly higher, allowing significant rotational mixing to occur. The primary age has increased to match that of the secondary as well, while the evolutionary mass, $\log g$ and T_eff are only slightly changed when comparing to the non-scaled results (middle column of Table 2.4). The major implication of this is that either the rotational axis and the normal to the orbital plane are heavily inclined, up to an estimated 68° to explain the observed projected rotational velocity, or the effect of rotational mixing is underestimated in the stellar-evolution models. We cannot exclude either that a mixing mechanism weakly dependent on rotation is operating on stars in this mass regime. Distinguishing between these scenarios however requires greater confidence in the quality of the CNO abundance measurements.

We summarize our results of Sects. 2.4 and 2.5 in an HRD and a Kiel diagram that is overplotted by several of the evolutionary tracks and isochrones of Brott et al. (2011), see Fig. 2.6. We note that while the location of the models on the HRD matches well, there is a mismatch of the spectroscopic mass inferred from the FASTWIND models and the evolutionary masses. In the Kiel diagram, there is a poorer match as expected from the mass discrepancy discussed in Sect. 2.4.2.

2.6 Conclusions

In this chapter, we have obtained disentangled spectra of 9 Sgr using a combination of high angular resolution astrometry and spectral grid disentangling with the `fd3` code. The astrometric measurements solidify the long period of 8.9 yr and have a near edge-on inclination of 86.5°. Our results confirm the presence of an O3V+O5V massive binary, which has inferred dynamical masses of about 53 and 39 $M_\odot$, making 9 Sgr one of the most massive galactic O+O binaries ever resolved. Furthermore, to our knowledge, this is only the second instance of a dynamical mass estimate of a galactic O3V star (the other from Mahy et al. 2018).

By re-deriving the semi-amplitudes of the RV curves through a spectral disentangling analysis, we clear up the contradictory results between the previous RV measurements of Rauw et al. (2012) and the high inclination of the interferometric orbit by Le Bouquin et al. (2017). Furthermore, the results show 9 Sgr is a member of the young open cluster NGC 6530. The combined dynamical, atmospheric, and evolutionary modeling shows 9 Sgr contains massive stars of roughly 53 $M_\odot$ and 37 $M_\odot$ for the primary and secondary, respectively. 9 Sgr is a unique system in the far top left corner of the HRD, and therefore provides an equally unique opportunity to use its stellar and systemic parameters to compare with massive star evolutionary models, as well as binary formation scenarios.

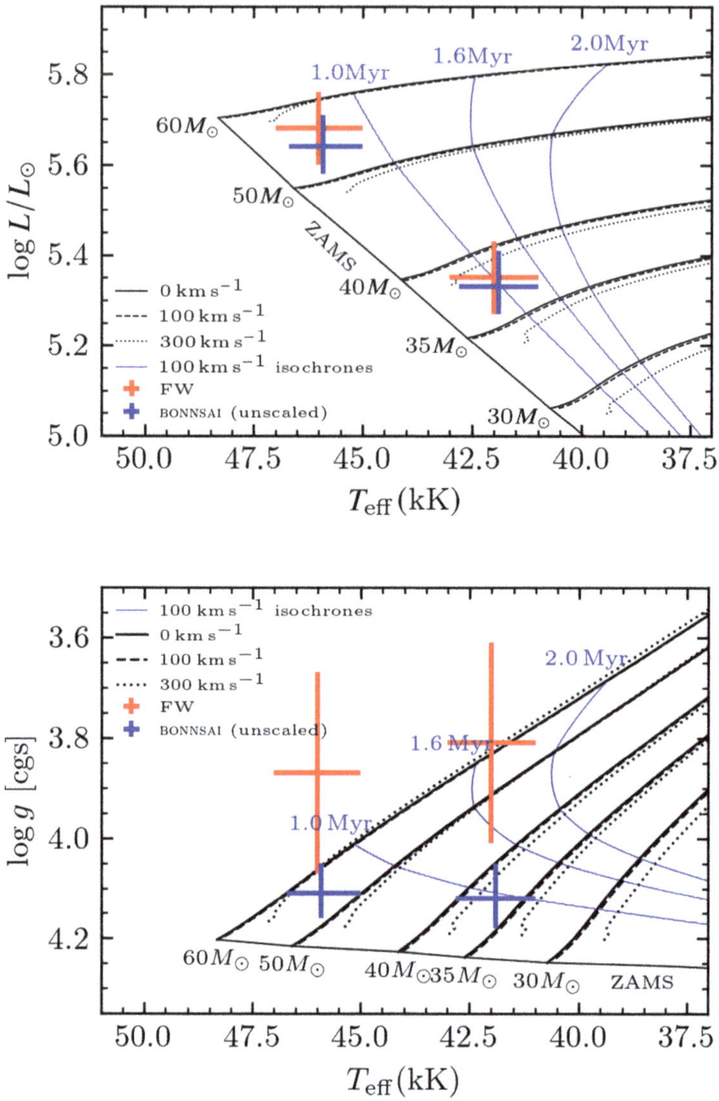

Fig. 2.6 *(Top)* HRD showing the location of the best-fit FASTWIND models (FW) and highest-likelihood evolutionary models from Brott et al. (2011) (BONNSAI). Overplotted are evolutionary tracks for different masses and initial rotation velocities (grey lines), along with isochrones for the 100 km s^{-1} initial rotation velocity models (blue lines). *(Bottom)* Kiel diagram with an equivalent legend as the HRD above. Using the scaled CNO abundances does not appreciably move the model of the primary in either diagram

Chapter 3
Modeling Tidal Distortion in 1D Stellar Structure

"A beginning is the time for taking the most delicate care that the balances are correct."

— Princess Irulan

This chapter is mainly based on:

Modeling overcontact binaries, I. The effect of tidal distortion
M. Fabry, P. Marchant and H. Sana
Astronomy & Astrophysics, 661, A123 (2022)

3.1 Introduction

In Sect. 1.4, we discussed how massive contact binaries play an important role in the formation of stellar mergers. They are also prime candidates for binary BH mergers in the CHE channel (Sect. 1.5.1). However, the theory of massive contact binaries, with radiative envelopes, required to produce evolutionary models, has not yet been well developed. Recently, new analysis techniques of observations have allowed for an improved characterization of the components of a massive contact binary (Palate et al. 2013; Abdul-Masih et al. 2020, 2021). Such observations give critical constraints to evolutionary models that try to explain the evolution of contact systems. However, to model these strongly interactive systems, important approximations need to be made (Marchant et al. 2016; Menon et al. 2021). Even though orbital evolution, tidal torques and mass transfer by RLOF is usually taken care of, the tidal deformation of the stellar structure from a binary companion or energy transfer in contact layers are not taken into account by these models.

Since centrifugal forces introduce a latitudinal dependence and tidal forces an additional azimuthal one, in order to properly compute stellar models for single rotating or binary stars, the stellar-structure equations would need to be recast and solved in 2D or 3D, respectively. From a computational standpoint, doing

M. Fabry, *Detailed Structure and Evolution Modeling of the Tightest Massive Binary Stars*, Springer Theses, https://doi.org/10.1007/978-3-032-28788-5_3

full 3D stellar evolution is not feasible with current resources, so simplifying assumptions need to be made in order to remain in 1D. Kippenhahn and Thomas (1970) introduced a methodology to correct the spherical stellar-structure equations, Eqs. (1.5), to account for deformation from the spherical geometry by rotation, which was later extended by Endal and Sofia (1976). This method is built on the premise that hydrostatic layers of stars fall on equipotential surfaces of the system. For single-star rotation, this method has been implemented in the stellar evolution code Methods for Experiments in Stellar Astrophysics (MESA, Paxton et al. 2011, 2013, 2015, 2018, 2019; Jermyn et al. 2023). Currently, these single-star corrections are also called upon when using the code to evolve binary stars. This means that while the centrifugal deformation from stellar rotation is accounted for, tidal deformation as a result of a nearby companion is not.

In this chapter, we apply the method of Kippenhahn and Thomas (1970) and Endal and Sofia (1976) to synchronized binary stars, as a first step in consistently modeling tidally deformed stars. The methods developed here are applicable to model twin contact binaries (binaries of mass ratio unity), where no energy is transferred between components, as well as for very close detached and semi-detached systems that are tidally synchronized. In Sect. 3.2, we calculate the modified 1D stellar-structure equations in a conservative potential and compute the structure correction factors f_P and f_T for a Roche binary. In Sects. 3.3 and 3.4, we consider the Eddington limit and boundary conditions as appropriate to the modified equations, respectively. Section 3.5 shows stellar models computed using the MESA code comparing currently implemented physics with the new methods in representative case studies. Lastly, Sect. 3.6 presents our final remarks. All the input files needed to reproduce the results of this chapter, and associated data products, can be downloaded from the Zenodo data repository.[1]

3.2 Methods

3.2.1 Modifications to the Spherical Stellar Structure Equations

In non-rotating stars, the equations of stellar structure and evolution are solved in a 1D space of the radius, r, or mass coordinate, M, and assume spherically symmetric stellar shells. As such, the equations of stellar structure are described by Eqs. (1.5). However, several physical effects can break the spherical symmetry that is assumed when calculating 1D stellar-evolution models. Centrifugal forces introduce a latitudinal dependence and a binary companion introduces an additional azimuthal dependence. Yet, if the total potential to which a star is subjected, Ψ, which is the sum of all gravitational and rotational contributions, is conservative,

[1] https://doi.org/10.5281/zenodo.5840879.

Kippenhahn and Thomas (1970) showed that such effects can be accounted for in 1D equations. The only assumption that remains is the one of shellularity: We assume that stellar isobars exactly coincide with equipotential surfaces of the given geometry. Therefore, the total pressure, P, is constant at constant Ψ. Von Zeipel (1924) showed that in hydrostatic equilibrium, a solidly rotating stellar shell has all its thermodynamical quantities, including the temperature, T, and density, ρ, remaining constant along isobars. However, he also showed that in this case, the stellar shell cannot be in radiative thermal equilibrium (see Sect. 1.1.5). Therefore one expects a deviation from shellularity coupled with large-scale meridional flows (Eddington 1925; Sweet 1950). Still, large horizontal flows are also present, keeping the rotation of isobars near to solid (Zahn 1992), so that the star is kept approximately shellular.

Instead of spherically symmetric stellar shells, we write the equations of stellar structure in terms of equipotential surfaces, which can be characterized by a single independent variable and thus be appropriate for 1D stellar evolution. Here, we summarize the model of Endal and Sofia (1976) (henceforth ES) that cast the equations of stellar structure in 1D given a conservative potential Ψ. This potential can, for example, be the Roche potential of a single rotating star (Sect. 3.2.2) or of a binary star (Sect. 3.2.3).

We start by defining r_Ψ as the equivalent radius of a sphere enclosing the same volume V_Ψ enclosed by an equipotential surface:

$$V_\Psi \equiv \frac{4\pi r_\Psi^3}{3}. \tag{3.1}$$

Hence, as the spherical relation from radius to volume is retained and the mass density, ρ, is assumed constant over an equipotential, the equation of continuity is unaffected with respect to its spherical equivalent (Eq. (1.5a)):

$$\frac{\partial M_\Psi}{\partial r_\Psi} = \rho \frac{\partial V_\Psi}{\partial r_\Psi} = 4\pi r_\Psi^2 \rho, \tag{3.2}$$

where M_Ψ is the total mass enclosed by the equipotential. The variable r_Ψ (or M_Ψ in a Lagrangian prescription) acts as the new independent variable to describe stellar models.

Next, we denote with g the magnitude of the effective gravity at each point in space:

$$g = |\nabla \Psi| = \frac{\mathrm{d}\Psi}{\mathrm{d}n}, \tag{3.3}$$

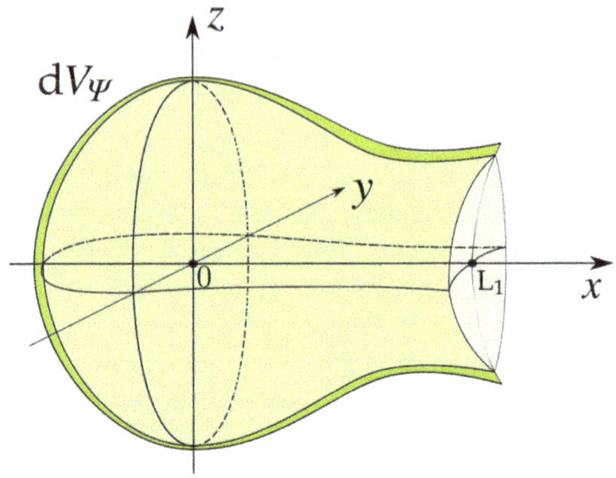

Fig. 3.1 Illustration of the volume of a shell in the Roche potential. The shell does not have uniform thickness, due to the effective gravity $d\Psi/dn$ varying across the equipotentials. The L_1 surface is also curved, see Sect. 3.2.4

with dn being the (perpendicular) distance between potential surfaces Ψ and $\Psi + d\Psi$. Considering then a change in equipotential volume, we have:

$$dV_\Psi = \int_\Psi dn\, dS \tag{3.4}$$

$$= d\Psi \int_\Psi g^{-1} dS = \langle g^{-1} \rangle S_\Psi d\Psi, \tag{3.5}$$

where S_Ψ is the surface area of the equipotential, and we defined, for convenience, the average of a quantity f over an equipotential as:

$$\langle f \rangle \equiv \frac{1}{S_\Psi} \int_\Psi f dS. \tag{3.6}$$

Figure 3.1 illustrates a differential volume, dV_Ψ, in the Roche potential for a shell lying above the equipotential of L_1 (see Sect. 3.2.4 for the definition of the L_1 surface). The volume element is thinner around the poles of the primary, as $d\Psi/dn$ is larger there than around the equator and near the neck of the contact shell.

Continuing from Eq. (3.5), we have

$$d\Psi = \frac{1}{\langle g^{-1} \rangle S_\Psi} dV_\Psi = \frac{1}{\langle g^{-1} \rangle S_\Psi \rho} dM_\Psi, \tag{3.7}$$

which, along with the statement of hydrostatic equilibrium $\nabla \boldsymbol{P} = -\rho \nabla \boldsymbol{\Psi}$ gives the modified momentum equation:

$$\frac{\partial P}{\partial M_\Psi} = -\frac{G M_\Psi}{4\pi r_\Psi^4} f_P, \tag{3.8}$$

with the correction to the standard spherical equation (Eq. (1.5b)) contained in the factor

$$f_P = \frac{4\pi r_\Psi^4}{GM_\Psi S_\Psi} \frac{1}{\langle g^{-1} \rangle}. \tag{3.9}$$

This factor thus measures the strength of the pressure gradient acting on a deformed shell (as a result of the non-spherical potential) with respect to a spherical shell enclosing the same volume and mass. Note that, in general, $S_\Psi \neq 4\pi r_\Psi^2$ and $\langle g^{-1} \rangle \neq r_\Psi^2/GM_\Psi$, so that $f_P \neq 1$.

Similarly to the mass-continuity equation, the equation of energy conservation retains its spherical form (Eq. (1.5c)), as equivalent radius, and enclosed mass and volume relate to each other spherically. Hence,

$$\frac{\partial L_\Psi}{\partial M_\Psi} = \varepsilon - \frac{\partial U}{\partial t} - P\frac{\partial(1/\rho)}{\partial t}, \tag{3.10}$$

where L_Ψ is the energy flowing through the equipotential per unit time, ε is the (specific) nuclear energy generation rate, U the specific internal energy, and t is time.

Considering lastly the equation of energy transport through radiative layers of the star, we have for the radiative flux:

$$\mathcal{F} = -\frac{4acT^3}{3\kappa\rho}\boldsymbol{\nabla T} = -\frac{4acT^3}{3\kappa\rho}\frac{\partial T}{\partial \Psi}\boldsymbol{\nabla\Psi}, \tag{3.11}$$

by the chain rule. Equation (3.11) is the von Zeipel theorem (von Zeipel 1924), stating that the radiative flux $\mathcal{F}$ is proportional to the local effective gravity $g = |\boldsymbol{\nabla\Psi}|$. Taking then the integral over an equipotential surface:

$$L_\Psi = \int_\Psi \mathcal{F}\cdot \mathbf{dS} = -\frac{4acT^3}{3\kappa}\langle g^{-1} \rangle \langle g \rangle S_\Psi^2 \frac{\partial T}{\partial M_\Psi}, \tag{3.12}$$

using the previous results, namely, Eqs. (3.3), (3.6), and (3.7). Rewriting using (3.8) and (3.9) transforms this into the usual spherical-like form of radiative energy transport (cf. Eq. (1.20)):

$$\frac{\partial \ln T}{\partial \ln P} = \frac{3\kappa}{16\pi acGM_\Psi}\frac{P}{T^4}L_\Psi\frac{f_T}{f_P} \equiv \nabla_{\text{rad}}\frac{f_T}{f_P}, \tag{3.13}$$

with the second correction factor to the spherical equations being

$$f_T = \left(\frac{4\pi r_\Psi^2}{S_\Psi}\right)^2 \frac{1}{\langle g \rangle \langle g^{-1} \rangle}. \tag{3.14}$$

Similarly to f_P (Eq. (3.9)), the correction f_T/f_P measures the strength of the radiative gradient of a deformed stellar shell relative to that of a spherical shell of the same volume and enclosed mass.

3.2.2 Single Star Rotation

A stellar shell rotating with angular velocity Ω exterior to a mass M has a total potential defined by

$$\Psi(r,\theta) = \Phi(r,\theta) - \frac{r^2\Omega^2\sin^2\theta}{2}, \tag{3.15}$$

where θ is the polar angle and Φ is the gravitational potential due to the mass interior to the shell. We noted in the introductory chapter that stars, especially massive ones with radiative envelopes, are heavily condensed toward their centers, and so significant deformation from spherical symmetry is only expected in the outermost layers. This justifies that the gravitational potential Φ of the star is taken as $\Phi = -GM/r$ of a point mass, with M as the total mass of the star.

Paxton et al. (2019) calculated analytical approximations (power-series expansions) to the corrections, f_P and f_T, and provided fits for these as a function of the variable $\omega = \Omega/\Omega_{\rm crit} = \Omega/\sqrt{GM_\Psi/r_{\rm e}^3}$, with $r_{\rm e}$ the equatorial radius of the shell. In Appendix B.1, we provide new, more accurate fits without increasing the degree of the polynomial expansions.

3.2.3 Synchronized Roche Binaries

We expect that tidal torques in binary systems are very sensitive to the separation of components (Zahn 1975), causing stars that significantly fill their RL to be synchronized with their orbits. Therefore, the total potential of these systems is the well-known Roche potential and may be written in the corotating frame as follows:

$$\Psi(x,y,z) = \Phi(x,y,z) - \frac{G(M_1+M_2)}{2a^3}\left[\left(x - \frac{M_2}{M_1+M_2}a\right)^2 + y^2\right], \tag{3.16}$$

where Φ is the gravitational contribution due to the two component stars, a their separation, and the second term represents the centrifugal contribution of the synchronous rotation around the center of mass (cf. Eq. (1.38)). We employ the same reasoning as in Sect. 3.2.2 to approximate the gravitational potential Φ by point mass potentials of mass M_1 and M_2, that is, the total masses of the component stars. Finally, defining the mass ratio $q = M_2/M_1$, normalizing all distances $r \to r/a$

and shifting $\Psi \to \Psi a/GM_1 - q^2/2(1+q)$, we arrive at the dimensionless form of the potential in spherical coordinates:

$$\Psi(r,\theta,\phi) = -\frac{1}{r} - q\left(\frac{1}{r'} - r\sin\theta\cos\phi\right) - \frac{1+q}{2}r^2\sin^2\theta, \tag{3.17}$$

where $r^2 = x^2+y^2+z^2$, $r\sin\theta\cos\phi = x$, $r\sin\theta\sin\phi = y$ and $r\cos\theta = z$ relate to the Cartesian frame, and $r' = \sqrt{1-2r\sin\theta\cos\phi + r^2}$.

3.2.4 Contact Shells

In the context of 1D evolution models, we have to distinguish the two stellar components within the geometry of a Roche binary. For equipotential shells lying within their respective RLs, this is trivial, as such layers are physically separated, but shells in contact are shared between the two stars. We therefore construct three "splitting surfaces," one through each of the colinear Lagrangian points L_1, L_2, and L_3, separating the Roche geometry in three parts. In Fig. 3.2, we show an illustration of the Roche geometry in the equatorial plane and highlight the colinear Lagrangian point equipotentials. We construct them in such a way that the condition:

$$\nabla\boldsymbol{\Psi}\cdot\mathbf{dS} = 0 \tag{3.18}$$

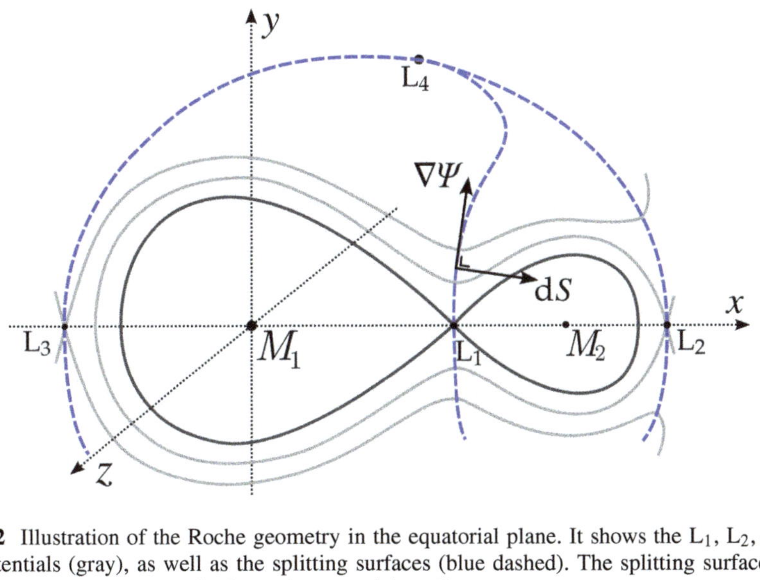

Fig. 3.2 Illustration of the Roche geometry in the equatorial plane. It shows the L_1, L_2, and L_3 equipotentials (gray), as well as the splitting surfaces (blue dashed). The splitting surfaces have their normals always perpendicular to the potential gradient

Fig. 3.3 Splitting surfaces for $q = 0.001$ at L_1. The color gradient follows the value of Ψ, starting at L_1 in black. *(Top)* Surface calculated by a uniform initial θ sampling around the Lagrangian point. The surface is not appropriately covered with this choice. *(Bottom)* Surface calculated with a denser sampling of the initial θ around $\pi/2$

holds on all points of these splitting surfaces (see Fig. 3.2, in blue dashed lines). In other words, the surface normal is always perpendicular to the gradient of the potential of Eq. (3.17).

We need to compute splitting surfaces for every mass ratio that we want to consider later. To construct them, we start at corresponding Lagrangian point, and take a tiny step away from it in the yz plane, as the gradient $\nabla\Psi$ is strictly zero at the Lagrangian point itself. Thus we take $\mathbf{r}_{\rm start} = (x_{\rm Lag}, \delta\cos\theta, \delta\sin\theta)$, for 2000 values of θ in the range $[0, \pi/2]$ (we make use of the symmetry $y \rightarrow -y$ and $z- \rightarrow -z$) and $\delta = 1 \times 10^{-14}$. Once the starting points are determined, new points of the surface are computed in sequences:

$$\mathbf{r}_{n+1} = \mathbf{r}_n + \delta_2 \boldsymbol{\nabla\Psi}|_{r_n}, \tag{3.19}$$

with $\delta_2 = 1 \times 10^{-4}$. The algorithm stops when $\Psi = 0.2\Psi_{\rm L_3} + 0.8\Psi_{\rm L_4}$, i.e., when the tracer has computed enough points away from the Lagrangian point to cover all shells we want to split (in actual stellar evolution calculations, shells lying beyond L_2 are unstable, see Sect. 5.2.3).[2] For most mass ratios, taking for θ a simple uniform sampling of the polar angle is sufficient, but for very low values of q, the gradient quickly takes the points away from the equatorial plane, see the top panel of Fig. 3.3. In that case, sampling closer to $\theta = \pi/2$ is required and produces a better covered

[2] *Beyond* in this context means at higher equipotentials.

surface (bottom panel of Fig. 3.3). In extreme cases (especially at the L_2 and L_3 surfaces), another strategy is needed still. First we compute the equatorial points, with $\theta = \pi/2$ and thus $z = 0$ for all $\mathbf{r}_n$ of that sequence. Then, for each of those points we take a tiny step in the z direction, of size $\delta = 1 \times 10^{-14}$, and the algorithm proceeds normally from there.

We note that by this construction of the splitting surfaces, the scalar product $\mathbf{dn} \cdot \mathbf{dS}$ is zero. This ensures that Eq. (3.4) still holds, and if the contact layers are shellular, there is no net flux of radiative energy between the components following Eq. (3.11). However, since the condition of shellularity cannot be held exactly, there will exist a flow $L_{\Psi,trans} = \int_{d\Psi} \mathcal{F} \cdot \mathbf{dS_{split}}$ across equipotentials, representing energy transfer from one component to the other. In this work, we assume this contribution is zero and defer the modeling of energy transfer to later work. We note also that for all cases $q \neq 1$, the L_1 splitting surface is *not* a vertical plane through L_1 (cf. Figs. 3.2 and 3.3), which differentiates our results of equipotential volumes and surface areas from those of Mochnacki (1984) and Marchant et al. (2021).

3.2.5 Numerical Calculations

In order to use the modified stellar-structure equations in a stellar-evolution code, we require for each shell in the geometry the quantities V_Ψ, S_Ψ, $\langle g \rangle$, and $\langle g^{-1} \rangle$ which specify r_Ψ, f_P, and f_T of Eqs. (3.1), (3.9), and (3.14). Additionally, the specific moment of inertia, $i_{\rm rot}$, is needed to calculate the rotational velocity of the equipotential shell from its angular momentum: $\Omega_\Psi = j_{\rm rot}/i_{\rm rot}$. Since the Roche potential does not have analytic formulae to describe equipotential surfaces, we approach this problem using numerical methods.

First of all, as noted above for the splitting surfaces, the problem is symmetric under $y \rightarrow -y$ and $z \rightarrow -z$, so we need only explicitly do calculations in the quadrant of positive y and z. Then, given a set of mass ratios, q, and target equipotential values, Ψ_0, we calculate, in a spherical coordinate system on rays of constant (θ, ϕ), the points $r = r(\theta, \phi)$ that lie on the corresponding equipotential surface, such that:

$$\Psi_0 - \Psi(r, \theta, \phi) = 0. \tag{3.20}$$

When solving Eq. (3.20), care has to be taken in order to avoid roots beyond the splitting surfaces (Sect. 3.2.4), as these belong to a part the geometry not associated to the considered component star, and would instead belong to the other component or to the region beyond the outer Lagrangian points. We therefore restrict the solver to values r smaller than the value $r_{\rm split}(\theta, \phi)$ on the splitting surface that the ray (θ, ϕ) crosses. If it finds no roots in the interval $(0, r_{\rm split})$, $r_{\rm split}$ is recorded instead as the splitting surfaces mark the maximal extent of the components.

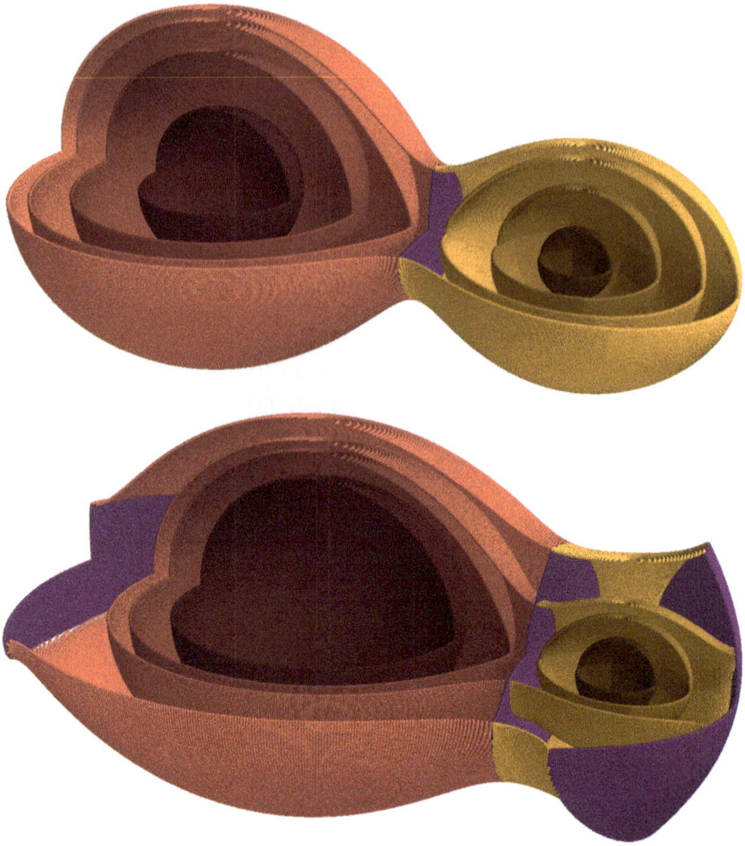

Fig. 3.4 Three-dimensional point clouds of Roche equipotentials. *(Top)* The $q = 0.3$ case that includes a moderately overflowing shell, with the splitting surface between the red and yellow components marked in purple. *(Bottom)* The $q = 0.1$ case that includes several heavily overflowing shells, exceeding even the L_2 and L_3 equipotentials. While for L_1 it is not immediately obvious, the L_2 and L_3 splitting surfaces are clearly curved (cf. Fig. 3.2 for the equatorial plane). Note that shells above L_2 are not necessarily physical in a hydrostatic star, as heavy mass loss and a dynamical merger is expected once a star (moderately) overflows the L_2 equipotential

We show 3D point clouds of some of our computations for $q = 0.3$ in Fig. 3.4, where we took the points in the $(+y, +z)$ quadrant and mirrored them to fill out two other quadrants. We see that shells well within the RL look spherical, which is expected as the potential is well approximated with $\Psi \approx 1/r$ there. Near the RL, the equipotentials are teardrop shaped, whereas overflowing layer follow the peanut-like shape as introduced in Sect. 1.4. The splitting surface separating the two components is colored purple. The bottom panel of Fig. 3.4 show heavily overfilling shells for $q = 0.1$, so much so that they cross the L_2 and L_3 splitting surfaces. It should be clear that these surfaces are not flat, but that they curve and follow the gradient $\nabla\Psi$ as explained in Sect. 3.2.4. Note however that the situation in the bottom panel is likely not to occur in nature. Once a contact binary overflows the L_2 equipotential, mass loss is expected (Shu et al. 1979), and a dynamical-timescale inspiral could ensue. Once the components reach an overflow commensurate with what is plotted in Fig. 3.4, co-rotation is likely lost, and the Roche potential stops being the appropriate geometry of he system.

Next, for these points $r(\theta, \phi)$, we numerically integrate the following quantities:

$$V_\Psi = \int_\Psi \frac{r^3}{3} \sin\theta \mathrm{d}\theta \mathrm{d}\phi, \tag{3.21}$$

$$S_\Psi = \int_\Psi \frac{1}{\hat{\mathbf{n}} \cdot \hat{\mathbf{r}}} r^2 \sin\theta \mathrm{d}\theta \mathrm{d}\phi, \tag{3.22}$$

$$\langle g \rangle = \frac{1}{S_\Psi} \int_\Psi \frac{g}{\hat{\mathbf{n}} \cdot \hat{\mathbf{r}}} r^2 \sin\theta \mathrm{d}\theta \mathrm{d}\phi, \tag{3.23}$$

$$\left\langle g^{-1} \right\rangle = \frac{1}{S_\Psi} \int_\Psi \frac{1}{g\hat{\mathbf{n}} \cdot \hat{\mathbf{r}}} r^2 \sin\theta \mathrm{d}\theta \mathrm{d}\phi, \tag{3.24}$$

$$\begin{aligned} i_{\mathrm{rot}} &= \frac{\int_\Psi (x^2 + y^2) \mathrm{d}m_\Psi}{\int_\Psi \mathrm{d}m_\Psi} = \frac{\int_\Psi r^2 \sin^2\theta g^{-1} \mathrm{d}\Psi \mathrm{d}S}{S_\Psi \left\langle g^{-1} \right\rangle \mathrm{d}\Psi}, \\ &= \frac{1}{S_\Psi \left\langle g^{-1} \right\rangle} \int_\Psi \frac{r^2 \sin^2\theta}{g\hat{\mathbf{n}} \cdot \hat{\mathbf{r}}} r^2 \sin\theta \mathrm{d}\theta \mathrm{d}\phi, \end{aligned} \tag{3.25}$$

where $\hat{\mathbf{n}} \cdot \hat{\mathbf{r}} = \frac{\mathrm{d}\Psi}{\mathrm{d}r} \Big/ |\nabla \boldsymbol{\Psi}|$ and we used the notation $\mathrm{d}m_\Psi = \rho \mathrm{d}n \mathrm{d}S$ to signify a small mass element on an equipotential shell, such that $\mathrm{d}M_\Psi = \int_\Psi \mathrm{d}m_\Psi$. From these, f_P and f_T can be computed for each equipotential shell with Eqs. (3.9) and (3.14).

We consider the component of a Roche binary at the origin with mass M_1 and a companion with mass M_2, and we computed all the necessary integrals of Eqs. (3.21)–(3.25) in the two-dimensional parameter space of mass ratio $q \equiv M_2/M_1$ and Roche equipotentials, Ψ_0. We emphasize here that we calculate solely the properties of the component with mass M_1 and that the properties of the companion M_2 can be obtained by inverting the mass ratio. We sampled 280 equally spaced mass ratios in logarithmic space in the interval $\log q \in [-7, 7]$, as well as the equal mass ratio $\log q = 0$ case while in potential space, we sampled 158 values in several logarithmic ranges from $\Psi_0 = 50\Psi_{\mathrm{L}_1}$ to $\Psi_0 = \frac{1}{2}(\Psi_{\mathrm{L_{out}}} + \Psi_{L_4})$ of different densities. With $\mathrm{L_{out}}$ we denote the outer Lagrangian point of the considered component of the binary (always having $x_{\mathrm{L_{out}}} < 0$, which is L_3 if $q \leq 1$ and L_2 if $q \geq 1$). As quantities exhibit important variations around the overflow potentials Ψ_{L_1} and $\Psi_{\mathrm{L_{out}}}$, those will be the regions we sample more densely. We note that $\Psi_{\mathrm{L}_1} < \Psi_{\mathrm{L}_2} \leq \Psi_{\mathrm{L}_3} < \Psi_{\mathrm{L}_4} = \Psi_{\mathrm{L}_5}$, as in our convention Ψ is strictly negative.

Figure 3.5 shows the integration results of f_P for various mass ratios as a function of their radius relative to the RL radius r_Ψ/R_{RL}, where we defined $r_{\Psi_{\mathrm{L}_1}} \equiv R_{\mathrm{RL}}$. We observe discontinuities at the Lagrangian overflow points, which is a result of the splitting surfaces limiting the considered volume to one component (Fig. 3.2).

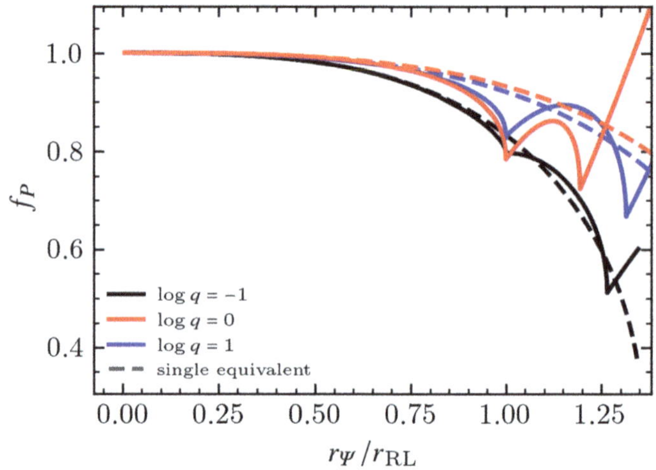

Fig. 3.5 Structure correction factor f_P as a function of fractional RL radius $r_\Psi / R_{\rm RL}$ for various mass ratios. The discontinuities in all graphs correspond to the crossing of the Lagrangian points L_1 (at $r_\Psi / R_{\rm RL} = 1$, by construction) and $L_{\rm out}$ at larger radii. For comparison, the equivalent single-rotating star f_P curves are overplotted in dashed lines

For a comparison with the single-rotating star deformation model, we overplot the equivalent f_P values as if the binary star was interpreted as a single-rotating star with the same volume and rotational velocity (see Appendix B.2 for how the translation is done).

3.2.6 Comparison to Literature

As verification of our integrations, we compare our results to calculations performed by Mochnacki (1984). As this study considered vertical splitting surfaces through L_1 at all mass ratio's, for contact layers we are limited to comparing the mass ratio unity case. Figure 3.6 shows excellent agreement between our f_P values and those computed via Eq. (3.9) using the results from Mochnacki (1984), with the relative difference being less than about 1 part in 10^3. A resolution convergence test of our integrations show our calculations converge to the 10^{-5} level at eight times our original resolution. The final computations of all integrals for use in stellar evolution instruments are done with this eight-fold resolution increase, and corresponds to dividing the interval $\phi = [0, \pi]$ in $n_\phi = 5280$ parts and $\cos\theta = [0, 1]$ in $n_\theta = n_\phi/2$ parts so that $\Delta\phi = \pi/n_\phi$ and $\Delta\cos\theta = 2/n_\phi$.

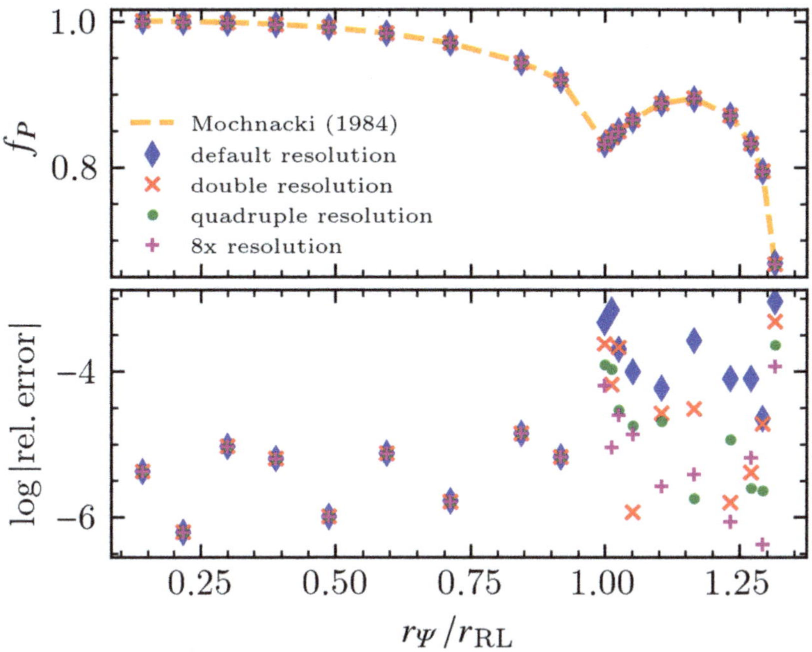

Fig. 3.6 Comparison of the calculations of the structure correction factor f_P from Mochnacki (1984) and those in this work, as a function of the fractional RL radius r_Ψ / R_{RL}. We performed a resolution convergence test with double, quadruple and octuple the default spatial resolution, and conclude our integrations have well converged at an eight-fold resolution increase

3.3 The $\Omega\Gamma$ Limit

Thanks to the radiation pressure of the photon flux that transports energy outward, a star gains radiative support against its self-gravity. In the situation where these effects are canceled out exactly, i.e., the effective gravity is zero, the star has reached the Eddington limit, which is given as (for a single, non-rotating and thus spherical star):

$$L = L_{\mathrm{Edd}} \equiv \frac{4\pi c G M}{\kappa}. \tag{3.26}$$

Defining then the Eddington factor Γ as:

$$\Gamma \equiv \frac{L_{\mathrm{rad}}}{L_{\mathrm{Edd}}}, \tag{3.27}$$

with L_{rad} the radiative luminosity, equal to L in a radiative envelope, $\Gamma = 1$ is then an equivalent statement of the Eddington limit. If $\Gamma > 1$, the sum of radiative and gravitational acceleration is directed outward. In the interior of the star this can in principle be compensated by an inversion of the gas pressure (Joss et al. 1973), but if a star approaches $\Gamma = 1$ at the surface, strong outflows are expected to develop (e.g., Gräfener et al. 2011).

In general, stars rotate however, and after adding a centrifugal contribution to the force balance, we have:

$$\boldsymbol{g}_{\mathbf{tot}} = \boldsymbol{g}_{\mathbf{grav}} + \boldsymbol{g}_{\mathbf{rot}} + \boldsymbol{g}_{\mathbf{rad}} = \boldsymbol{g} + \boldsymbol{g}_{\mathbf{rad}} = 0, \quad (3.28)$$

where we write $\boldsymbol{g} = \boldsymbol{\nabla \Psi}$ as the effective gravity. Langer (1997) then considered the Eddington limit on the equator of a rotating (but non-deformed) star, and, assuming that the radiation field is isotropic, the radiative flux is $\mathcal{F} = L/4\pi R^2$, so that the force balance reduces to

$$1 - \omega^2 - \Gamma = 0, \quad (3.29)$$

where we have defined $\omega = \Omega/\Omega_{\text{crit, class}}$ as the fractional classical critical rotation rate $\Omega_{\text{crit, class}} = \sqrt{GM/R^3}$. It is important to make the distinction here as the formal critical velocity is lowered to:

$$v_{\text{crit}} = \sqrt{\frac{GM}{R}(1 - \Gamma)} = \sqrt{v_{\text{crit, class}}(1 - \Gamma)}. \quad (3.30)$$

Therefore, the closer a star is to the Eddington limit, the lower the critical velocity will be.

Maeder and Meynet (2000) later noted that this treatment does not take von Zeipel's theorem into account, namely that the radiative flux is dependent on local effective gravity, Eq. (3.11). Expressing this for a rotating star to lowest order in rotational velocity Ω gives

$$\boldsymbol{g}_{\text{tot}} = \boldsymbol{g}\left(1 - \Gamma_\Omega(\vartheta)\right), \quad (3.31)$$

where ϑ signifies a polar dependence and $\Gamma_\Omega \equiv \Gamma \times f(\Omega, \vartheta)$ is now a rotation-dependent Eddington factor. Since the strength of the radiative flux is a function of the effective gravity through the von Zeipel effect, Maeder and Meynet (2000) showed that, neglecting the ϑ dependence, Eq. (3.31) bifurcates when $\Gamma = 0.639$. Below this, $\boldsymbol{g} = 0$ is the only solution, so that, independent of luminosity, the break up velocity is the classical velocity $v_{\text{crit,class}} = \sqrt{GM/r_e}$, with r_e the equatorial radius of the rotating star. Above the bifurcation point, both $\boldsymbol{g} = 0$ and, importantly,

$$1 - \Gamma_\Omega = 0, \quad (3.32)$$

determine the critical velocities, the second of which is lower than the classical critical velocity. The lowest of these two velocities then determines the physical break-up velocity of a rotating star.

We further expand on the notion of break-up limits by considering the case of deformed stars due to an arbitrary, but conservative, potential. As a starting point, we consider the same criterion of zero net acceleration, $\boldsymbol{g}_{\mathbf{tot}} = 0$, as the stability boundary. The contributions to this acceleration are the effective gravity, $\boldsymbol{g}$, as a result of the potential, and the radiative acceleration. Using von Zeipel's expression for radiative flux (Eq. (3.11)), combined with the equation of hydrostatic equilibrium results in:

$$\boldsymbol{g}_{\mathbf{rad}} = -\frac{4aT^3}{3}\frac{\mathrm{d}T}{\mathrm{d}P}\boldsymbol{g}, \tag{3.33}$$

which, upon using Eq. (3.13) as the radiative gradient in the non-spherical geometry, gives the stability criterion:

$$\boldsymbol{g}\left(1 - \Gamma\frac{f_T}{f_P}\right) = 0. \tag{3.34}$$

We recognize that von Zeipel's law of radiation can break down close to critical velocity due to significant baroclinicity, that is, departure from shellularity (see, e.g., Espinosa Lara and Rieutord 2011). Continuing, however, on our assumption of shellularity, Eq. (3.34) is of a very similar form as that of Maeder and Meynet (2000), namely, $\boldsymbol{g}(1 - \Gamma_\Psi) = 0$, where $\Gamma_\Psi = \Gamma \times f$ is again a product of the classical Eddington factor with a function dependent on the geometry. For the single rotating star deformation, $f = f(\omega)$, dependent on the fractional critical rotation rate, while in the tidal deformation case the function f is of the form $f = f(r_\Psi/R_{\mathrm{RL}}, q)$, dependent on the mass ratio and degree of Roche filling. In Eq. (3.34), the classical solution $\boldsymbol{g} = 0$ is present, of course, where the effective gravity vanishes as a result of rotational (and tidal) support only. As above, in this case there is no corresponding Eddington luminosity as it is independent of any radiative acceleration due to the von Zeipel effect. The other solution is given by:

$$1 - \Gamma\frac{f_T}{f_P} = 0, \tag{3.35}$$

which translates to a modified expression of the Eddington luminosity:

$$L_{\mathrm{Edd},\Psi} = L_{\mathrm{Edd}}\frac{f_P}{f_T}, \tag{3.36}$$

or the Eddington factor (see also Sanyal et al. 2015):

$$\Gamma_\Psi = \frac{L_\Psi}{L_{\mathrm{Edd},\Psi}} = \frac{\kappa L_\Psi}{4\pi c G M_\Psi}\frac{f_T}{f_P}. \tag{3.37}$$

These expressions, as in the case of the equations derived in Sect. 3.2.1, are applicable in any conservative potential Ψ.

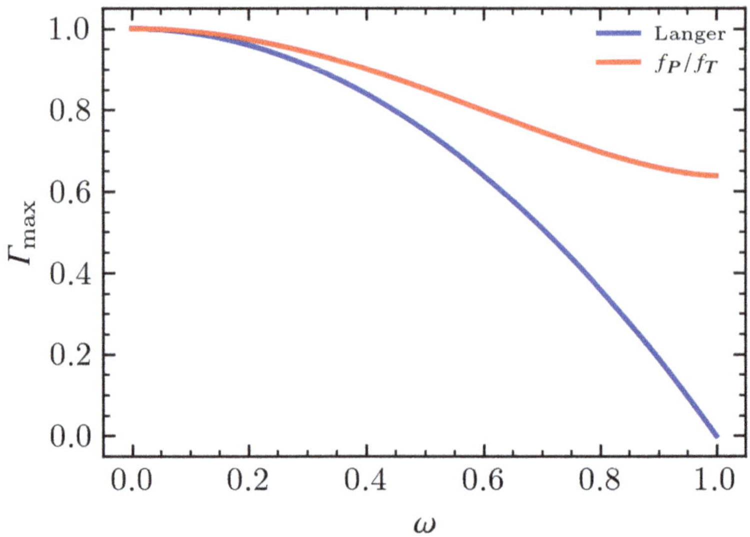

Fig. 3.7 Maximal Eddington factor $\Gamma_{\max}$ that a rotating star can have as a function of fractional classical critical rotation rate of $\omega = \Omega/\Omega_{\text{crit, class}}$. Our model of shellular stars has a maximal reduction to $\Gamma_{\max} = 0.639$

As the criteria for instability, Eqs. (3.29), (3.32), and (3.35) all express a departure of the Eddington limit of a star from the classical Eddington limit of $\Gamma = 1$. In the model developed here, it amounts to a reduction of the maximal Eddington factor by the ratio f_P/f_T, which is smaller than one in both the single rotating star or synchronized binary case. In Fig. 3.7, we show the comparison of this reduction from the results of Langer (1997) to this work in the case of a single rotating star. Moreover, consistent to Maeder and Meynet (2000), we find a bifurcation in the critical rotation velocity at $\Gamma = 0.639$. Below this number, only the classical critical velocity, $\omega = 1$, will make the star unstable, while above it, this velocity is reduced. The case of a synchronized Roche binary is shown in Fig. 3.8. For example, a binary with a mass ratio of unity, overflowing to its outer Lagrangian point, has its Eddington limit reduced to about 84% its classical value.

3.4 Atmospheric Boundary Conditions

The equations of stellar structure must be supplied with appropriate boundary conditions (BCs) at the center and atmosphere of the star. In the center, the conditions are such that $r = M = L = 0$, which transform to $r_\Psi = M_\Psi = L_\Psi = 0$ in our equipotential shell model. At the surface, an atmosphere model is integrated to supply the pressure P_{surf} and temperature T_{surf}. A common choice is the so-called gray, Eddington atmosphere, which is a plane-parallel model integrated from

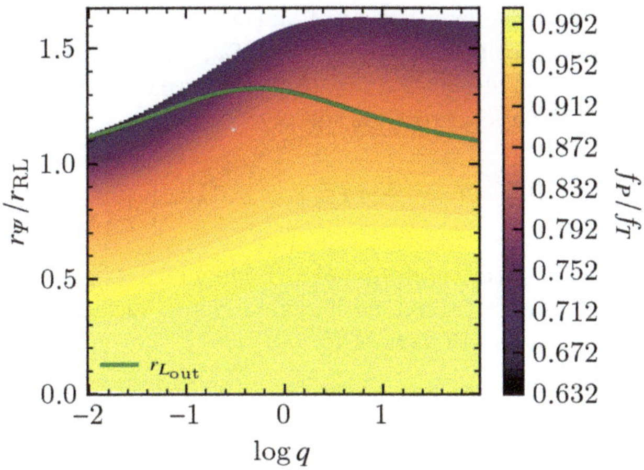

Fig. 3.8 Reduction, f_P/f_T, of the maximal Eddington factor as a function of mass ratio, q, and fractional overflow radius, $r_\Psi / R_{\rm RL}$. The equivalent radius of a star filling up to its outer Lagrangian point is marked with the green line

$\tau = 0$ to some predetermined atmospheric optical depth $\tau_{\rm surf}$, and is presented in, for example, Cox and Giuli (1968, henceforth CG).

In spherical stars, the surface optical depth $\tau_{\rm surf}$ is reached at the same physical perpendicular depth, z, at all points in the star. For deformed stars, however, the optical depth τ across an equipotential depends on local effective gravity, as follows:

$$d\tau = \kappa\rho dz = -\kappa\rho \frac{dn}{d\Psi} d\Psi = -\kappa\rho g^{-1} d\Psi, \tag{3.38}$$

with κ and ρ as the opacity and density of the equipotential, respectively. Consequently, equipotential surfaces do not coincide with surfaces of constant optical depth. For example, in single rotating stars, a fixed $\tau_{\rm surf}$ is reached at a deeper equipotential on the poles compared to on the equator, resulting in a different observed effective temperature between those regions (which can further be modulated by a gravity darkening law). The same is true for highly deformed binary stars like contact systems. The atmosphere properties output by stellar evolution codes should therefore also take into account this effect of position dependent atmospheric depth.

Within the treatment of deformed stars in a conservative potential, we proceeded as follows. We start by defining the global effective temperature $T_{{\rm eff},\Psi}$ of an equipotential by the law of Stefan-Boltzmann:

$$L_\Psi = \int_\Psi \mathcal{F} \cdot \mathbf{dS} \equiv S_\Psi \sigma T_{{\rm eff},\Psi}^4, \tag{3.39}$$

that is, the global effective temperature of an equipotential shell is that of a black body radiating a luminosity L_Ψ over an area S_Ψ, which need not be spherical. Next, using the equation of radiative flux (Eq. 3.11), we can define a local effective temperature $T_{\mathrm{eff},\ell}$ that varies across the equipotential:

$$\mathcal{F} = \frac{g}{\langle g \rangle} \sigma T_{\mathrm{eff},\Psi}^4 = \sigma T_{\mathrm{eff},\ell}^4. \tag{3.40}$$

We now consider a gray, plane-parallel, Eddington-approximated atmosphere to construct appropriate BCs. At each point on the surface equipotential, the temperature profile can be written as (CG):

$$T^4(\tau) = \frac{1}{2} T_{\mathrm{eff},\ell}^4 \left(1 + \frac{3}{2}\tau\right). \tag{3.41}$$

This expression is derived using two standard assumptions that can be similarly applied to the case of deformed stars. First, it is assumed that the radiation pressure at all optical depths can be written as $P_r(\tau) = \frac{1}{3} a T^4(\tau)$. Second, the intensity of radiation on top of the atmosphere is assumed to be isotropic in the outward directions, meaning that we ignore limb darkening, so that the pressure at the $\tau = 0$ surface is computed to be (CG):

$$P_r(0) = \frac{2\mathcal{F}}{3c}. \tag{3.42}$$

The first BC we consider constrains the temperature T_{surf} of the outermost boundary by requiring that $T_{\mathrm{surf}} = T_{\mathrm{eff},\Psi}$. To find the second BC on the surface pressure P_{surf}, we consider the equation of hydrostatic equilibrium:

$$\frac{dP}{d\tau} = \frac{g}{\kappa}. \tag{3.43}$$

Taking a point on the surface equipotential where $T_{\mathrm{eff},\ell} = T_{\mathrm{eff},\Psi}$, then by Eq. (3.40), we have that $g = \langle g \rangle$. As we also require that $T_{\mathrm{surf}} = T_{\mathrm{eff},\Psi}$, Eq. (3.41) implies that the surface optical depth is $\tau_{\mathrm{surf}} = 2/3$. The surface pressure can then be integrated at this point of the equipotential surface by assuming a thin atmosphere such that g and κ are constants.

$$P_{\mathrm{surf}} = \int_0^{\tau_{\mathrm{surf}}} \frac{\langle g \rangle}{\kappa} \mathrm{d}\tau + P_r(0) = \frac{2}{3} \frac{4\pi G M_\Psi}{\kappa S_\Psi} \frac{f_P}{f_T} + \frac{2}{3} \frac{\sigma T_{\mathrm{surf}}^4}{c}, \tag{3.44}$$

where we substituted $\langle g \rangle$ with the definitions of f_P and f_T in Eqs. (3.9) and (3.14), respectively.

In summary, the atmospheric BCs are represented by the following system of equations:

$$\begin{cases} P_{\mathrm{surf}} = \dfrac{2}{3}\dfrac{4\pi G M_{\Psi}}{\kappa(T_{\mathrm{surf}}, P_{\mathrm{surf}}) S_{\Psi}}\dfrac{f_P}{f_T} + \dfrac{2}{3}\dfrac{\sigma T_{\mathrm{surf}}^4}{c}, & (3.45) \\ T_{\mathrm{surf}} = \sqrt[4]{\dfrac{L_{\Psi}}{\sigma S_{\Psi}}}. & (3.46) \end{cases}$$

We emphasize that this surface temperature T_{surf} of the stellar models is not equal to the local effective temperature across the whole surface equipotential. Of the latter we can observe the variations (cf. Fig. 1.25). However, it is possible to relate the surface temperature output by stellar models using these BCs to an observed local effective temperature at any point on the surface equipotential by applying Eq. (3.40). Finally, we mention that going forward, whenever the effective temperature is referenced, we mean the global effective temperature as defined by Eq. (3.39).

3.5 Stellar Model Comparison

We assess the effect of including tidal distortion on the structure of the star as well as the modified boundary conditions by computing stellar-evolution models using the 1D `MESA` code, version 15140, with the MacOS `MESAsdk` v.21.2.1. Starting at the zero-age main sequence (ZAMS), we compare the evolution of stars in three cases: (1) a 5 $M_\odot$ single rotating star, (2) a 5 $M_\odot$ detached star in orbit with a compact object, and (3) a 32 $M_\odot$ twin binary in contact.

The next subsection lists the physical assumptions made when computing stellar models. The following subsections detail and discuss the results of the respective case studies.

3.5.1 Physical Ingredients

3.5.1.1 Mixing, Microphysics, and Stellar Winds

In all stellar models, we used the Ledoux criterion (Ledoux 1947) for determining convective regions. Within such regions, convection itself is modeled via the mixing length theory of Böhm-Vitense (1958) as described by CG, with the mixing length parameter $\alpha = 2$. We used efficient semiconvective mixing (Langer et al. 1983), with $\alpha_{\mathrm{sc}} = 100$ as proposed by Schootemeijer et al. (2019). Thermohaline mixing is applied as prescribed in Kippenhahn et al. (1980), with an efficiency of $\alpha_{\mathrm{th}} = 1$. During the main sequence, the convective core is allowed to overshoot into the radiative envelope. We use a step overshoot as calibrated by Brott et al. (2011) with $f = 0.345$, $f_0 = 0.01$, meaning that the diffusion coefficient is taken as constant from f_0 pressure scale heights into the convective region out to $f - f_0$ pressure scale heights away from the (formal) boundary. All other extra mixing processes, including rotational mixing from, for example, Eddington-Sweet type circulations are ignored in these sample models so as to isolate the effects of including the different forms of rotational deformation.

In terms of microphysics, we use the basic nuclear network consisting of ^{1}H, ^{3}He, ^{4}He, ^{12}C, ^{14}N, ^{16}O, ^{20}Ne, and ^{24}Mg, which is sufficient to follow the main-sequence evolution (Paxton et al. 2011) and the reaction rates are taken from the JINA library (Cyburt et al. 2010). MESA uses a blend of different equations of state (EOS) from Saumon et al. (1995), Timmes and Swesty (2000), Rogers and Nayfonov (2002) and Potekhin and Chabrier (2010), as described in Paxton et al. (2019). Radiative opacities are similarly determined from a blend of opacity tables from Iglesias and Rogers (1996) and Ferguson et al. (2005). The solar metallicity, $Z_\odot = 0.0142$, as well as the relative metal fractions from are taken from Asplund et al. (2009).

Wind mass loss of individual stars is included following the prescription of Brott et al. (2011). Mass loss through winds is accounted for by following the prescription of Brott et al. (2011). If the surface hydrogen fraction is $X > 0.7$, the mass loss rate is taken either from Vink et al. (2001), for temperatures above the iron bi-stability jump (calibrated also by Vink et al. 2001), or the maximum of the rates from Vink et al. (2001) and Nieuwenhuijzen and de Jager (1995), below this temperature. For $X < 0.4$, the wind prescription of Hamann et al. (1995) is used, albeit decreased by a factor of ten. When $0.4 < X < 0.7$, the wind is linearly interpolated between the above results.

3.5.1.2 Rotation and Binary Physics

At all times during the evolution of all models, angular momentum is continuously diffused throughout the whole star so as to enforce rigid-body rotation during the evolution. This, as with the omission of rotational mixing, is a way to reduce differences between the different rotational distortion models. When modeling single-star rotation, the structure correction factors f_P and f_T that we use are the ones presented in Appendix B.1. In binary models, the angular velocity is always synchronized to the orbital period, so that the Roche potential (Eq. (3.17)) describes the geometry of the system, and the timescale of this synchronization is the orbital period itself. The structure corrections f_P and f_T are determined by interpolating the results of Sect. 3.2.5 on a grid of mass ratio and fractional RL radius ($q = M_2/M_1, r_\Psi/R_{\mathrm{RL}}$), where the RL sizes are determined according to

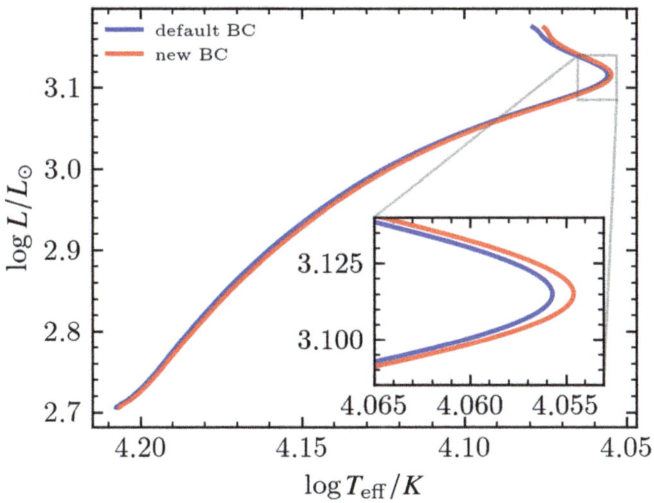

Fig. 3.9 HRD of the main-sequence evolution of 5 $M_\odot$ rotating stars in MESA using the default and new BCs. The inset zooms in on the blue hook near the TAMS

the Eggleton approximation (Eggleton 1983).[3] We do not model mass nor energy transfer in these models, as this is not needed for detached or equal mass binaries.

3.5.2 Main-Sequence Evolution of a Single Rotating Star

We compare the main-sequence evolution of a 5 $M_\odot$ star rotating initially at around 70% critical velocity at Solar metallicity with and without the modified atmospheric BCs developed in Sect. 3.4. In MESA, the default BCs are integrated from a standard spherically symmetric, plane-parallel, gray Eddington atmosphere (Eddington 1926). Our new BCs, however, take into account the modified pressure and area of the outer surface of the star, as represented by the correction factors, f_P, f_T, and the non-spherical area, S_Ψ, in Eqs. (3.45)–(3.46).

Figure 3.9 shows the evolutionary track of the single rotating stars in an HRD. We see that the change in BC has little impact on the overall evolution, with the largest difference of the surface properties occurring near hydrogen depletion ($X_c = 0.001$), where, at constant L, the corresponding $T_{\rm eff}$ varies by about 0.7%. Since the spherical BC calculates $T_{\rm eff}$ from $L = 4\pi r^2 \sigma T_{\rm eff}^4$, comparing with Eq. (3.46) at constant luminosity and radius results in the temperatures differing by a factor

[3] We note that during the calculation of this grid, RL sizes are computed exactly to construct the quantity $r_\Psi/R_{\rm RL}$. These values are then interpolated by MESA with fractional RL sizes computed with the Eggleton approximation of $R_{\rm RL}$.

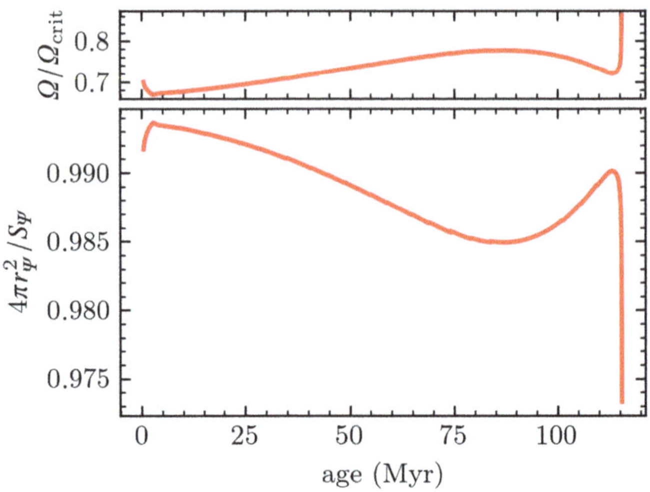

Fig. 3.10 Fraction of equivalent spherical area to true area at the surface as a function of age of the single rotating 5 $M_\odot$ model. A lower fraction indicates a further departure from spherical symmetry due to rotation. The top panel shows the fractional critical rotation rate for comparison

$\sqrt[4]{4\pi r_\Psi^2/S_\Psi}$. We plot the quantity $4\pi r_\Psi^2/S_\Psi$ as a function of stellar age in Fig. 3.10. As the star loses thermal equilibrium nearing hydrogen exhaustion, it contracts and tends toward critical rotation at the surface (recall that we are enforcing solid-body rotation). At this limit, the area ratio reaches its minimum of about 0.945, which, per Eq. (3.39) and at constant luminosity, leads to an effective temperature ratio of 0.986, meaning a 1.4% difference. In the bulk of the main sequence, however, we have an area ratio of around 0.99, which corresponds to just a 0.25% difference in effective temperature.

3.5.3 Evolution of a Detached B-type Star

We modeled the evolution of a 5 $M_\odot$ solar metallicity B-type star in a binary with a point mass companion of 3 $M_\odot$ ($q = 0.6$) and with an initial period of three days until the star reaches its RL. With this setup we mimic the situation of intermediate-mass stars found in short-period eclipsing binaries. We varied the rotation distortion model between (1) no rotation, (2) single star rotation where only centrifugal forces are taken into account, and (3) tidal deformation that accounts for both centrifugal and tidal forces. For the latter two, we also varied the atmospheric BCs between the new atmospheric BCs (Sect. 3.4) and the default spherical ones. For the non-rotating model, the new BCs reduce to the default ones, as there is no departure from spherical symmetry.

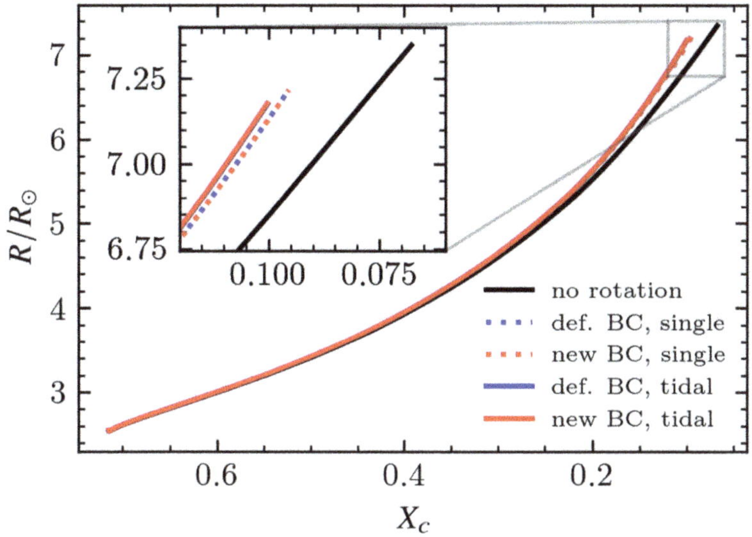

Fig. 3.11 Radius evolution of a 5 $M_\odot$ star starting on a three-day orbit around a 3 $M_\odot$ point mass, as a function of central hydrogen fraction X_c. The difference between the rotation models is small while the one resulting from the different BCs is imperceivable

Figure 3.11 shows the radius evolution of the stars as a function of the central hydrogen fraction, X_c. At the onset of evolution, where the star is well within its RL, we see negligible difference between the various assumptions on BCs or rotation. However, as the star approaches its RL size, at constant X_c, the family of rotating tracks differ from the non-rotating model by up to 5%. This is a significant difference, and is resolvable by very precise observations of eclipsing binaries, whose errors on measured radii can be better than 3% (Torres et al. 2010; Graczyk et al. 2018; Pavlovski et al. 2018; Tkachenko et al. 2020, and references therein). Therefore, modeling stars in eclipsing binaries as non-rotating in evolution codes could induce a systematic mismatch when fitting measured radii to inferred radii from such models. Finally, we note that the difference in radii between the tidal and single rotating star models, which is at most 1%, is only resolvable by the most accurate of observations, if at all. Furthermore, the difference in radii between considered BC is on the order of 0.1% and is not perceivable on the figure.

3.5.4 Evolution of a Twin Contact Binary

We evolved a 32 $M_\odot$ star with a metallicity of $Z = Z_\odot/2$ in a binary system with its twin, meaning all properties of the stars are identical. In particular, this means that the mass ratio is unity, and we can ignore binary interaction effects like mass or energy transfer. We start the evolution at the ZAMS with a 1.5 day orbit and stop the

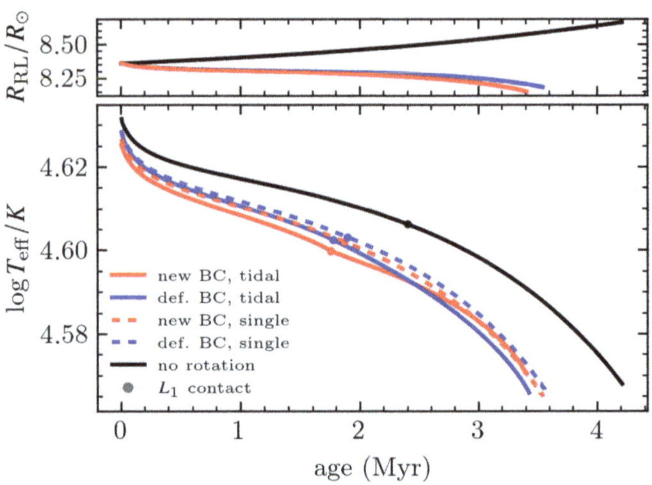

Fig. 3.12 Effective temperature evolution of the 32 $M_\odot$ contact model with the different assumptions on BC and deformation model. We notice that the models including rotation are up to 5% cooler during contact. Between the different rotation models and BCs, we find only minor differences. The top panel shows the RL evolution, which explains the difference in age at onset of overflow of the Lagrangian points through spin-orbit coupling

simulation once the stars reach the outer Lagrangian points. This setup is the closest we can currently and consistently model massive contact binaries with. VFTS 352 in the Large Magellanic Cloud (Almeida et al. 2015) is the prime example of a massive, twin contact binary.

In Fig. 3.12, we show the effective temperature evolution of the 32 $M_\odot$ model under the various rotation model and BC assumptions. We notice again, like in the detached case, that the observable stellar parameter (in this plot $T_{\rm eff}$) can vary up to 5% in the late stages of the contact configuration. The difference between the rotation models are minor, on the order of 1%, and the tidally distorted model with the new BCs of Sect. 3.4 shows a rather complex evolution. It kinks at the RL filling point and subsequently crosses the other rotating tracks as the degree of contact increases. This behavior may be related to the crossing of the structure correction factors, f_P and f_T, as r_Ψ keeps rising (see Fig. 3.5). Moreover, there is a ~25% age difference between the rotating and non-rotating models. This difference can be understood in the context of spin-orbit coupling. In the non-rotating model, the star can store no angular momentum, so that tidal synchronization has no effect (at least with the current implementation in MESA) and the RL size slowly increases due to wind mass loss (see top panel Fig. 3.12).[4] However, when the star is modeled with angular momentum, wind mass loss causes an efficient spindown of the star.

[4] One can incorporate a form of spin-orbit coupling even in non-rotating models by keeping track of the moment of inertia of the models, and assuming co-rotation with the orbit.

Tidal synchronization then keeps the star rotating close to the orbital velocity at the cost of decreasing the orbital separation, thus reducing the RL size and accelerating the points of contact and L_2 overflow. This effect is also visible in the detached binary of Sect. 3.5.3. We note that in the extreme case where the orbital angular momentum is lower than three times the spin angular momentum of the stars, this process causes the so-called Darwin instability (Darwin 1879; Hut 1980), where the stars catastrophically spiral in on each other on the orbital timescale.

3.6 Discussion and Conclusions

We developed the methodology needed to account for tidal deformation in binary stars in 1D stellar evolution codes. They are represented in the structure-correction factors of f_P and f_T that multiply the equations of hydrostatic equilibrium and radiative energy transport, respectively. Additionally, modified expressions of the Eddington limit and atmospheric boundary conditions as a result of the deformation have been calculated.

We show that the radii predicted by stellar-evolution models can shift by $\sim$5% if centrifugal deformation is included, while also including tidal deformation leads to a smaller $\sim$1% effect. This means comparing observed stellar parameters to non-rotating models could result in a mismatch and this could have implications for high-precision astrophysics. In a study where observed surface properties with uncertainties on the order of 5% are used, centrifugal deformation has to be included in the stellar models. In binaries, if the precision of measurement is better than 1%, then the tidal deformation has to be taken into account. We note that the effect observed here is only due to geometrical considerations, that is, the equator bulges out due to rapid rotation or the presence of a companion. Mixing as a result of rapid rotation was not included and will further affect the radius evolution (and age). Generally, the effect of rotational mixing is to keep the stars compact as fresh fuel is introduced into the core, so this may somewhat cancel the geometrical effect. However, the exact interplay and outcome of these competing effects is far from established.

The methods developed here have a wide range of applicability as they can be used for any (semi-)detached binary of arbitrary mass ratio. This can be applied, for instance, in studying the effect of tidal forces in mass transfer stability, which can have a potential impact on the formation of gravitational-wave sources through stable mass transfer (van den Heuvel et al. 2017; Marchant et al. 2021). There are, however, limitations to our model, most notably that the structure-correction factors computed from the binary Roche potential only apply to fully tidally locked systems. The general case, where the rotation of the components is free, cannot be described with a conservative potential, and is thus not suitable to the treatment of Kippenhahn and Thomas (1970) and Endal and Sofia (1976) used here. Developing a global model of free rotation including tides is a complex problem, but a potential workaround would be to consider free rotation for stars well within their RLs and use the centrifugal deformation corrections of single stars, and then implementing a switch to the synchronized binary corrections once a star fills an appreciable amount of its RL.

For contact configurations, so far only the mass ratio of unity can be considered consistently, as energy transfer in the contact layers, is zero. When we include the processes of mass and energy transfer in contact layers, which is the subject of the next chapter, the methods developed here enable the consistent modeling of all contact binary systems.

Chapter 4
Modeling Energy Transfer in Contact Binaries

"E.T. phone home..."

— E.T., the Extra-Terrestrial

This chapter is mainly based on:

Modeling contact binaries, II. The effect of energy transfer
M. Fabry, P. Marchant, N. Langer and H. Sana
Astronomy & Astrophysics, 672, A175 (2023)

4.1 Introduction

The modeling of energy transfer (ET) goes back to the 1960s in an effort to resolve Kuiper's paradox. Lucy (1968) first considered ET in common convective envelopes, and computed the first approximated structure models for W UMa stars. Later, Lucy (1976), Flannery (1976), Hazlehurst (1985) and Kähler (1989) calculated models that are out of thermal equilibrium and show cyclic behavior (see also, Chap. 6). Shu et al. (1976, 1979) and Lubow and Shu (1977, 1979) (collectively SLA) constructed models of contact binaries where they dropped the requirement of a continuous structure at the layer coinciding with the equipotential surface of the first Lagrangian point, L_1. These models resolve Kuiper's paradox regardless of the thermal structure of the envelopes, namely, whether they are radiative or convective. Despite much criticism (see e.g., Hazlehurst 1993; Kähler 1989), this is the simplest model of ET in radiative envelopes available in the literature.

Accurate modeling of massive, long-lived contact systems has been attempted in the past. Marchant et al. (2016) computed detailed evolution model grids of massive binaries with initial periods down to 0.5 d, which includes the regime of contact binaries at the zero-age main sequence. Those models however did not include ET between contact components, as the models concerned binaries of mass ratio close to unity, $M_2/M_1 = 0.8$–1, and the effect was thought to be minimal.

M. Fabry, *Detailed Structure and Evolution Modeling of the Tightest Massive Binary Stars*, Springer Theses, https://doi.org/10.1007/978-3-032-28788-5_4

Sen et al. (2022) used the models of Marchant (2018) to study the semi-detached Algol systems, although these models also included contact phases. Menon et al. (2021) computed models with initial mass ratios down to $M_2/M_1 = 0.6$ with the study of massive, long-lived contact binaries in mind. They found, also without including ET in contact phases, a strong correlation between observed mass ratio and period in contact systems that is broadly in agreement with observations. However, the mass-ratio distribution they derive is heavily skewed toward values close to unity, which is not supported observationally. These authors suggested that including ET in contact phases of unequal mass components could alleviate this discrepancy.

In this chapter, we apply the ET model of SLA in common stellar layers to modern stellar structure models to further advance our evolutionary modeling of massive contact binaries. In Sect. 4.2, we describe and discuss the theory of ET. Section 4.3 describes the physical setup of the stellar evolution code, along with our ET implementation. In Sect. 4.4, we compare models computed with and without ET in contact layers and discuss the difference in the observable properties of the models. Lastly, in Sect. 4.5, we provide our concluding remarks.

4.2 Theory of Energy Transfer

4.2.1 Simple Considerations

Theoretical modeling of contact binaries started with Kuiper (1941), who stated that stable contact systems of uniform composition with unequal masses cannot exist as a result of differing mass-radius relations. For the galactic ZAMS models of Brott et al. (2011), we derive the mass-radius relation of single stars to be approximately:

$$\left.\frac{R_2}{R_1}\right|_{\text{ZAMS}} = \left(\frac{M_2}{M_1}\right)^{0.57} = q^{0.57}. \tag{4.1}$$

However, the condition of contact in a binary following the Roche geometry constrains the surface of the stars to the same equipotential, which leads to:

$$\left.\frac{R_2}{R_1}\right|_{\text{Roche}} \approx \left(\frac{M_2}{M_1}\right)^{0.46} = q^{0.46}, \tag{4.2}$$

for stars not overflowing their RL too much. Clearly, Eqs. (4.1) and (4.2) cannot be satisfied simultaneously, unless $M_1 = M_2$. However, while the contact condition needs to be satisfied from dynamical arguments, the stellar structure does not, a priori, need to follow a single-star model. Even though the dense stellar core will be largely unperturbed due to the companion, the outer layers of contact components are highly distorted, causing variations in the total radii with respect to spherical models, as seen explicitly in Chap. 3. This result had not yet taken the possibility of ET into account and we expect further changes under the consideration that a hotter gas under similar pressure takes up more volume. Therefore, the ZAMS mass-radius relation in Eq. (4.1) is not expected to be satisfied under general contact conditions, and so Kuiper's paradox can be resolved by providing alternative stellar models that have a mass-radius relation closer to the contact condition in Eq. (4.2).

Following the Roche-lobe geometry, Lucy (1968) finds the ratio of the surface areas S_2/S_1 of two stars in contact to be proportional to $(M_2/M_1)^{\beta}$, with $\beta = 0.96$. Using the approximation $\beta \simeq 1$, combined with von Zeipel's theorem of gravity darkening, $T_{\rm eff}^4 \propto g$ (von Zeipel 1924), and the Stefan-Boltzmann law, $L \propto S T_{\rm eff}^4$, results in the simple expectation that in contact binaries, the luminosity ratio follows the mass ratio (Lucy 1968; Tassoul 2000):

$$\frac{L_2}{L_1} \simeq \frac{M_2}{M_1} = q. \tag{4.3}$$

Single main-sequence stars, on the other hand, follow the well-known mass-luminosity relation:

$$\frac{L_{\rm s,2}}{L_{\rm s,1}} \simeq \left(\frac{M_2}{M_1}\right)^{\alpha} = q^{\alpha}, \tag{4.4}$$

with $\alpha \simeq 2$–3 for the upper main sequence (Gräfener et al. 2011; Köhler et al. 2015). We use the symbol "$\simeq$" to denote these relations are approximations to simple power laws. We see that this leads to a difference between the luminosity of a single star $L_{\rm s,1}$ and the luminosity L_1 of a star of the same mass in a contact binary of

$$\Delta L_1 = L_1 - L_{\rm s,1} = -f L_1, \tag{4.5}$$

and for the companion

$$\Delta L_2 = L_2 - L_{\rm s,2} = f L_1 \simeq \frac{f}{q} L_2, \tag{4.6}$$

since $L_2 \simeq q L_1$ and we require $\Delta L_1 + \Delta L_2 = 0$ to conserve energy. This defines f as

$$f \simeq \frac{q - q^{\alpha}}{1 + q^{\alpha}}. \tag{4.7}$$

Therefore, the two stars in a contact binary can fulfill the single-star mass-luminosity relation in Eq. (4.4) in their cores and the contact binary mass-luminosity condition in Eq. (4.3) at their surfaces if the amount of energy per time, given by Eqs. (4.5) and (4.6), is transferred from the more massive to the less massive star in their common envelope.

4.2.2 Models of Energy Transfer

The general solution to Kuiper's paradox is to consider detailed stellar models with the inclusion of ET between the binary components. Several models of ET are given in the literature.

Lucy (1968) and Biermann and Thomas (1972) provided a first solution by adjusting the adiabatic constants of convective envelopes in contact components. However, this is an unsatisfactory solution, since this setup requires the stars to be burning hydrogen through different nuclear chains or cycles in the case of Lucy (1968) or that the models exhibit inaccurate light curves, as in Biermann and Thomas (1972). Other models, such as those of Lucy (1976) or Flannery (1976), relaxed the requirement of thermal equilibrium and constructed models of W UMa stars that exhibited thermal cycles. Kähler (1989) presented a detailed model that required turbulent motions in the common envelope to explain early-type (radiative) W UMa binaries.

Meanwhile, SLA presented the contact discontinuity model of contact binaries, by relaxing the requirement of continuous structural quantities across the RL. This is the only model that treats the common envelope as a single volume of the binary structure, at the price of hiding a heat engine in a very thin region around the RL. Schematically, their model can be represented as shown in Fig. 4.1. The components generate luminosities L_1 and L_2 respectively in their cores, and have this energy reorganized at the base the common envelope, just above the RL. The bulk of the common envelope, and in particular the surface, then uniformly radiate the combined luminosity, $L_1 + L_2$. Locally, gravity darkening is still applied, which, for radiative envelopes, is the von Zeipel theorem, $L \propto S\langle g\rangle$.

One peculiar feature is that this model necessitated a temperature inversion at the RL layer in one of the components as otherwise they would not be able to construct thermally stable contact models of uniform composition (in order to model binaries at zero age). This feature has received criticism in that the proposed heat engine violates the second law of thermodynamics and cannot be stable over thermal timescales (see Hazlehurst 1993; Kähler 1989, and references therein). Eventually, Kähler (2004) drew the following conclusion based on all collected theoretical arguments: internal circulation currents must exist in the less luminous component to reduce the radiative temperature gradient, since the luminosity carried by radiation is reduced by the circulation luminosity.

Given the complexity of the theoretical problem of the structure of contact binaries, especially with radiative envelopes, it is beyond the scope of this thesis to further develop the analytic theory. Instead, we apply an ET model in modern stellar-structure calculations. Using shellularity as our base assumption of the stellar structure (see Sect. 4.3.2 for the precise definition), the model of SLA is a natural choice, although we recognize that this comes with the apparent thermodynamical problems stated above. However, we believe we avoid the most fundamental one, as we do not explicitly require a temperature inversion in our models. We only used the model of SLA to compute the amount of energy transferred (see Sect. 4.2.3).

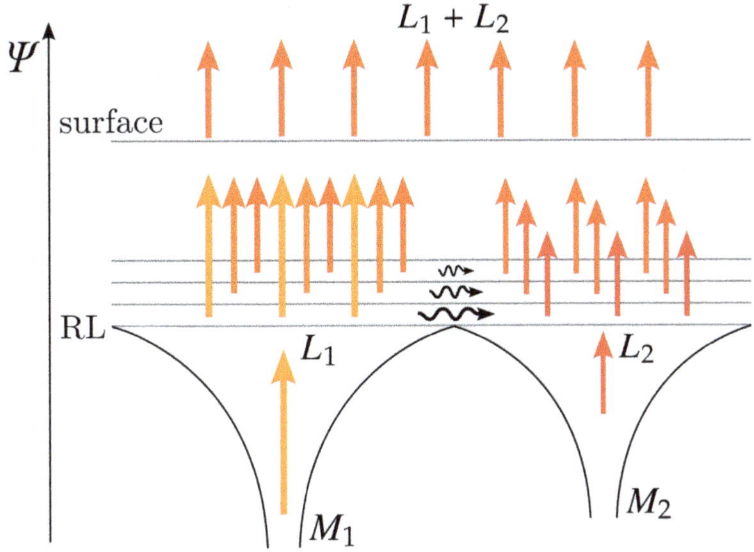

Fig. 4.1 Illustration of the energy-transfer process as modeled by Shu et al. (1979). The y-axis is the Roche potential, while the (omitted) horizontal axis represents an arbitrary horizontal direction along the equipotentials (which are disconnected in the wells of the stellar cores). A heat engine at the base of the common envelope moves energy from the more luminous to the less luminous component, so that in the common envelope, and also at the surface, the star radiates uniformly. Gravity darkening (not shown) can, however, still produce variations of the fluxes

4.2.3 *Energy Transfer in the Roche Geometry*

The work of SLA provides a general model of ET by introducing the notion of an energy flow at the base of the common envelope. If the contact layers are shellular, and they satisfy von Zeipel's gravity darkening, conservation of energy at the RL implies:

$$L_1' = (L_1 + L_2)\frac{S_1\langle g\rangle_1}{S_1\langle g\rangle_1 + S_2\langle g\rangle_2}, \tag{4.8a}$$

$$L_2' = (L_1 + L_2)\frac{S_2\langle g\rangle_2}{S_1\langle g\rangle_1 + S_2\langle g\rangle_2}. \tag{4.8b}$$

Here, S is the surface area of the RL, $\langle g\rangle$ is the surface averaged effective gravity at the RL, and the primed quantities specify the state just above the ET layer, while unprimed those just below. This equation specifies that the fraction of the total luminosity that each component radiates is proportional to $S\langle g\rangle$. The transferred luminosity then equals

$$L_{\rm trans} = L_1 - L_1' = \frac{L_1 S_2 \langle g \rangle_2 - L_2 S_1 \langle g \rangle_1}{S_1 \langle g \rangle_1 + S_2 \langle g \rangle_2}. \tag{4.9}$$

Comparing Eqs. (4.8) against Eqs. (4.5)–(4.6), we find for the fraction f:

$$f = \frac{L_1 S_2 \langle g \rangle_2 - L_2 S_1 \langle g \rangle_1}{(L_1 + L_2) S_1 \langle g \rangle_1}, \tag{4.10}$$

which is consistent with Eq. (4.7) as $S_2 \langle g \rangle_2 / S_1 \langle g \rangle_1 \simeq q$ and $L_2/L_1 \simeq q^{\alpha}$.

4.3 Methods

To investigate the effect of ET on the evolution of contact binaries, we computed binary evolution models using the stellar-evolution code MESA (Paxton et al. 2011, 2013, 2015, 2018, 2019; Jermyn et al. 2023), version r22.11.1. These models are the first massive binary-evolution models that include ET in contact layers. We followed the evolution of binaries from the ZAMS until the less massive component overflows the second Lagrangian point.

4.3.1 Physical Assumptions in MESA

The microphysical setup of our stellar evolution models remains mostly the same as in Chap. 3, with some additions from the newer MESA version. We use the same nuclear microphysics as in Chap. 3. The EOS has an additional table available from the Skye project (Jermyn et al. 2021), its blend specified in Jermyn et al. (2023). Radiative opacities, are as in Chap. 3, with Compton scattering opacity added from Poutanen (2017), and electron conduction opacities are taken from Cassisi et al. (2007) and Blouin et al. (2020). Like Chap. 3, we set the metallicity of stars to the solar value, $Z_{\odot}$, where $Z_{\odot} = 0.0142$, with metal fractions as determined from Asplund et al. (2009).

Wind-mass loss is determined similarly as in Chap. 3, and we allow part of the wind launched by a star to be accreted by its companion using the Bondi-Hoyle mechanism (Bondi and Hoyle 1944) as implemented by Hurley et al. (2002).

For internal mixing and rotation, we use the same setup as in Sect. 3.2.

The treatment of mass transfer (MT) is explained below in Sect. 4.3.3, while the implementation of energy transfer (ET) is shown in Sect. 4.3.4.

4.3.2 Shellularity and Roche Lobe Geometry

Since we deal with highly tidally deformed stars, we use the modifications to the stellar-structure equations from Sect. 3.2.1 to incorporate the RL geometry into a one-dimensional stellar-evolution code. The stars are therefore modeled as hydrostatic structures living in the Roche potential Ψ of a fully synchronized binary. Furthermore, we assume stellar layers to be shellular. Shellularity is reached when all intensive quantities (in particular the temperature, pressure, and mass density) are constant along a stellar layer and when such a layer coincides with a unique equipotential surface.

It should be emphasized that in the context of 1D models, the full shellularity of layers of a (single) star is an assumption—not a self-consistently modeled feature—as, of course, there is no 3D structure to explicitly test the shellularity of a stellar layer in the Roche potential. However, overflowing layers of components in a contact binary can be tested whether they are shellular *with respect to each other* if the temperature, density, and so on at the 1D cells are equal for equal values of Roche potential. Given the method we used to split the common envelope of the contact binary (Sect. 3.2.4), the computation of the tidal deformation corrections based on equipotential surfaces, and the usage of the corresponding outer boundary condition (Sect. 3.4), the final ingredient of ET in the common envelope developed here will ensure the shellularity of one component with the other—again, defined as having similar cell values of pressure, temperature, and so on—for similar values of potential.

In contact binaries, the outer stellar layers are shared between the components and a choice must be made as to what part of the envelope belongs to each component. To this end we constructed what we call a splitting surface emanating from the Lagrangian point L_1, by following the gradient $\nabla\boldsymbol{\Psi}$ outward in all directions (see Sect. 3.2.4). With this setup, the volume of both components is uniquely defined, and in Sect. 3.2.5 we computed integrals in the Roche potential necessary to represent distorted shells in 1D. The volume-equivalent radius then acts as the new independent variable of a stellar shell, and is defined as in Eq. (3.1). Given a certain splitting strategy, and the requirement of the stellar surface being on a common equipotential, a relation of the form:

$$r_{\Psi,2} = F(q; r_{\Psi,1}), \tag{4.11}$$

with $q = M_2/M_1$, constrains the radii of the overflowing components 1 and 2. This is a statement of the contact condition and it is at the basis of Kuiper's paradox (Sect. 4.2). For vertical splitting surfaces through L_1, Marchant et al. (2016) found the following fit:

$$\frac{r_{\Psi,2} - R_{\mathrm{RL},2}}{R_{\mathrm{RL},2}} \approx q^{-0.52} \frac{r_{\Psi,1} - R_{\mathrm{RL},1}}{R_{\mathrm{RL},1}}. \tag{4.12}$$

With our setup of the splitting surface however, this approximation is no longer accurate for significantly overflowing shells of mass ratios away from unity. Therefore, in this thesis, we evaluate the function F in Eq. (4.11) by interpolating the results of the Roche integrations obtained in Sect. 3.2.5.

Retaining the form of the fit of Marchant et al. (2021), for the radius of a component overflowing to its outer Lagrangian point, L_{out}, which is L_2 for the less massive and L_3 for the more massive component, we constructed a new fit from the integration results of Sect. 3.2.5, which is:

$$\frac{R_{L_{\mathrm{out}}} - R_{\mathrm{RL}}}{R_{\mathrm{RL}}} = \frac{3.3752}{1 + \left(\frac{\ln q + 1.0105}{\sigma}\right)^2} \cdot \frac{1}{9.0087 + q^{-0.4022}}, \tag{4.13}$$

$$\sigma = \frac{62.9237}{15.9839 + q^{0.2240}}.$$

This fit has an error smaller than 0.1% in the range $-7 \leq \log q \leq 7$, compared to 1% if we used the fit of Marchant et al. (2021). We use this fit to determine when the least massive component overflows to L_2, after which we stop the evolutionary simulation.

4.3.3 Mass Transfer

The components of a contact binary are, per definition, in mechanical contact with each other. Thus, stellar material can be exchanged throughout the evolution of such a system. However, MT cannot occur at an arbitrary rate, since the surface of both components are constrained by the contact condition in Eq. (4.11). If this were not satisfied, pressure gradients would arise across the surface of the contact binary, and we expect this would induce strong, horizontal flows equilibrating the surface.[1] Thus, in our simulations, we implicitly adjust the MT rate either to keep the donor just below its RL radius in a semi-detached configuration, or so the component radii satisfy the contact condition of Eq. (4.11) in a contact configuration (see also Marchant et al. 2016).

[1] Horizontal in this context means perpendicular to the local effective gravity, $\boldsymbol{g} = \nabla\boldsymbol{\Psi}$. Conversely, vertical means parallel to $\nabla\boldsymbol{\Psi}$.

4.3.4 Energy Transfer

Not only are the contact layers of both stellar components in mechanical contact, facilitating MT, there is also thermal contact between these layers, allowing for ET. As elaborated in Sect. 4.2, we model ET in contact binaries as the transfer of luminosity of one component to the other (Eq. (4.9)), at the location of the RL. It should be mentioned that no further corrections to the ET rate need to be computed since the splitting surface we use is parallel to the local effective gravity and thus so is the radiative flux too. Therefore, since the regular energy transport by radiation (or by convection) in the interior of either component is vertical, no vertical energy flux crosses over toward the other component. Our splitting strategy thus nicely decouples horizontal and vertical energy flows and in this approximation, and the ET rate computed here is not affected by the vertical energy transport present in the components.

The stability and performance of numerical simulations are severely impacted by introducing sharp discontinuities at the RL radius; in this case, that of luminosity. To remedy these issues, we used the following approach.

First, for numerical stability over evolutionary steps, we smooth the amount of ET by computing:

$$\tilde{L}^{n}_{\text{trans}} = pL^{n}_{\text{trans}} + (1-p)\tilde{L}^{n-1}_{\text{trans}}, \tag{4.14}$$

where $\tilde{L}^{n}_{\text{trans}}$ is the smoothed transferred luminosity at an evolutionary step, n, L_{trans} is the luminosity to be transferred as per the Roche geometry (Eq. (4.9)), and p is the smoothing factor. In all simulations including ET, we use a moderate smoothing of $p = 0.5$.

Then, we implement the luminosity transfer as a constant extra source of specific heat $\varepsilon_{\text{RL},1,2}$ in both stars, occurring in the cells with a radius within 1.00 and 1.01 times their respective RL radii, a choice meant to reflect the estimate of Shu et al. (1979) (see Eq. (4.17) below and also Appendix C.1 for further details). The magnitude of $\varepsilon_{\text{RL},1,2}$ is determined from the required luminosity:

$$\varepsilon_{\text{RL},1}\Delta m_1 = \tilde{L}_{\text{trans}}, \tag{4.15a}$$

$$\varepsilon_{\text{RL},2}\Delta m_2 = -\tilde{L}_{\text{trans}}, \tag{4.15b}$$

where $\Delta m_{1,2}$ is the mass of the cells we put the extra heat in.

Since MT is modeled as removing stellar material from the top of the donor and putting it on top of the accretor, during MT, the common layers are out of thermal equilibrium due to expansion or compression. To correct for this effect and regain shellularity, especially at the surface, we transport additional heat in the common envelope according to Eq. (4.9), where now we evaluate $S\langle g\rangle$ at the surface layer of both components instead of the RL. The constant heat source $\varepsilon_{\text{surf},1,2}$ is then computed similarly to Eq. (4.15), only now Δm encompasses all shells from 1.01 times the RL up to the surface.

As a final approximation, we linearly scale the necessary ET for shallow contact systems, defined as $0 \leq \frac{R-R_{\rm RL}}{R_{\rm RL}} \leq 0.01$ for either star:

$$\tilde{L}_{\rm trans,\,shallow} = \tilde{L}_{\rm trans} \frac{\min\left(\frac{R_1-R_{\rm RL,1}}{R_{\rm RL,1}}, \frac{R_2-R_{\rm RL,2}}{R_{\rm RL,2}}\right)}{0.01}. \tag{4.16}$$

We recognize that using our ET prescription in fast MT phases is the weakest element of our simulations. When the MT timescale is shorter than the thermal timescale, the star falls out of thermal equilibrium, which is characterized by strong (vertical) gradients in the luminosity profile. During such phases, the energy budget of the star is redistributed vertically in order to regain as best as possible thermal equilibrium. Adding to this process the horizontal energy transport in the layers near the RL could induce complex interactions between both energy flows. Even though we mention the horizontal and vertical energy flows to be decoupled above, this is an idealization and the approximation may break down in quickly evolving phases. Furthermore, while our computation of how much energy is transferred depends only on the geometry and the theorem of von Zeipel, which does not require thermal equilibrium of the layers, hydrodynamical effects break the validity of von Zeipel's theorem (von Zeipel 1924), which can be significant near the Lagrangian point L_1.

4.4 Stellar Models

4.4.1 Detailed Example of ET Evolution

Here, we detail the evolution of a 25 $M_\odot$ primary and a 20 $M_\odot$ secondary (so $q_{\rm init} = 0.8$) in an initial orbit of 1.4 days. This initial configuration is chosen such that, at the onset of the long lasting contact phase, this model mimics the observed galactic contact binary V382 Cyg studied in Abdul-Masih et al. (2021). We simulate both cases where ET, as implemented in Sect. 4.3.4, is included and excluded.

In Fig. 4.2, we show the HRD broken up into four distinct phases of the evolution, while in Fig. 4.3, we show the MT rate and mass-ratio evolution. First, we have the pre-interaction phase (leftmost column) followed by a short-lived, fast case A MT phase from the primary to the secondary when the donor experiences RLOF (second column of Fig. 4.2). During this MT phase, the components briefly enter into contact, as the accretor loses thermal equilibrium and swells beyond L_1. After thermal relaxation of the accretor, the system detaches, after which slow case A MT is initiated (third column of Fig. 4.2), still from the primary to the secondary, although the mass ratio has inverted by this point. This phase of nuclear-timescale MT occurs in a semi-detached configuration and is referred to as an Algol phase. Finally, the evolution of the now more massive secondary catches up to that of the

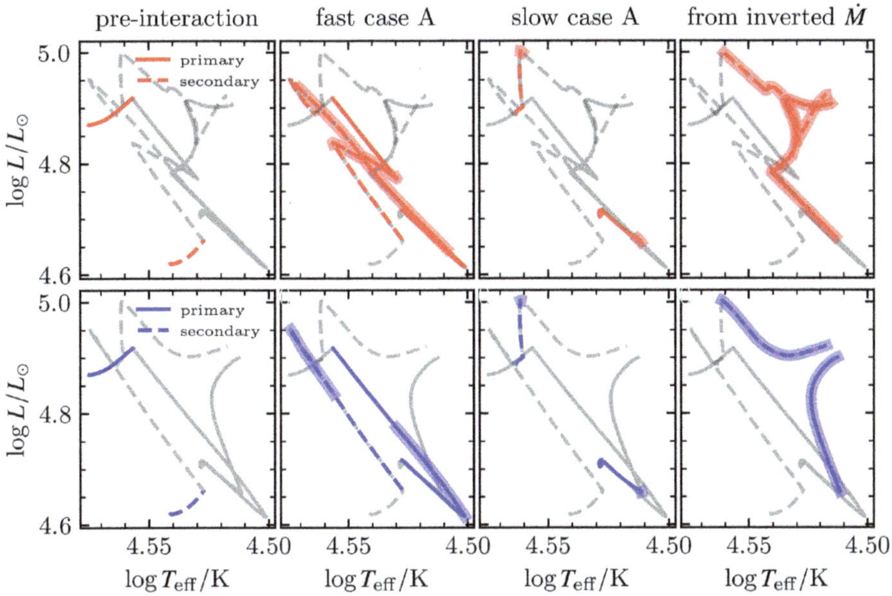

Fig. 4.2 HRD of a $25\,M_\odot + 20\,M_\odot$ binary with an initial period of 1.4 d. Red lines indicate a simulation with the inclusion of ET, while blue lines ignore ET. Moving left to right, the columns highlight four subsequent phases in the evolution (see Sect. 4.4.1 for details). Contact phases are highlighted with thick outlines

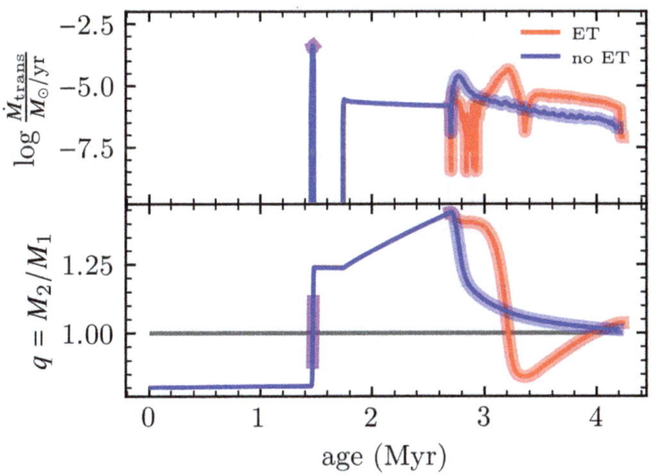

Fig. 4.3 MT rate (top panel) and mass-ratio evolution (bottom panel) of the $25\,M_\odot + 20\,M_\odot$ binary studied in Sect. 4.4.1. Contact phases are highlighted with thick outlines

stripped primary, contact is engaged again and MT inverts (rightmost column of Fig. 4.2), moving material from the secondary back to the primary (however see below). This contact configuration evolves on the nuclear timescale and ends when L_2 overflow occurs.

Qualitatively, the evolution of this system is similar with and without inclusion of ET. Up to and including the slow case A MT phase, the evolution of this system is nearly identical. One exception is the different HRD tracks during the fast MT case. In the second column of Fig. 4.2, the red tracks exhibit a "hook," while the blue tracks do not. This is a direct consequence of ET in the short contact phase changing the outer structure of the components. The secondary donates energy to the primary and cools a bit, jumping to the right of the HRD (and vice versa for the primary). The point at which the primary and secondary tracks touch (at roughly $\log L/L_{\odot} = 4.8$ and $\log T_{\rm eff}/{\rm K} = 4.545$) corresponds to the situation where $M_1 = M_2$. This is consistent with the theory since we expect that for $M_1 = M_2$ we have similar luminosities per Eq. (4.3) and also similar effective temperatures, following the law of Stefan-Boltzmann. When contact is disengaged, the secondary (at this point more massive than the primary) jumps back to the left as the energy sink to the primary disappears. Given the timescale of this MT phase, however, it is unlikely to observe contact binaries in this phase and thus to constrain the effects of ET.

Still, we do observe quantitative differences between the energy-transferring and non-energy-transferring models during the final, inverted MT phase, which is long-lived and therefore more likely to be observed. In particular, from the $q(t)$ curves in the bottom panel of Fig. 4.3, we see that the ET model spends more time at mass ratios $q \gtrsim 1.4$ versus the no-ET model. Furthermore, the no-ET model monotonically approaches a mass ratio of unity without further inverting the mass ratio or the MT direction. In contrast, the ET model experiences two inversions of the mass ratio at around $\tau = 3.2$ and 4.0 Myr and the MT direction is inverted at around $\tau = 3.3$ MYr.

In Fig. 4.4 (top panel), we show the differential duration of the long-lived contact phase, plotted against $\bar{q} = \min(M_1/M_2, M_2/M_1)$, while the cumulative duration is shown in the bottom panel.[2] The top panel shows three distinct peaks in the red $\bar{q}$ distribution, two of which correspond to the periods where the ET model is inverting its MT direction (at $q \approx 1.4$ and $q \approx 0.85$) and the last corresponds to a mass ratio close unity $\bar{q} \approx 0.97$ at the end of the simulation. Comparing the two former peaks to the no-ET curve at those $\bar{q}$ shows that the ET model spends more time at these mass ratios than the no-ET model. As a cumulative distribution, from the bottom panel of Fig. 4.4, we deduce that it is about four times more likely to observe the ET model in a configuration more extreme than $\bar{q} = 0.8$ than the no-ET model, even though the total length of the contact phase is similar to within 0.05 Myr.

When comparing the profiles of the components in both models, we note a considerable difference in the structure of the outer layers of the components. In

[2] We use $\bar{q}$ here since in an observed system, one has no a-priori information on what the most massive component is.

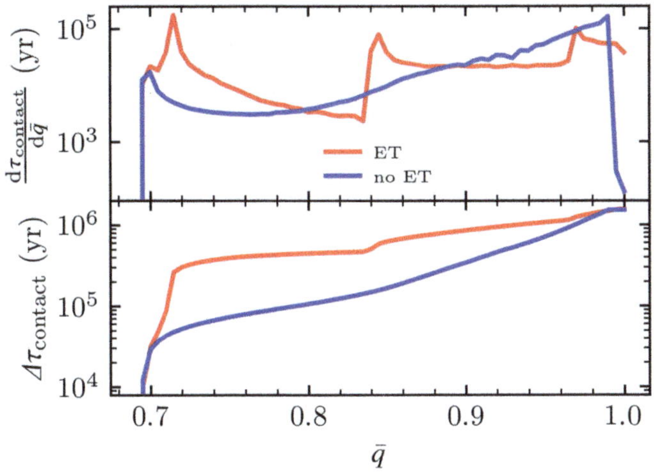

Fig. 4.4 Contact duration distributions as function of observed mass ratio $\bar{q} = \min\left(\frac{M_1}{M_2}, \frac{M_2}{M_1}\right)$. *Top:* Differential duration (in bins of $\Delta\bar{q} = 0.005$). *Bottom:* Cumulative duration in configurations more extreme than $\bar{q}$

Fig. 4.5, the temperature, density, and luminosity profiles of the models at the start of inverted MT phase ($\tau = 2.7\,\mathrm{Myr}$) are shown. The independent coordinate is chosen to be $\tilde{r} = \frac{r - R_{RL}}{R - R_{RL}}$, so as to map the overflowing layers to the interval [0, 1]. At this point in the evolution, in both cases of ET and no ET, the stars are in a contact configuration and the primary is stripped to $M_1 = 18.3\,M_\odot$, while the secondary accreted mass to $M_2 = 26.4\,M_\odot$. We note that these masses are close to the component masses Abdul-Masih et al. (2021) found for V382 Cyg. We see that in the system without ET, the overflowing shells do not satisfy our definition of shellularity since the temperature and density profiles of the primary and secondary do not match for constant Roche potential (which approximately corresponds to constant $\tilde{r}$). The ET models however match significantly better, as seen in the left column of Fig. 4.5 by the joining of the temperature and density profiles of the primary and secondary. In particular, at the surface, the density of the ET components agree to within 0.2%, while the temperature to within 0.5%, whereas the surface properties of the no-ET components vary more than 10% (as expected for models of differing mass). While the former differences are hard to constrain from measurement, current data-analysis techniques are able to measure surface temperatures with accuracies better than 10%, so that the no-ET model can be ruled out observationally.

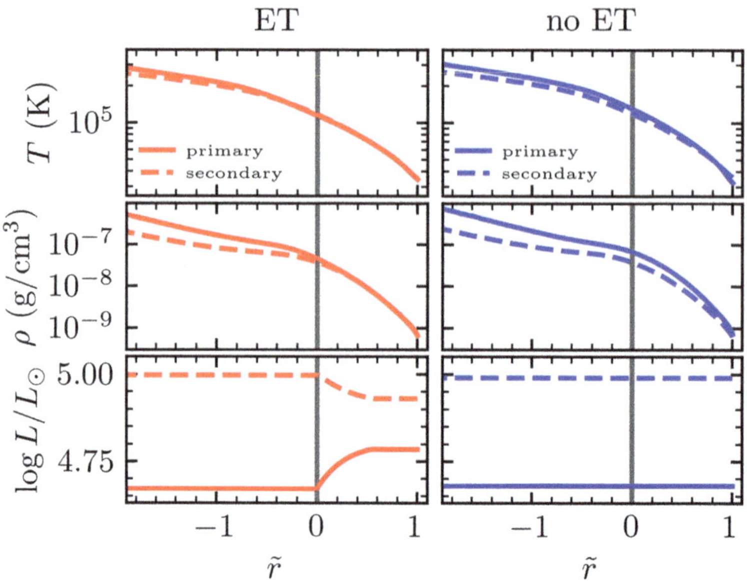

Fig. 4.5 Profiles of outer layers of the binary components at the onset of inverse MT. It shows (top to bottom) the temperature, density, and luminosity profiles for both components in the ET (red, left column) and no ET (blue, right column) cases, all as a function of the scaled radius coordinate $\tilde{r} = \frac{r-R_{\rm RL}}{R-R_{\rm RL}}$. The gray vertical lines show the location of the RL

A significant shortcoming of our model is the assumed thickness of the ET layer. Shu et al. (1979) argued that the thickness d of the ET layer is on the order of:

$$\frac{d}{a} \sim \delta^{0.4}, \quad \delta = \frac{L_{\rm trans}/a^2}{\rho h c_s}, \tag{4.17}$$

with a as the binary separation, and ρ, h, and c_s as the density, specific enthalpy and local sound speed evaluated at the RL, respectively. Lubow and Shu (1979) computed this number to be $d/a \sim 10^{-2}$ for stars of masses around 4–8$M_\odot$ which justifies SLA in modeling the layer as a discontinuity in the stellar profile, located at the RL radius.

However, direct computation of Eq. (4.17) for our 25 $M_\odot$ + 20 $M_\odot$ model show that this estimation breaks down for higher masses, see the red line in Fig. 4.6. This suggests that the energy redistribution flow, modeled as a discontinuity by SLA, is not sufficient in these higher mass stars. Except when the binary has reached considerable overflow, where $d/a \lesssim 10^{-2}$, we find that the thickness needed can be a significant fraction of the binary separation, even surpassing it in the early stages of the contact phase.

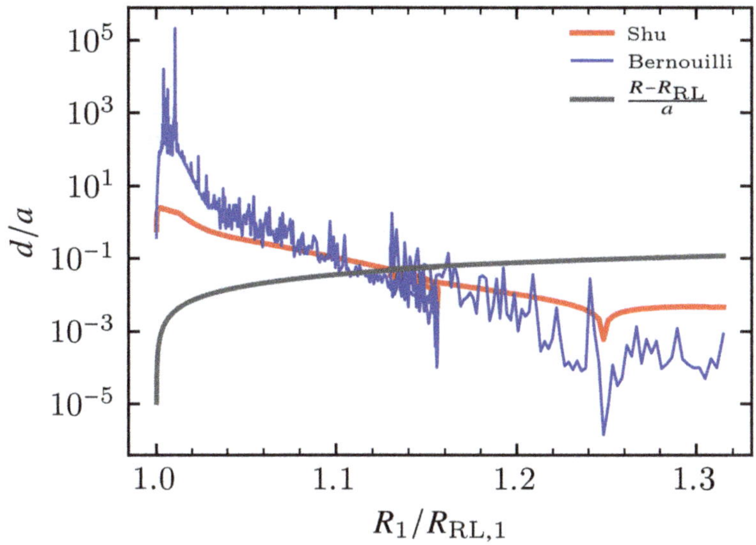

Fig. 4.6 Thickness of the ET layer, d, with respect to the binary separation, a, as a function of the radius of the primary star during the long lived contact phase. The gray line gives the physical size of the overflowing layers, and corresponds to the maximal width the ET layer can assume

Another way to compute the thickness of the ET layer is to use Bernouilli's equation. If we consider fluid motion from far away from L_1 on the primary star (location i) toward L_1 (location f), we have

$$\frac{1}{2}v_f^2 + \int_i^f \frac{dP}{\rho} = 0, \tag{4.18}$$

where we have already canceled the potential terms Ψ_i, Ψ_f since we move along an equipotential surface, and the initial velocity, v_i, is assumed to be negligible. Making the estimation:

$$\int_i^f \frac{dP}{\rho} \approx \left(\frac{1}{\rho_f} + \frac{1}{\rho_i}\right)\left(P_f - P_i\right), \tag{4.19}$$

and assuming that the transferred energy through the binary neck of width, b, by a mass flow of $\dot{M} = \rho_i v_f b^2$ is:

$$L_{\rm trans} = \rho_i v_f b^2 (c_{p,1} + c_{p,2})(T_f - T_i), \tag{4.20}$$

we compute for the minimal thickness of the ET layer:

$$b \approx \sqrt{\frac{L_{\rm trans}}{\rho_i (c_{p,1} + c_{p,2})(T_f - T_i)\sqrt{\left(\frac{1}{\rho_f} + \frac{1}{\rho_i}\right)(P_f - P_i)}}}. \tag{4.21}$$

Finally the thickness of the layer at the neck is related to the thickness far away from $\mathrm{L_1}$ via:

$$\frac{d}{a} \approx \left(\frac{b}{a}\right)^2, \tag{4.22}$$

since the Roche potential varies quadratically near $\mathrm{L_1}$ and linearly elsewhere. Equation (4.21) gives a lower limit on the width of ET layer so that a balanced mass flow, $\dot{M}$, in the contact binary can carry a to be transferred luminosity, $L_{\rm trans}$. Conversely, it can be interpreted as $L_{\rm trans}$ being the maximal luminosity the mass flow can carry in a layer of fixed width, b.

We plot the Bernouilli-computed thickness of Eq. (4.22) in Fig. 4.6. Similarly, we see that the required thickness, d/a, is much larger than what the radius of the primary star allows room for. It is only at later times, when the mass ratio has equilibrated and deeper contact is engaged, that the estimated width becomes smaller than the overflow rate, $R - R_{\rm RL}$, of the primary.

4.4.2 Mass Versus Luminosity Ratios

During the nuclear-timescale, inverted MT phase, contact is engaged so that our ET scheme acts to move luminosity from one component to the other. As mentioned in Sect. 4.2, we expect the luminosity ratio of contact binaries to follow the mass ratio, $L \propto M$, as opposed to detached stars following a single-star mass-luminosity relation, $L \propto M^\alpha$, with $\alpha \simeq 2$–3. Figure 4.7 shows the evolution of the luminosity ratio as function of the mass ratio during the slow MT phase of the 25 $M_\odot$ + 20 $M_\odot$ explored in Sect. 4.4.1. In this graph, the models evolve from the top right at $q \approx 1.4$ to near-equal mass ratio on the left. We see that the model not including ET follows closely a $q^{2.2}$ relation, appropriate for single stars in the mass range of 10–30$M_\odot$. As the models including ET engage into deep contact however, their mass-luminosity ratio changes drastically from the $q^{2.2}$ line in near-contact to the q^1 relation in full contact.

We give (overplotted on Fig. 4.7) the measurements of several observed massive contact binaries of Abdul-Masih et al. (2021), Yang et al. (2019) and Lorenzo et al. (2014) (see also Fig. 2 of Langer 2022). Curiously, the luminosity ratios from Abdul-Masih et al. (2021), while they do follow an $L \propto M$ trend, are offset to lower L_2/L_1 than predicted. Either the luminosity of the primary is overestimated or the

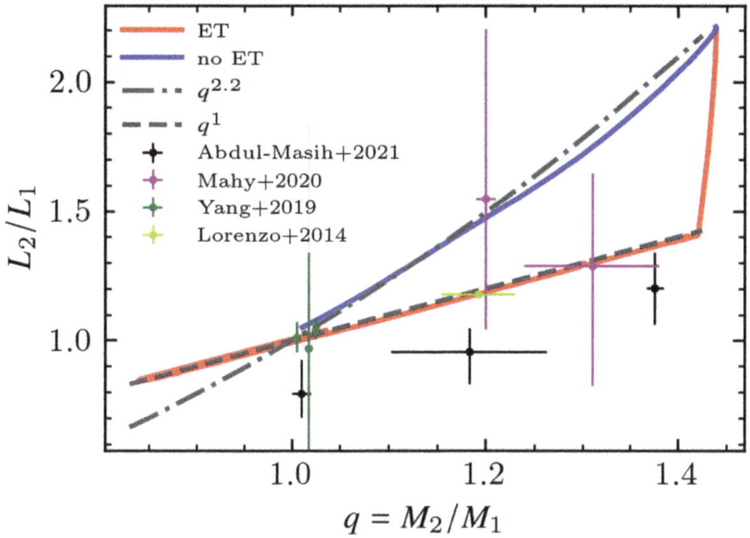

Fig. 4.7 Luminosity ratio versus mass ratio during the nuclear-timescale MT of the 25 $M_\odot$ + 20 $M_\odot$ system of Sect. 4.4.1. Measurements from observed massive (near-)contact systems from Abdul-Masih et al. (2021), Yang et al. (2019), and Lorenzo et al. (2014) are overplotted. The systems from Mahy et al. (2020a) were classified as 'uncertain configurations'

uncertainties are underestimated. The systems included from Mahy et al. (2020a) were categorized as 'uncertain configurations' since the measurement of the radius was consistent with being above as well as below the RL. In the context of the mass-luminosity relation however, we expect that the system from Mahy et al. (2020a) at $q \approx 1.3$ (VFTS 563) is a true contact system, while the one at $q \approx 1.2$ (VFTS 217) is not (although within 1σ it could be either).

4.5 Conclusions

In this chapter, we have taken a step forward in the detailed modeling of contact binaries, by implementing a model of ET in detailed stellar-structure and evolution models. From Fig. 4.5, we see that this relatively simple model (i.e., the inclusion of a heat source or sink in the stellar model as proxy for the ET) is capable of making the common layers of stellar components in a contact configuration shellular. Models without such ET do not exhibit their common layers to be shellular, which, from theoretical arguments, would drive strong horizontal flows equilibrating all gradients, in particular pressure.

From Figs. 4.3 and 4.4, we see that the time spent in deep contact at mass ratios between $\bar{q}$ = 0.7–0.8 is extended when ET was included, versus when it was ignored. This is a promising result, in that if this trend persists across the parameter space of the total mass, initial mass ratio, and initial period, ET could provide an answer to the discrepancy between the observed mass-ratio distribution of massive contact systems and its predicted distribution. The computation and analysis of a full grid of models for a population synthesis study is the topic of the next chapter.

Chapter 5
Energy Transfer in a Population of Massive Contact Binaries

> *"I know my apprehensions might never be allayed, and so I close, realizing that perhaps the ending has not yet been written."*
>
> *—Atrus*

This chapter is based on:

Modeling contact binaries, III. Properties of a population of close, massive binaries
M. Fabry, P. Marchant, N. Langer and H. Sana
Astronomy & Astrophysics, 695, A109 (2025)
and:
Effects of limited core rejuvenation on the properties of massive contact binaries
J. Vandersnikt & M. Fabry
Astronomy & Astrophysics, 695, A223 (2025)

5.1 Introduction

In Chaps. 3 and 4, we have developed the methodology to consistently take into account tidal deformation of close binary components and energy transfer in the common layers of contact binaries. In the introduction, we have also established that the state-of-the-art models have trouble reproducing the observed mass-ratio distribution of massive contact systems. While the models indicate that, once nuclear-timescale contact is engaged, the masses should equalize rapidly (Menon et al. 2021), long-term monitoring of the period derivative shows observed systems are stable on the main-sequence lifetime (Abdul-Masih et al. 2022). Still, in Chap. 4, we showed that for particular systems, energy transfer in radiative envelopes can

M. Fabry, *Detailed Structure and Evolution Modeling of the Tightest Massive Binary Stars*, Springer Theses, https://doi.org/10.1007/978-3-032-28788-5_5

alter the mass-ratio evolution of massive contact systems, and has the possibility to relieve the discrepancy between models and observations.

In this chapter, we compute a grid of short-period, massive binary-evolution models and investigate the effect of energy transfer on the mass-ratio distribution of contact systems. In Sect. 5.2, we describe the modifications to the stellar evolution code with respect to the previous chapters, the initialization of the binary models, termination conditions, and we detail the calculations used for making population predictions. Next, Sect. 5.3 presents and discusses the results obtained from the population synthesis, while Sect. 5.5 gives concluding remarks.

5.2 Methodology

We use the detailed binary evolution code MESA, version 22.11.1 (Paxton et al. 2011, 2013, 2015, 2018, 2019; Jermyn et al. 2023), to compute grids of binary models that undergo case A mass transfer, from which we assemble a synthetic population. The parameters to be varied in each grid are the initial primary mass, $M_{1,\mathrm{init}}$, initial mass ratio, $q_{\mathrm{init}} = M_{2,\mathrm{init}}/M_{1,\mathrm{init}}$, and the initial period, p_{init}. For initial primary mass, we sample $M_{1,\mathrm{init}} = \{8, 9, 10, \ldots, 19\} \cup \{20, 22, 24, \ldots, 48\} \cup \{50, 52.5, 55, \ldots, 70\}\, M_{\odot}$. The initial mass ratio is $q_{\mathrm{init}} = 0.6$–0.975, spaced uniformly with $\Delta q_{\mathrm{init}} = 0.025$, while the initial period is $p_{\mathrm{init}} = 0.5$–$8\,\mathrm{d}$, spaced logarithmically with $\Delta \log(p/\mathrm{d}) = 0.04$. We compute two grids with these parameter variations: one that includes the effect of energy transfer (ET), while the other ignores it. This amounts to a total of 35,712 detailed binary-evolution models. Throughout the discussion, we will refer to the models that include ET as the "ET models," while those that do not the "no-ET models."

5.2.1 *Physical Assumptions in MESA*

We use a similar setup of the MESA code as in Chap. 4, i.e., we use the same microphysics, convection and overshoot parameters (with two exceptions, see below), wind prescription, deformation geometry, and mass- and energy-transfer (MT, ET) calculations. In contrast to Chaps. 3 and 4 we make the following modifications to the modeling setup:

- We do not assume rigid-body rotation throughout the evolution, so that differential rotation is allowed. However, we do include fast tidal synchronization by uniformly modifying the angular momentum of the star on the timescale of the orbital period, p,

$$\Delta j_{\mathrm{tides}} = \left(1 - \exp\left(\frac{-\Delta t}{p}\right)\right) (\omega_{\mathrm{orb}} i_{\mathrm{rot}} - j)\,. \tag{5.1}$$

Here Δt is the timestep of the evolution, j and i_{rot} are the current angular momentum and the moment of inertia of the stellar layer, respectively, and $\omega_{orb} = 2\pi/p$ is the orbital angular velocity. We model spin-orbit coupling to conserve total angular momentum, so that the required angular momentum to synchronize both stars in this fashion is then subtracted from the orbital angular momentum. Given that Δt is much longer than the orbital period in evolutionary calculations (years to thousands of years versus days), Eq. (5.1) specifies efficient tidal synchronization. We thus expect little departure from solid-body rotation in our very close binary simulations, since processes that cause departures from it operate on longer timescales than the orbital one.

- We also allow for rotational mixing in our models. We include Eddington-Sweet circulation, the Goldreich-Schubert-Fricke instability and both the dynamical- and secular-shear instability, all following the implementation of Heger et al. (2000).
- We reduce the parameters of MLT to $\alpha_{MLT} = 1.5$, and the semiconvective efficiency to $\alpha_{sc} = 1$, to be in line with Menon et al. (2021).

5.2.2 *Binary Initialization*

Once initial masses and a period are selected, the binary model is initialized as follows. Non-rotating, single-star models of mass $M_{1,\mathrm{init}}$ and $M_{2,\mathrm{init}} = qM_{1,\mathrm{init}}$ are interpolated from pre-computed zero-age main sequence (ZAMS) models of solar metallicity ($Z_\odot = 0.0142$, with metal fractions following Asplund et al. 2009) and a helium content of $Y = 0.2703$ and loaded into a binary model. We set the eccentricity to zero as we expect circular orbits for close binaries (Zahn 1975). The single-star models are then spun up to the appropriate Keplerian angular velocity ω_{orb}. Since the models are brought out of thermal equilibrium due to the spin up, what follows is a thermal relaxation period onto the true ZAMS of the models, during which MT, ET or any other mass loss is turned off. We also keep the period constant to the initial value and enforce rigid body rotation at the orbital angular velocity. We define the ZAMS as the point where the luminosity of both stars is within 1% of their nuclear luminosity. When this condition is reached, we make a decision depending on the rate of overflow of the stars.

First, if both stars are within their RL, the simulation proceeds as normal, without further intervention. Second, if at least one star overflows its RL, this means that MT must have started in the pre-main sequence (PMS) phase. To simulate this, we artificially boost the orbital period to put the components in a detached configuration, and then, on a thermal timescale of the most massive star (defined here as $\tau = 0.75GM_1^2/R_1L_1$), drain the added angular momentum from the system to again reach the initial period. During this process, the components reach the RL, and MT and ET are allowed. Finally, if one of the stars overflows the second Lagrangian point (L2OF) at the ZAMS, we terminate the simulation immediately as we assume a merger would have happened on the PMS.

5.2.3 Outcomes and Termination

The outcome of our simulations broadly falls in three categories.

- The binary reaches L_2 in a contact configuration. In this situation, we assume mass loss will start from L_2, carrying away angular momentum, which we expect results in a merger on a dynamical timescale (see, e.g., Pejcha 2014). We stop the simulation at this point and classify these systems as "mergers." A variant of this outcome is when the MT calculation determines that $\dot{M}$ is a very high value. We choose to take a limiting value of $\dot{M} = 10^{-1} M_{\odot}\mathrm{yr}^{-1}$ across the whole grid. MT rates higher than this value are at least two orders of magnitude above the thermal timescale MT rates of massive main-sequence stars (except perhaps for the highest masses in our grid, where $\dot{M}_{\mathrm{thermal}} \approx 5 \times 10^{-3}\, M_{\odot}\mathrm{yr}^{-1}$), and we expect MT to evolve toward dynamical timescale MT.
- One of the stars (most likely the initially more massive one) reaches core hydrogen exhaustion, without triggering a merger. At this point, the star will start undergoing rapid evolution, and we stop the simulation calling these "(main-sequence) survivors."
- The simulation can encounter numerical difficulties, by failing to reach convergent models during the evolution. In this case, we halt the simulation and signal this system encountered an "error."

5.2.4 Population Synthesis Computations

When doing population synthesis calculations, each model in our grid needs to be assigned a relative weight to accurately predict the occurrence rate of a certain system in a population. Because the probability, $\mathcal{P}$, of finding systems in a certain configuration ϑ_0 is proportional to the time the system spends in that configuration, we have

$$\mathcal{P}(\vartheta_0) \propto \int \delta(\vartheta - \vartheta_0) \mathcal{W}_{\mathcal{N}} \, \mathrm{d}\mathcal{N}, \tag{5.2}$$

where we sum the time each of the models, $\mathcal{N}$, spends at the configuration ϑ_0, using the Dirac delta function, δ. The integrand is then multiplied with its statistical weight, $\mathcal{W}_{\mathcal{N}}$, coming from the likelihood of this model being created from the star and binary formation processes. Since we vary the initial primary mass, initial mass ratio and initial period, and take into account the star-formation rate (SFR) history, the total (relative) weight is going to be,

$$\mathcal{W}_{\mathcal{N}} \mathrm{d}\mathcal{N} = \mathcal{W}_{M_{1,\mathrm{init}}} \mathrm{d}M_{1,\mathrm{init}} \mathcal{W}_{q_{\mathrm{init}}} \mathrm{d}q_{\mathrm{init}} \mathcal{W}_{p_{\mathrm{init}}} \mathrm{d}p_{\mathrm{init}} \mathcal{W}_{\mathrm{SFR}} \, \mathrm{d}t. \tag{5.3}$$

The individual weights are taken from models of observed young populations. We take

$$\mathcal{W}_{M_{1,\mathrm{init}}}\,\mathrm{d}M_{1,\mathrm{init}} = M_{1,\mathrm{init}}^{-2.35}\,\mathrm{d}M_{1,\mathrm{init}}, \tag{5.4a}$$

$$\mathcal{W}_{q_{\mathrm{init}}}\,\mathrm{d}q_{\mathrm{init}} = q_{\mathrm{init}}^{0}\,\mathrm{d}q_{\mathrm{init}} = \mathrm{d}q_{\mathrm{init}}, \tag{5.4b}$$

$$\mathcal{W}_{p_{\mathrm{init}}}\,\mathrm{d}p_{\mathrm{init}} = (\log p_{\mathrm{init}})^{0}\,\mathrm{d}\log p_{\mathrm{init}}, \tag{5.4c}$$

$$\mathcal{W}_{\mathrm{SFR}}\,\mathrm{d}t = \mathrm{d}t, \tag{5.4d}$$

coming from the Salpeter initial-mass function (Salpeter 1955; Kroupa 2001; Bastian et al. 2010) and the modeled mass-ratio distributions found from a sample of O-type stars from Sana et al. (2012), which are shown to be similar for B-type stars (Villaseñor et al. 2021; Banyard et al. 2022). No sample of observed binaries covers periods shorter than 1 d, so we adopt Öpik's law (Öpik 1924), $\mathcal{W}_{p_{\mathrm{init}}} = 1/p_{\mathrm{init}}$, which is close to the findings of Almeida et al. (2017). Unless the index of the period-distribution power-law is positive and high, which is not supported by any observational sample of young, massive stars, short-period binaries are favored over longer ones. We use no twin binary excess in our setup, since high-mass stars have small twin fractions (Sana et al. 2012; Moe and Di Stefano 2017). The age prior indicates we assume a constant star-formation history.

Given we have a grid of a discrete number of models, each grid point represents a finite area of the birth distributions in Eqs. (5.4). In particular, we compute:

$$\int_{\mathrm{lower}}^{\mathrm{upper}} M_{1,\mathrm{init}}^{-2.35}\,\mathrm{d}M_{1,\mathrm{init}} \propto M_{1,\mathrm{init,upper}}^{-1.35} - M_{1,\mathrm{init,lower}}^{-1.35}, \tag{5.5}$$

for the contribution of the model of initial primary mass $M_{1,\mathrm{init}}$, and the upper and lower bounds are taken in the middle between our grid points. For points on the edge of the grid, we linearly extrapolate the spacing outward and then use Eq. (5.5) as usual. In this way, the model with initial primary mass $8\,M_\odot$ has $M_{\mathrm{upper}} = 8.5\,M_\odot$ and $M_{\mathrm{lower}} = 7.5\,M_\odot$, while the model of highest initial mass has $M_{\mathrm{upper}} = 71.125\,M_\odot$ and $M_{\mathrm{lower}} = 68.875\,M_\odot$. A similar computation for the bins of q_{init} and p_{init} is performed.

With this setup, we can compute the probability of finding a contact binary within a bin of a certain observed period and mass ratio:

$$\mathcal{P}_{\mathrm{contact}}(q_{\mathrm{obs}}, p_{\mathrm{obs}}) = \frac{\int \delta(q - q_{\mathrm{obs}})\delta(p - p_{\mathrm{obs}})\mathcal{W}_{\mathcal{N}}\,\mathrm{d}\mathcal{N}}{\int \mathcal{W}_{\mathcal{N}}\,\mathrm{d}\mathcal{N}}, \tag{5.6}$$

where we defined

$$q_{\mathrm{obs}} = \min(M_2/M_1, M_1/M_2) \tag{5.7}$$

as the observed mass ratio. Is is understood that we filter the models only to times where they are in contact.

5.3 Results

After computing the binary-evolution models according to the methods described in Sect. 5.2, we obtain the following outcomes for our models. Figures 5.1 and 5.2 show the termination conditions of all of our models. We observe a good completion rate, with less than 9% of models with ET failing and 5% of models without ET failing. Furthermore, the lower-mass models in the range $p_{\rm init} = 1.7$–2.1 d that fail do so very close to the TAMS. Hence we do not expect them to contribute much to the time spent in contact.

5.3.1 *Roche-Lobe Overflow at the Zero-Age Main Sequence*

First, we discuss the models that experience RL overflow (RLOF) at the ZAMS. These are marked with the backward hashing (\\) in Figs. 5.1 and 5.2. Recall from Sect. 5.2.2, we expect these systems to strongly interact on the PMS, and to account for this possibility, we perform an artificial shrinking of the orbit toward their initial period. Figure 5.3 shows the orbital period and mass-ratio evolution of the 20 $M_{\odot}$ model with initial mass ratio of $q = 0.8$ and initial periods between 0.8–1.05 d. Note the logarithmic scale on the x-axis. The models reach our definition of the ZAMS at around 0.02 Myr, after which we boost the period and initiate the shrinking on the thermal timescale of the primary.

We emphasize that we do not treat the shrinking process fully. Sana et al. (2017) analyzed the RV-dispersion of young stars in M17, and inferred that the underlying period distribution fits best if there is a short-period cutoff at $p \approx 50$ d. Presumably then, the stars start their migration inward toward a distribution that follows an Öpik-like law, Eq. (5.4c), down to periods of around 1 d. Our toy migration model thus only covers the final 1% of the migration into a RL filling configuration (from $\sim$1.2 d to $\lesssim$ 1 d). It is however uncertain how and on what timescale migration happens, but the RV-dispersion analysis of Ramírez-Tannus et al. (2021) suggests it might be on the nuclear timescale of massive stars. Therefore, there is a possibility that short-period binaries evolve significantly before they are observed as $\sim$1 d contact binaries. In contrast, we operate the shrinking on a (short) thermal timescale, so that the two stellar models do not have time to build a significant composition difference before they enter contact.

During the shrinking process, all models start MT before they reach their initial period again, signaling that significant interaction on the PMS could be important for very short binary systems. We note that, with or without ET, all models equalize to mass ratios close to unity within the first 0.1 Myr or so. The ET models show some

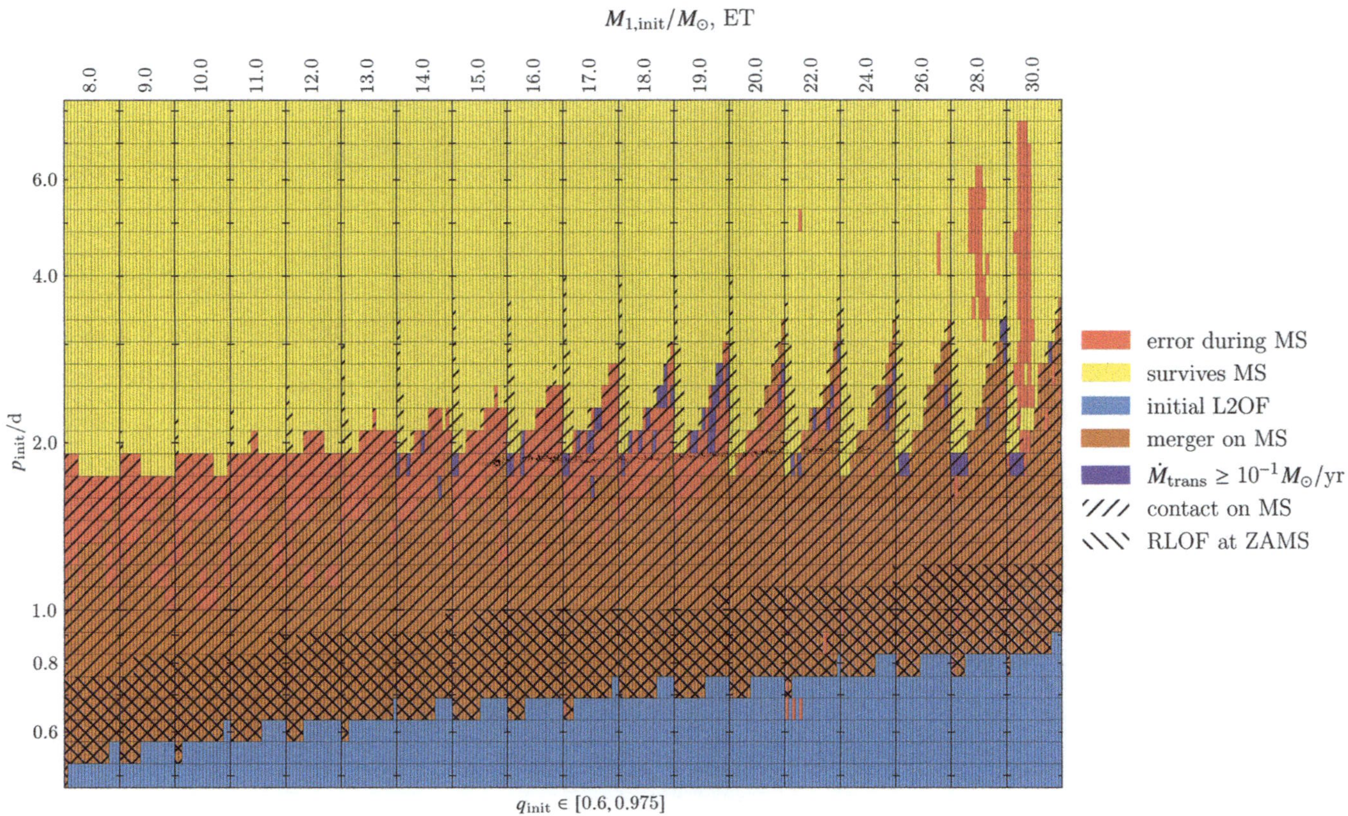

Fig. 5.1 Outcomes of our grid of models that include ET. Each wider column contains the models for the indicated initial primary mass $M_{1,\mathrm{init}}$. The columns, in turn, are subdivided horizontally by initial mass ratio q_{init}, increasing from left to right from 0.6 to 0.975, and vertically by initial period, P_{init}. MS = main sequence

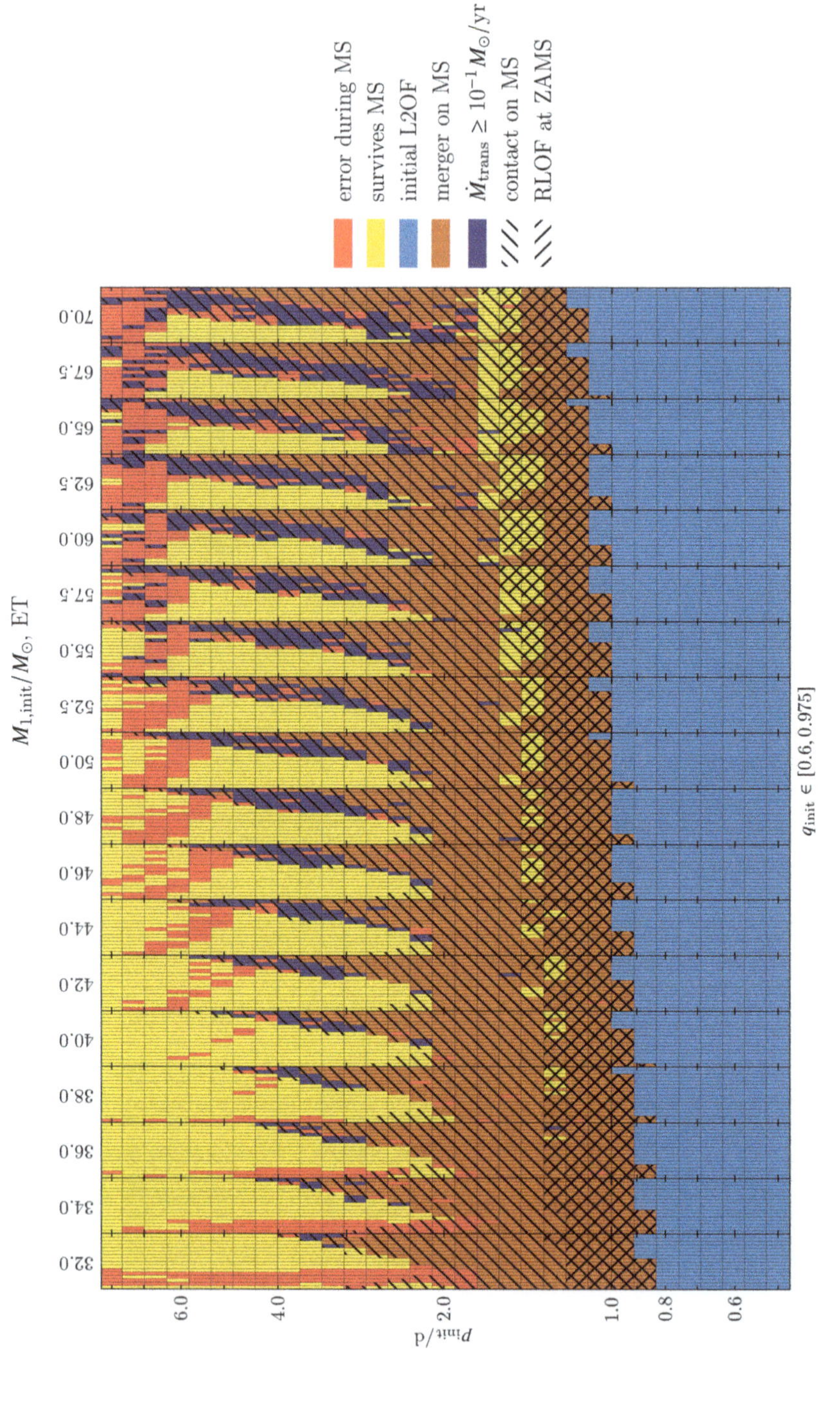

Fig. 5.1 (continued)

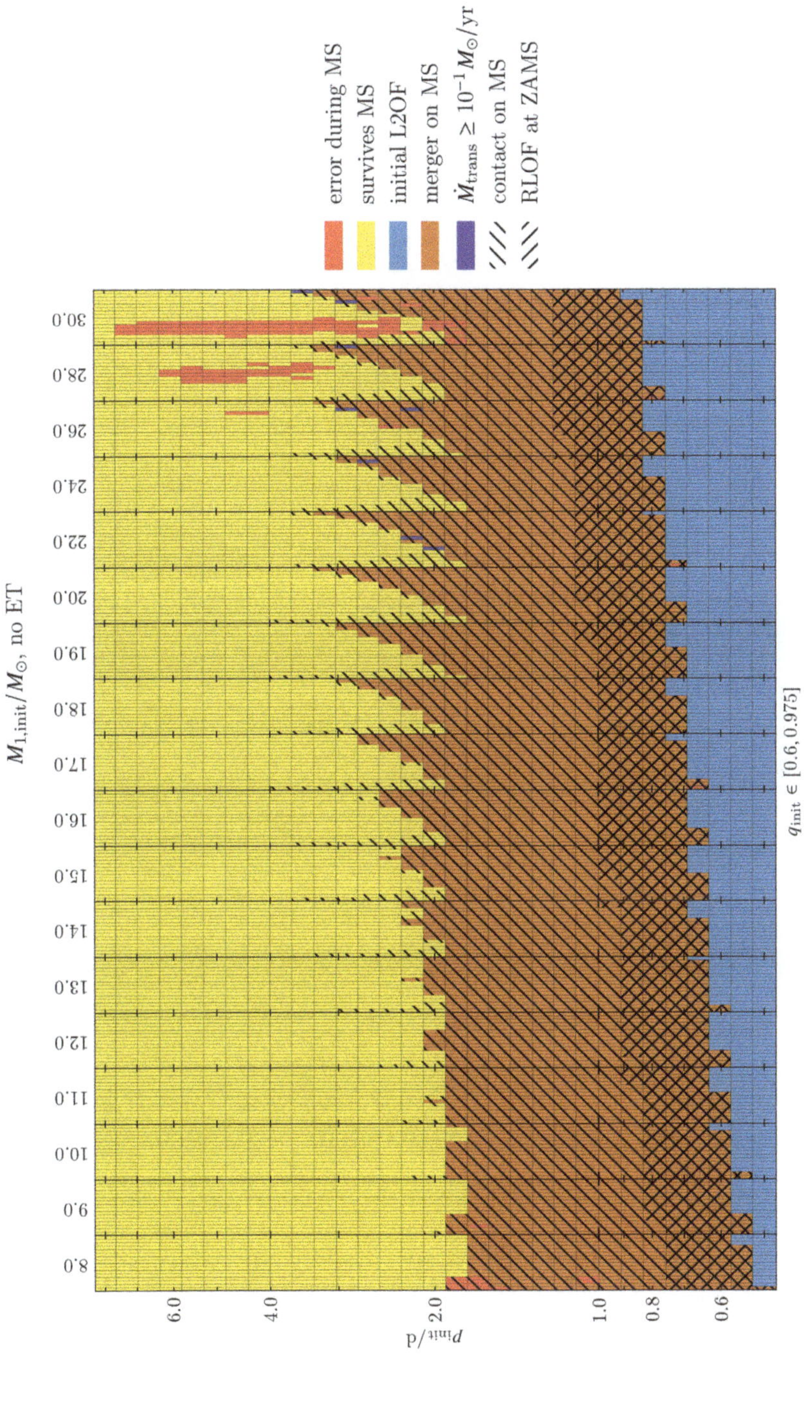

Fig. 5.2 Outcomes of our grid of models that do not include ET. The axes and legend are similar to Fig. 5.1. MS = main sequence

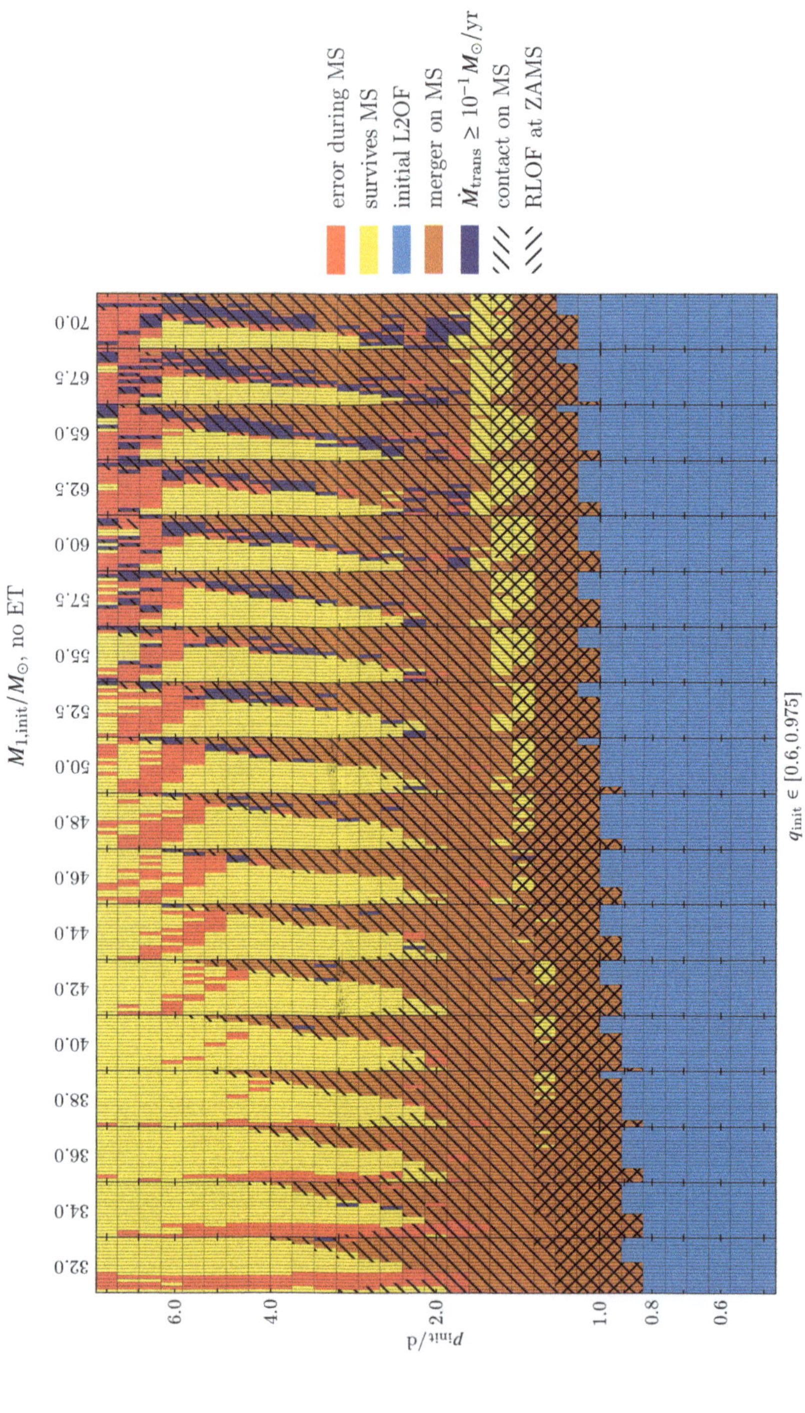

Fig. 5.2 (continued)

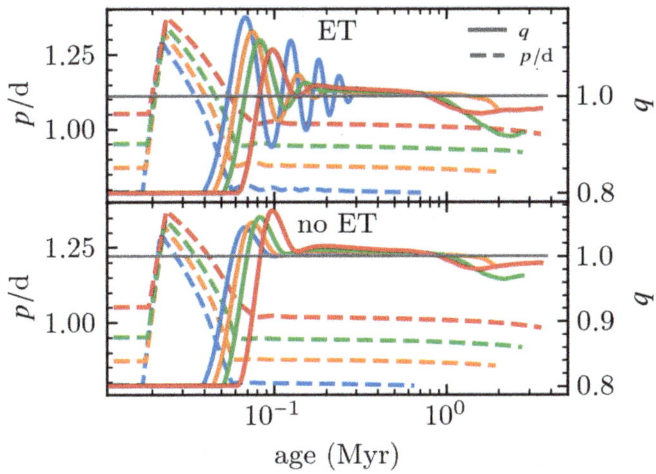

Fig. 5.3 Period and mass-ratio evolution of the $M_{1,\rm init} = 20\,M_\odot$ models with $q_{\rm init} = 0.8$ that experience RLOF at the ZAMS. The initial period of these models can be read off from the intersection with the left y-axis. All models interact before reaching their assumed initial period, meaning strong interaction on the PMS is expected

oscillations because ET is actively regulating MT. In cases where the energy donor is the mass accretor, MT is invigorated, and this is generally the case in thermal-timescale MT events as the accretor gains significant luminosity as the accreted material thermally relaxes. Finally, it has not yet been unambiguously shown theoretically that homogeneous stellar models allow for a contact configuration with unequal masses. However, all (numerical) evolution models, with or without ET, indicate that a composition difference is needed to create stable contact binaries with mass ratios away from unity. Since we chose to perform the tidal shrinking on the thermal timescale, preventing composition gradients from building, we find further evidence here that ET cannot create stable contact models of homogeneous stars.

Despite some qualitative differences in their evolution, the ET and no-ET models show no significant difference in the mass-ratio distribution of contact binaries that experience RLOF at the ZAMS. From Fig. 5.4 (see Fig. D.1 for the 2D probability density), we infer it is very unlikely to observe a contact binary that interacted on the PMS at mass ratios far from unity. From our models, there is only around 1% chance a binary with a mass ratio more extreme than $q = 0.8$ will be observed. Conversely, over 95% of all PMS-interacting binaries should be observed with $q > 0.95$.

Both observationally (Sana et al. 2017; Ramírez-Tannus et al. 2021) and theoretically (Tokovinin and Moe 2020), it has been suggested that hardening of binaries through disk migration is an ingredient in the formation of close binaries, such as contact binary progenitors. Given that these systems need a proper treatment of the binary formation process, from now on, we limit ourselves to the models that do not strongly interact on the PMS.

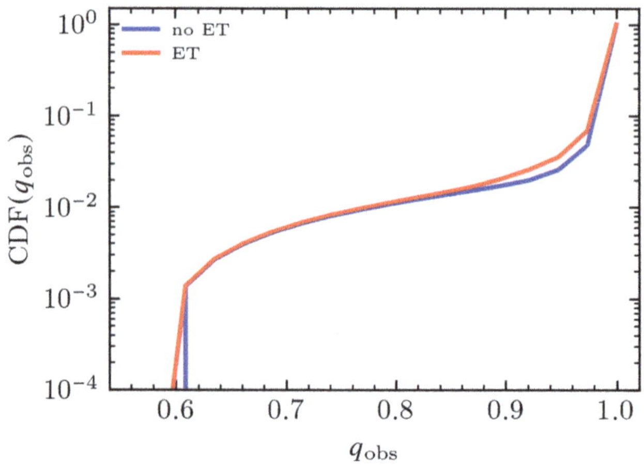

Fig. 5.4 Cumulative probability distribution of the mass ratio of contact binaries that interact on the PMS. Our models indicate almost no unequal-mass contact binaries should be observed

5.3.2 Interaction on the Main Sequence

5.3.2.1 Example Models

Next, we consider the models that interact only on the main sequence. We identify four broad categories of evolution for our case A interacting binaries.

The first group are the models of low to moderate mass ($M_{1,\mathrm{init}} \lesssim 45\,M_\odot$) that interact early on the main sequence, which were dubbed "System 1" by Menon et al. (2021). Here, the more massive star starts fast case A MT before an appreciable fraction of the main sequence is traversed. Contact is engaged immediately, and as soon as both stars regain thermal equilibrium, MT reverses and the system promptly evolves to mass ratios close to unity. We show an example of this type of evolution in Fig. 5.5, which has initial conditions $M_{1,\mathrm{init}} = 24\,M_\odot$, $p_{\mathrm{init}} = 1.15\,\mathrm{d}$ and $q_{\mathrm{init}} = 0.7$. We note that ET has little impact on the long-term evolution of this system. This kind of evolution is qualitatively similar to what we see for the PMS-interacting systems discussed in the previous section.

The second class of systems are those of low to moderate mass that start MT after an appreciable fraction of the main sequence has been traversed. Such systems were dubbed "System 2" by Menon et al. (2021). Here, we clearly distinguish fast case A from slow case A, after which MT reverses that eventually equalizes the masses of the components. This kind of evolution we studied in detail in Sect. 4.4. We show a further example model of this type in Fig. 5.6 (cf. Fig. 4.3). We see that ET has a significant effect for systems like these, because it moderates the MT rate at the point the nuclear timescale contact is engaged. In both the example given here and in Sect. 4.4, ET has the effect of delaying the mass equalization, so that, with ET, the model is more likely to be observed at unequal mass ratios.

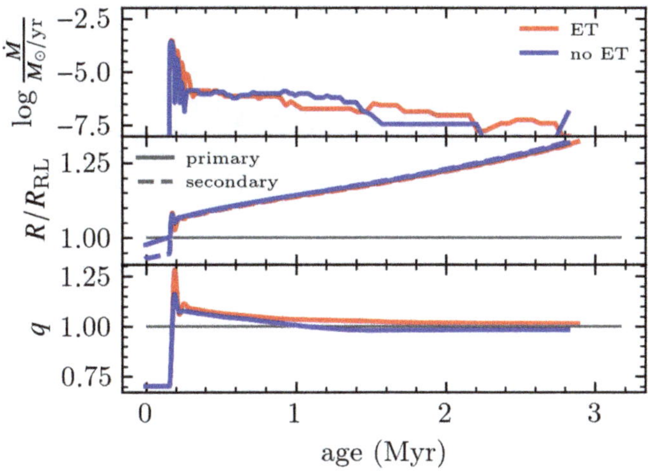

Fig. 5.5 Evolution of the system with initial parameters $M_{1,\mathrm{init}} = 24\,M_\odot$, $p_{\mathrm{init}} = 1.15\,\mathrm{d}$ and $q_{\mathrm{init}} = 0.7$. The top panel shows the MT rate, the middle panel the relative RL radii, and the bottom panel the mass ratio. This model behaves like "System 1" of Menon et al. (2021), and equalizes rapidly irrespective of whether ET is used or not

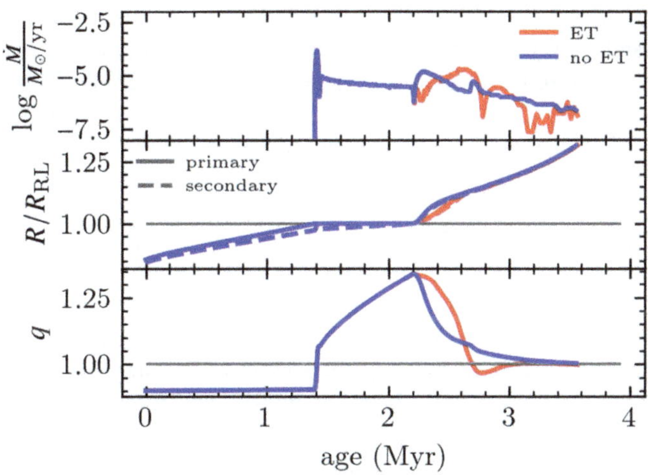

Fig. 5.6 Evolution of the system with initial parameters $M_{1,\mathrm{init}} = 30\,M_\odot$, $p_{\mathrm{init}} = 1.52\,\mathrm{d}$ and $q_{\mathrm{init}} = 0.9$, with similar panels as Fig. 5.5. This system behaves like "System 2" of Menon et al. (2021), and to what we studied in Sect. 4.4. Note that ET delays mass equalization somewhat

The third class of systems are those that avoid long-lived contact phases. They occur at all masses when case A MT starts late on the main sequence, corresponding to initial orbital periods $p_{\mathrm{init}} \gtrsim 2 - 3\,\mathrm{d}$. Most models in this regime even avoid contact during fast case A (the exceptions being at low mass and low mass ratio, $M_{1,\mathrm{init}} \lesssim 40\,M_\odot$, $q_{\mathrm{init}} \lesssim 0.7$). At high initial period, and/or low initial mass

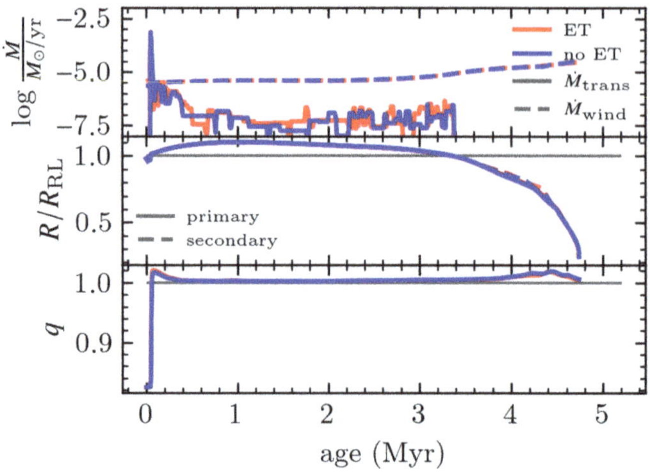

Fig. 5.7 Evolution of the system with initial parameters $M_{1,\mathrm{init}} = 67.5\,M_\odot$, $p_{\mathrm{init}} = 1.66\,\mathrm{d}$ and $q_{\mathrm{init}} = 0.825$. This model undergoes (near-)CHE, and its orbit widens due to significant wind mass loss

ratio, the primary of the system reaches core-hydrogen exhaustion (yellow boxes in Figs. 5.1 and 5.2), while at higher initial mass and higher initial mass ratio, the system merges before the masses are equalized during inverted MT (orange boxes in Figs. 5.1 and 5.2). The models in this category are most likely observable as pre-interaction or classical Algol systems, and were studied in detail recently by Sen et al. (2022) (see also, e.g., Pols 1994; Vanbeveren et al. 1998; Wellstein et al. 2001).

The fourth and final class of system we distinguish are the (very) massive models ($M_{1,\mathrm{init}} \gtrsim 45\,M_\odot$) that interact early on the main sequence. The models in this regime suffer significant mass and angular momentum loss due to high stellar winds (recall we model galactic binaries), and they also rotate quickly. Both these properties have the effect of keeping the radius to RL size small, because these models evolve nearly chemically homogeneously, and the wind mass loss causes orbital widening. An example of this evolution is shown in Fig. 5.7. Models in this class engage in nuclear-timescale contact, but eventually the orbit widens so the stars detach, and they avoid a main-sequence merger. For the specific model shown in Fig. 5.7, the secondary reaches core-hydrogen exhaustion first.

5.3.2.2 Population Properties

From our model outputs, we compute the probability of finding contact systems with a certain observed period and mass ratio. This amounts to evaluating Eq. (5.6) for our grids. Figure 5.8 shows the result. First we note that both the ET and no-ET models heavily cluster toward the parameters $(q_{\mathrm{obs}}, p_{\mathrm{obs}}) = (1, 1\,\mathrm{d})$, similarly to the

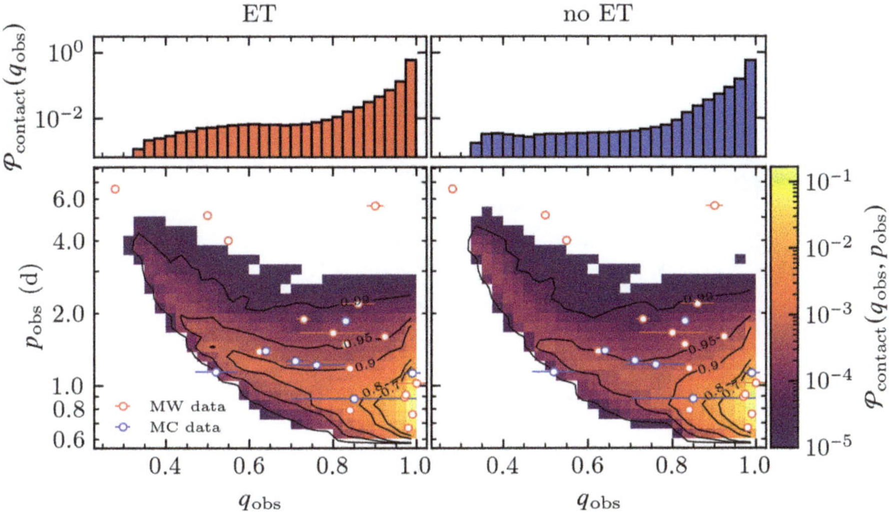

Fig. 5.8 Probability density in $(q_{\rm obs}, p_{\rm obs})$ (lower panels) and $q_{\rm obs}$ space (upper panels) for the models that experience contact during the MS. The thin black lines are HDIs of the indicated percentiles. ET has a measurable effect for about 10% of the systems. The parameters of observed systems from the Milky Way (MW) and Magellanic Clouds (MC) are listed in Tables D.1 and D.2

findings of Menon et al. (2021). Kolmogorov-Smirnov (KS) tests of the observed sample with respect to the ET and no-ET model distributions give p-values of 4.7×10^{-7} and 5.3×10^{-8} respectively, indicating that it is extremely unlikely these distributions correctly model the observed sample. To clarify the difference between the ET and no-ET distributions, we have drawn Highest Density Intervals (HDIs) on both panels of Fig. 5.8 (shown by black lines) to indicate in what regions a certain fraction of models would be found. Only in the 90th percentile we see a measurable shift toward lower observed mass ratios. This means that, with ET, the mass-ratio evolution of at most one in ten observed contact binaries will be significantly affected.

When we consider only the models with initial primary mass higher than 20, the situation is more pronounced, see the cumulative distributions in Fig. 5.9 (The 2D density plot is deferred to Fig. D.2). First we note that both the ET and no-ET models are less clustered toward mass ratios of unity, at mass ratios larger than $q = 0.5$, the dashed lines are above the full lines (compare also, e.g., the 70th percentile HDIs in Fig. D.2 with Fig. 5.8). However, the distributions indicate that ET will have a perceivable effect on up to 30% of the observed systems (note Fig. 5.9 has a logarithmic y-axis), which is a larger effect than for the full massive-star population. Despite this shift in the right direction, we still over-predict mass ratios close to unity, because the nuclear evolution of the models drive them to equal masses. Performing a KS test on the ET models of mass $M_{1,\rm init} > 20\,M_\odot$ returns a p-value of 0.002, indicating that, unless we require a confidence higher than 99.8%, we conclude that the current observed sample is not distributed according to the

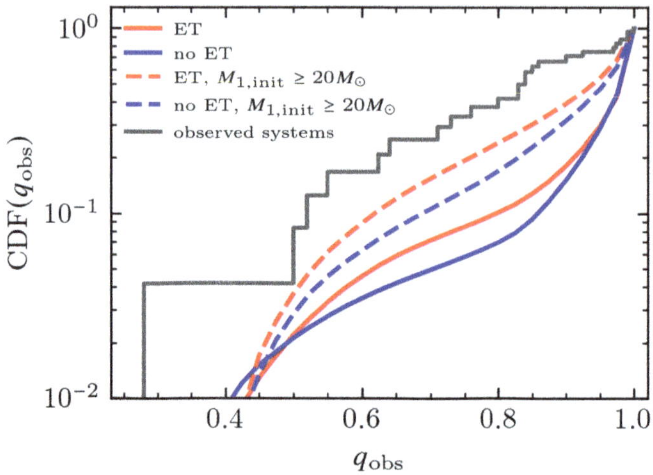

Fig. 5.9 Cumulative distributions as a function of observed mass ratio for the full synthetic populations (full lines), the populations considering only masses above 20 $M_\odot$ (dashed lines), and the observed sample of Tables D.1 and D.2 (gray line). Note the logarithmic y-axis, compared to Fig. 1.26

models. A KS-test on the no-ET models of high mass only gives a p-value of 1×10^{-4}.

5.3.3 Luminosity Ratios

Despite not significantly affecting the evolution of massive contact binaries, ET does affect the observable properties of the models. Since ET effectively changes the outer boundary condition of both stars, by raising the luminosity of the energy gainer, and reducing it for the energy donor, we expect a significantly different luminosity ratio of models in contact that include ET versus those that do not.

In Fig. 5.10, we plot the probability density as function of the observed mass- and luminosity-ratio of the whole population. Like in Fig. 4.7, we see a very clear difference between the ET and no-ET models. When ET is applied, the models follow closely a $L \sim q$ relation, as expected from the theory of Sect. 4.2. The no-ET models instead follow the single-star mass-luminosity relation $L \sim q^{2.5}$. Comparing with the observed sample, we find that the ET models match better, but refrain from making a formal goodness-of-fit estimation because some near-contact systems are also included, and the completeness of the sample is doubtful. We can still confidently say our implementation of ET naturally reproduces the similar effective temperatures of both components of a contact binary, signaling that ET must occur in the common layers.

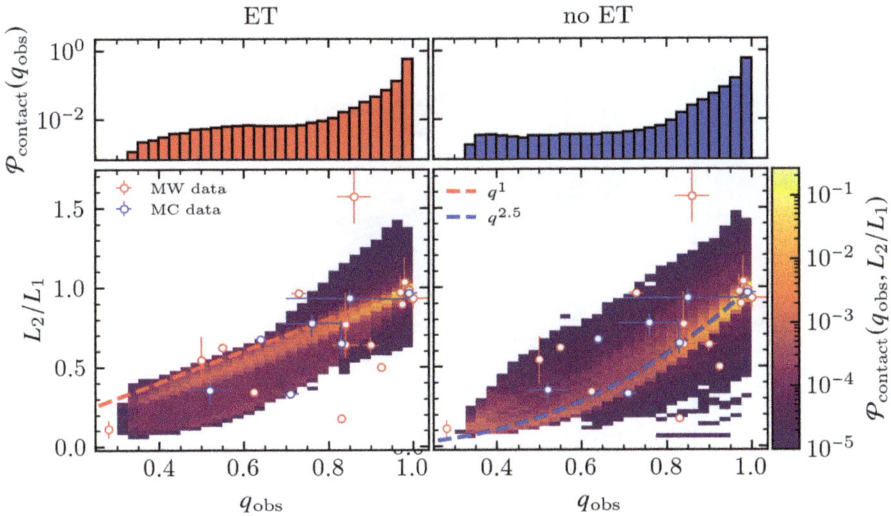

Fig. 5.10 Probability densities in observed mass- and luminosity-ratio space. L_2/L_1 here means the ratio of the luminosity of the less massive star to that of the more massive star. We observe a clear difference when ET is applied

5.4 Discussion

5.4.1 Rejuvenation

From the above results, it seems ET alone is insufficient to explain the mass-ratio distribution of observed massive contact binaries. ET has an effect on the evolution at the start of the contact phase, as seen in the previous chapter (Fig. 4.3), but after several thermal timescales, nuclear evolution dominates and the masses equalize.

To appreciably affect the mass-ratio evolution of contact binaries, we need to significantly alter the mass-radius relation of the components. Another process that heavily influences this is the rejuvenation process of the accretor during fast case A MT. Rejuvenation is the effect that an accretor experiences when growing its convective core, thereby allowing fresh hydrogen to mix in, which raises the central hydrogen fraction. The degree to which its core is allowed to grow depends on several factors. Of course the amount of mass transferred is the primary parameter, but there is little control over it as it is a quantity set by the initial masses and period. Instead, parameters such as the overshoot parameter, $\alpha_{\rm ov}$, the semiconvective efficiency $\alpha_{\rm sc}$, and the Ledoux parameter, B, can be adjusted. In stellar-evolution codes, overshooting can be forcibly inhibited if the Ledoux term (cf. Eq. (1.24)) is higher than some value, B, the Ledoux parameter:

$$\frac{\phi}{\delta}\nabla_\mu > B. \tag{5.8}$$

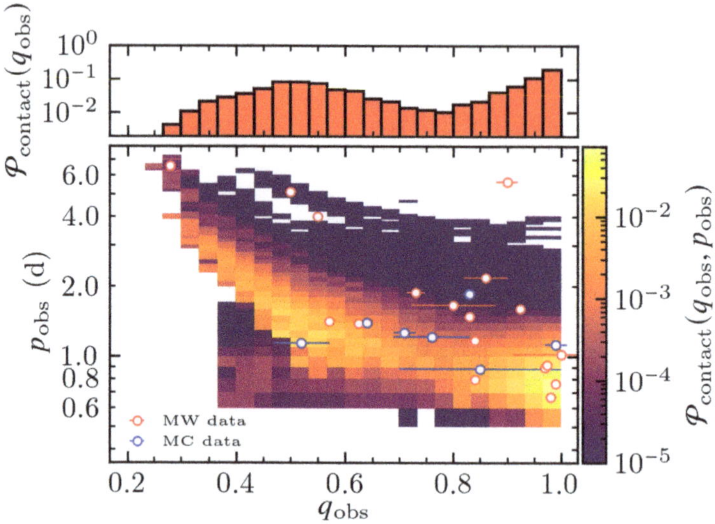

Fig. 5.11 Probability density of the population that has its rejuvenation strength reduces by a limited overshooting. The mass-ratio distribution is markedly affected, and is slightly bimodal

In the regions where overshooting is turned off this way, the semiconvective instability will tend to set in, since the high molecular gradient ∇_μ stabilizes the otherwise unstable region as per the Schwarzschild criterion. We note that this effect will only modify the models that interact only after an appreciable fraction of the main-sequence lifetime has passed, because the creation of a molecular weight barrier is a requirement.

In the dissertation of Förster (2023), it is shown that varying α_{ov}, α_{sc} and B has significant effects on the mass-ratio evolution of contact binaries. In this section, we do a similar study, but we include tidal deformation and ET. Förster (2023) found that varying the overshoot parameter has the largest effect on the observed mass ratio. Here, instead of directly varying α_{ov} to affect the size of the convective core, we effectively constrain its growth by setting $B = 10^{-4}$ instead, and recompute a sub-sample of the grid above. The models we consider here have initial parameters $M_{1,\mathrm{init}} = 8$–$40\,M_\odot$ with a spacing of $2\,M_\odot$, $q_{\mathrm{init}} = 0.4$–0.95 with a step of 0.05 and $p_{\mathrm{init}} = 0.4$–4.159 d logarithmically spaced with $\Delta \log p = 0.04$. We continue to use the tidal deformation corrections and ET for these models.

The resulting synthetic population that has a limited rejuvenation strength is shown in Fig. 5.11. First we note that the correlation between period and mass ratio is still very strong. However, compared to Fig. 5.8, mass-ratio distribution now is significantly different. Instead of one large cluster at $p{\sim}1$ d and $q{\sim}1$, now we see a slightly bimodal distribution at $q = 0.5$ and $q = 1$. In fact, comparing the CDFs this population over predicts low-mass ratio contact binaries, see Fig. 5.12.

We note that using a limiting B value to inhibit overshooting is a rather artificial way to manage the extent of the overshooting region. Furthermore it is highly

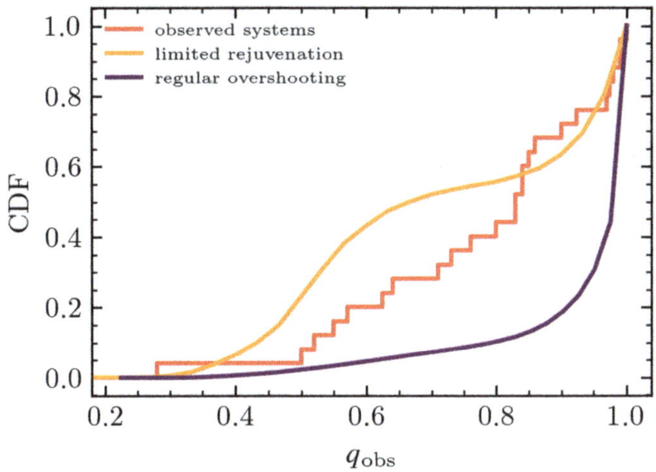

Fig. 5.12 CDF of the mass-ratio distribution of the observed contact-binary sample (red), the synthetic population that has its rejuvenation limited (orange), and the "regular" population from Sect. 5.3 (purple)

dependent on numerical aspects: the mesh and mesh-refining strategy that is used, whether smoothing is applied over composition boundaries, and also the timestep used for the integration. Another option is to use calibrations of convective-core overshooting coming from asteroseismological analysis to directly adapt α_{ov} (which may be rotation, age and/or mass dependent). Pedersen et al. (2021) inferred both the near-core and the envelope mixing profile in a sample of 26 slowly pulsating B stars. They find that stellar-evolution models with step overshooting (as opposed to diffusive, exponentially decaying overshoot) fits best for the largest fraction of their sample (17 of 26), using values $\alpha_{ov} \in [0.05, 0.4]$. Applying asteroseismic techniques to binary interaction products (such as Algols or merger products), with the intent of measuring the internal mixing properties, is currently in full development (Guo 2021; Aerts and Tkachenko 2023; Henneco et al. 2024a).

Given the results of Förster (2023), and those presented here, a thorough investigation of the near-core mixing parameters across the whole population of massive contact binaries, is urgently needed. In such a study, special care should be taken with respect to the numerical implementation of the B parameter (if used), since timestep-to-timestep variations at the core boundary could induce numerical instabilities. This work is unfortunately beyond the scope of this thesis. It would however be very fruitful, because if we have a complete set of observed contact binaries, then with appropriate model grids we can constrain the near-core mixing physics via the observed mass-ratio distribution. This is possible thanks to the geometric constraint set by the Roche geometry, forcing a fixed mass-ratio to radius-ratio relation, and the requirement of shellular contact layers. Stated differently, nuclear-timescale contact phases at a mass ratio away from unity require a specific rejuvenation process so that the mass-radius relation of both stars, post-case A interaction, approaches the Roche mass-radius relation. On top of this, ET can then play a role regulating the MT rate during the contact phase.

5.4.2 Uncertain Population Distributions

The KS tests performed above indicate that, from a statistical point of view, we must reject the hypothesis that the observational sample is distributed per our model distributions. Still, there could be observational biases that skew the sample. It could be that the sample of contact binaries is heavily biased toward detecting unequal-mass systems. However, owing to their short periods and large radial-velocity semi-amplitudes, all systems, regardless of mass ratio, should be picked up in long-term spectroscopic monitoring programs (such as VFTS, Evans et al. 2011 and TMBM, Almeida et al. 2017). Photometric characterization is not dependent on mass ratio either. There thus seem no obvious observational biases in terms of mass ratio in the current sample of massive contact binaries, apart from its completeness. In terms of primary mass however, characterization of individual systems seems biased toward higher mass. From Tables D.2 and D.1, the amount of O+O binaries that were studied far outnumbers the B+B contact systems when compared to a Salpeter IMF. Comparing the observational sample to models that reach down to initial primary masses of 8 $M_{\odot}$ might thus skew the results, since the sample contains an underrepresented number of B+B contact binaries compared to the distribution used in the population modeling.

A final consideration is the initial period distribution of binaries. Taking into account the indications that binaries harden over nuclear timescales, the true ZAMS period distribution does potentially not follow an Öpik-like law (i.e., $\mathcal{W}_p \approx 1/p$, cf., Eq. (5.4c)) down to periods of one day. Hence, as we have done above by considering high-mass stars alone, here we test whether imposing a low-period cut on the ZAMS period distributions reconciles the models with the observations. In practice, this means we prevent the systems that interact on the ZAMS (or early on the main sequence) to contribute to the population predictions. However, since we would have to compute new grids, we cannot incorporate models that result in short-period binaries as a result of orbital shrinking on the nuclear timescale. These two factors will therefore also reduce the total number of predicted contact binaries from our models.

In Fig. 5.13, we show the KS-test p-values when imposing different short-period cut-offs p_{cut}. We emphasize that this means that the birth distribution probability density is Öpik's law (Eq. (5.4c)) down to p_{cut}, but zero for $p_{\mathrm{init}} \leqslant p_{\mathrm{cut}}$. We see that the population predictions of contact binaries are very sensitive to the assumed birth distribution. At $p_{\mathrm{cut}} = 1.26\,\mathrm{d}$, the KS statistic of the mass-ratio distribution reaches a minimum of 0.16 for the ET-models, while it is 0.19 for the no-ET models. This corresponds to p-values of 0.53 and 0.30, respectively, indicating that, to high confidence, we cannot reject the hypothesis that the observed sample is drawn from the distribution with this period cut. Figure 5.14 shows the

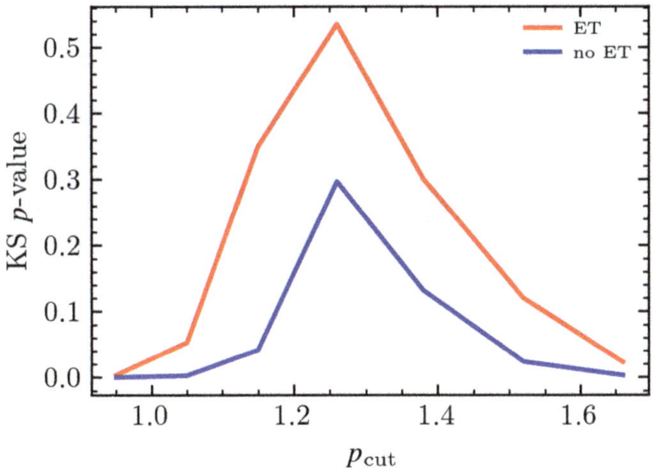

Fig. 5.13 KS-test p-values when comparing the empirical CDF of the observed sample of massive contact binaries (Tables D.1 and D.2) against Öpik initial period distributions with a low-period cut-off $p_{\rm cut}$

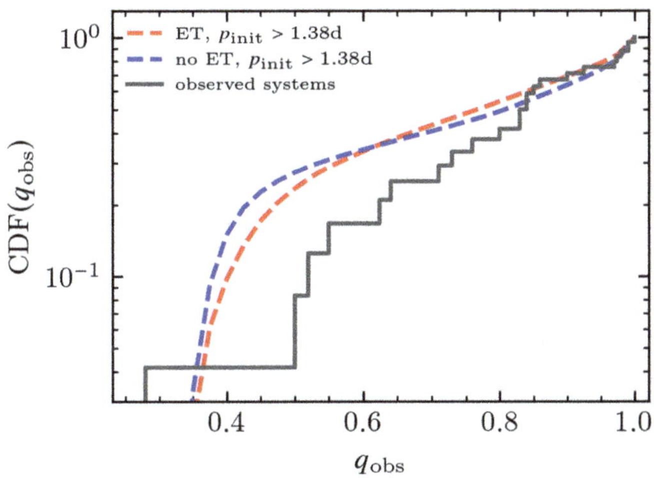

Fig. 5.14 CDFs of the population as function of observed mass ratio, using the low-period cut-off $p_{\rm cut} = 1.38$ d. From the KS statistic (Fig. 5.13), the observed sample is most likely drawn from this birth distribution

corresponding cumulative probability distributions, and points out that the models now overestimate the extreme-mass-ratio contact binaries. We note, however, that placing this cut without applying a hardening model cannot explain the period distribution of the sample, see Fig. 5.15. This suggests that the sample of contact binaries we observe today is likely coming from hardened systems that entered the contact regime later on the main sequence.

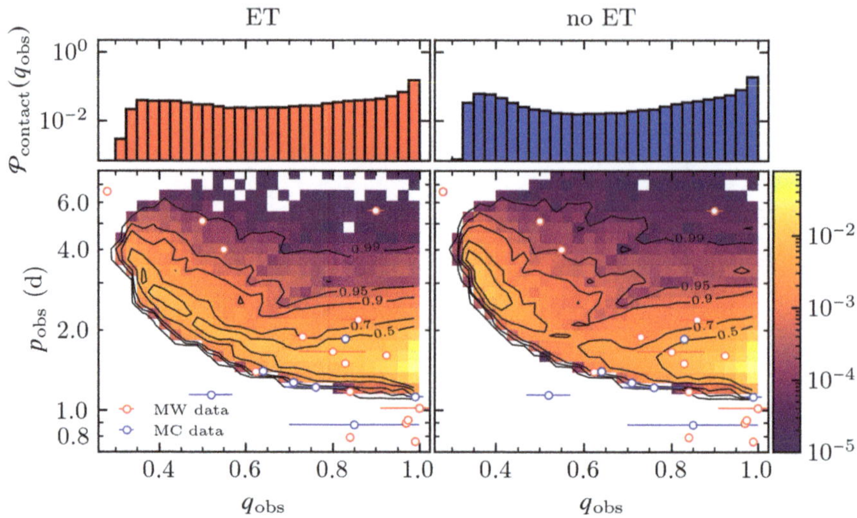

Fig. 5.15 Probability density distribution in $(q_{\rm obs}, p_{\rm obs})$ space, for the population with $p_{\rm cut} = 1.38$ d. With this period cut, we cannot predict the short-period contact binaries of $p \lesssim 1.2$ d, suggesting that period hardening operates for the observed contact binaries

5.5 Conclusions

In this chapter, we have applied our model of tidal deformation, developed in Chap. 3, and ET of Chap. 4 to the whole population of massive stars that undergoes a contact phase during the main sequence. We have then built a synthetic population of the models that include ET, and another that does not, and draw the following conclusions.

ET shifts the modeled mass-ratio distribution of massive contact-binary models. The impact is insufficient however to explain the observed mass-ratio distribution. While the MT rate is affected as ET changes the outer envelopes (and thus the boundary conditions) of the stellar models, nuclear evolution of the well-rejuvenated components dominates the long-term changes of the mass ratio. However, surface properties, in particular the luminosity and temperature, are heavily affected across the whole population of contact binaries. With ET, our population follows more closely the observed luminosity ratios of contact systems. It is thus very important to account for ET when comparing observed systems to models, and it is especially inappropriate to compare with single-star models, as long-lived contact binaries are most likely post-MT systems.

We suggest that the discrepancy between the predicted and observed mass-ratio distribution of contact binaries comes from the assumptions concerning near-core boundary mixing. In the models of this chapter, and those of Menon et al. (2021), very efficient rejuvenation was observed, thanks to overshooting regions being allowed to penetrate molecular weight barriers. Instead, Förster (2023) and

Sect. 5.4.1 both show that inhibiting overshooting using limiting values of the Ledoux term has important ramifications for the nuclear evolution, and thus the mass-ratio evolution of contact binaries. In this regard, asteroseismic modeling has calibrated in detail the mixing properties near the core boundary for single (rotating) stars (e.g., Pedersen et al. 2018, 2021; Michielsen et al. 2021), and can help constrain the near-core mixing parameters. A systematic study of these effects in mass accretors of binary systems is needed.

Finally, the observed sample of massive contact binaries is likely incomplete, as we miss many B+B contact binaries with respect to O+O ones, purely from IMF considerations. Furthermore, the period distribution that is commonly assumed, Öpik's law, could not be representative for the shortest-period binaries. Given that it is suggested migration of massive binaries occurs on the nuclear timescale (Sana et al. 2017; Ramírez-Tannus et al. 2021), the birth distribution at ZAMS of massive stars could be less skewed toward short periods than Öpik's law, or the distribution that, e.g., Sana et al. (2012) predicts. Instead, significant main-sequence evolution can occur before the components end up on their $\sim$1 d orbit and start to behave as a contact systems. Performing an orbital-period hardening on the nuclear timescale will thus have the effect of putting moderately evolved binaries in the contact regime where we are (most likely) observing them today.

Chapter 6
Low-Mass Contact Binaries

wie het kleine niet eert, is het grote niet weerd.

—Dutch proverb

This chapter contains work from:

Contact-binary Evolution with Energy Transfer and Saturated Magnetic Braking
M. Fabry and A. Prša
The Astrophysical Journal, 995, 19 (2025)

6.1 Introduction

The methods developed in this thesis were so far applied to massive stars only, the main reason being that the binary fraction is nearly unity for such stars. However, low-mass stars also occur in binaries, albeit less frequently (Fig. 1.14). Still, because the number of low-mass stars is orders of magnitude larger, tens of thousands of low-mass contact binaries have been identified (compared to only a couple dozen for the massive contact systems). This number will increase to over a million once the Vera C. Rubin observatory completes its survey (Prša et al. 2011b; Ivezić et al. 2019). However, while the photometric modeling of the surface of such systems is already very advanced (e.g., Prša et al. 2016; Conroy et al. 2020; Kobulnicky et al. 2022), detailed structure and evolution modeling of W UMa-type binaries is lacking, especially when all the physics of energy transfer, mass transfer, tidal friction and magnetic braking are concerned.

In this chapter, we briefly introduce the theory of low-mass contact binaries, discuss magnetic braking in the context of close binary evolution, and present the first results from a grid of low-mass contact-binary models.

M. Fabry, *Detailed Structure and Evolution Modeling of the Tightest Massive Binary Stars*, Springer Theses, https://doi.org/10.1007/978-3-032-28788-5_6

6.2 Theoretical Models

As mentioned in the general introduction (Sect. 1.4.2), initial modeling considerations were geared mostly toward understanding the W UMa-type contact binaries. Kuiper (1941) first deduced that, due to mismatching mass-radius relationships, contact binaries of unequal mass should not be stable. In response, many authors proposed a theory of ET in the common layers of the envelope. The first such model was that of Lucy (1967b), who proposed that if the adiabatic constants of the convective envelopes of both stars are the same (which they would not be if they were single stars), contact binaries of different mass could exist. In Lucy (1968), models were computed that indeed showed the Roche mass-radius relation of $R \propto M^{0.46}$ can be satisfied with an envelope of a single adiabatic constant. However, there were only a limited amount of configurations possible, because of the behavior of the envelope under different forms of nuclear burning (*pp*-chain vs. CNO cycle). Furthermore, this theory does not apply to the radiative envelopes of intermediate and high mass stars, of which we also observe contact binaries with unequal components.

Another model that was put forward was that of the contact discontinuity, which was mentioned in Sect. 4.2 (Shu et al. 1976, 1979; Lubow and Shu 1977, 1979). This model (as the name implies) relaxed the requirement of a continuous structure of the stars. At the RL, jump conditions connect the luminosity below the RL to that of the common envelope. Lubow and Shu (1977) then constructed two contact-binary models, one with a common convective envelope, and one with a common radiative one. However, this model was met with criticism by Papaloizou and Pringle (1979) and Hazlehurst and Refsdal (1980) of not being able to maintain the contact discontinuity over timescales longer than the thermal timescale.

A third set of models that was proposed are those that relax the requirement of thermal equilibrium of the components. Flannery (1976) and Lucy (1976) almost simultaneously produced models that include thermal relaxation oscillations (TROs). Flannery (1976) in particular showed that the system is unstable to MT occurring in a contact binary. Later, both Li et al. (2005) and Yakut and Eggleton (2005) also computed contact-binary models with TROs, see Fig. 6.1.

Despite being able to compute evolutionary models with ET (such as done in Chaps. 4 and 5), the actual mechanism of ET is not yet well understood. Differential rotation has been suggested by Yakut and Eggleton (2005), and Stępień (2009) developed an analytical 2D hydrodynamical model of a surface energy stream that is affected by the Coriolis force. Unfortunately, none of the theories outlined above have given a clear picture of contact-binary structure and evolution. In the future, perhaps 3D simulations of mass and energy transfer in a contact binary will allow us to gain insight in the circulation of material in contact-binary interiors.

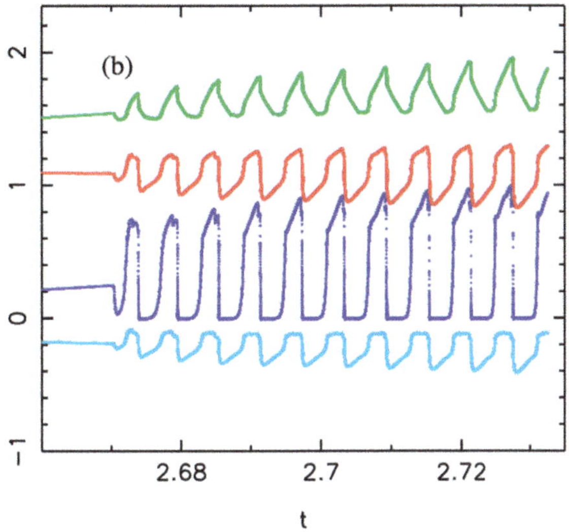

Fig. 6.1 Thermal Relaxation Oscillations in the model computed by Yakut and Eggleton (2005). The luminosities (in $L_\odot$) of the two components are given in red and green. The transferred luminosity (also in $L_\odot$) is in dark blue, while the light blue is the temperature difference of the components (in $\log T$). The x-axis is age in Gyr. *Figure adapted from Yakut and Eggleton (2005). © AAS, adapted with permission*

6.3 Magnetic Braking in Tidally Locked Binaries

Magnetic braking (MB) is the process low-mass stars undergo when their (low, but non-zero) stellar wind gets torqued by their magnetic field (e.g., Mestel 1968). Only much further from the surface of the star is the wind truly "free," and it can thus carry away much more angular momentum than the surface layers. Consequently, MB efficiently spins down low-mass stars with magnetic activity (including our Sun). By studying the rotational velocity of solar-type stars as a function of age, Skumanich (1972) found an empirical law (see also, Smith 1979):

$$v_{\mathrm{rot}} \propto t^{-1/2}. \tag{6.1}$$

Verbunt and Zwaan (1981) and Rappaport et al. (1983) translated the "Skumanich-law" into a MB prescription that steeply increases as the rotation period of the star decreases:

$$\frac{\mathrm{d}j}{\mathrm{d}t} \propto -MR^4 p_{\mathrm{rot}}^{-3}. \tag{6.2}$$

This MB law, which we dub the "classical MB" law has been widely used in the literate to study all kings of low-mass binaries, including cataclysmic variables (e.g., Patterson 1984; Knigge et al. 2011) and low-mass X-ray binaries (e.g., Podsiadlowski et al. 2002).

In close binaries, where tides are expected to be strong (Zahn 1977), the stellar rotation period is locked to the orbital period, and the angular-momentum loss caused by MB would thus correspond to a $\mathrm{d}j/\mathrm{d}t \propto p_{\mathrm{orb}}^{-3}$ relation. However, investigating the period distribution of detached eclipsing binaries from the Zwicky

Transient Survey, El-Badry et al. (2022) concluded that MB must saturate at a limiting rotational (and thus orbital, in tidally locked binaries) period. Many other studies that show evidence for saturated MB exist (e.g., Stauffer and Hartmann 1987; Keppens et al. 1995; Newton et al. 2017; Johnstone et al. 2021). In the following we will use the saturated MB prescription from Sills et al. (2000):

$$\frac{\mathrm{d}j}{\mathrm{d}t} \propto -MR^4 p_{\mathrm{crit}}^{-2} p_{\mathrm{rot}}^{-1}, \quad \text{if } p_{\mathrm{rot}} < p_{\mathrm{crit}}, \tag{6.3}$$

where p_{crit} depends on the convective turnover timescale of the star, and is in the range of 2.7–11 d for G- to M-type stars (Wright et al. 2011). For short enough period binaries then, the MB strength tapers off to a p_{orb}^{-1} relation, which El-Badry et al. (2022) needed to explain the many short-period binaries observed in their sample.

For the contact binaries studied in this thesis, we can make a similar argument as El-Badry et al. (2022). From the Roche geometry integrations of Chap. 3, we know that the volume-equivalent radius of the L_2 equipotential is maximally 30% larger than that of L_1 at the same separation a. In a circular binary, the angular momentum of the orbit J relates to a as:

$$J^2 = \frac{GM_1^2M_2^2}{M_1 + M_2} a. \tag{6.4}$$

Therefore, if we neglect any radius changes of the stars, or MT within the binary, only around 12% of the angular momentum needs to be lost for a binary that was just in contact to start overflowing the outer Lagrangian point L_2. The total lifetime of the contact binary can then be estimated by computing $J/\dot{J}$. For classical MB, the result is around 23 Myr for an equal mass binary of two 1 $M_\odot$ components, which is on the same order as the thermal timescale of solar-type stars. For that same binary, saturated MB instead predicts a lifetime of 460 Myr. This is longer than the thermal timescale, albeit still shorter than their nuclear one. Given we observe a lot of contact binaries, we are led to believe that MB must saturate. In fact, the population studied by El-Badry et al. (2022) contain the detached progenitors of contact binaries, and they already require saturation at rotational periods longer than those for contact binaries.

It should be noted that massive stars do not suffer from MB, because they have a radiative envelope in which no magnetic dynamo can operate. We therefore do not expect a strong surface magnetic field can be formed, which is why we ignored MB and its effects in the previous chapters on massive contact binaries.

6.4 A Low-Mass Contact Binary Model with ET and Saturated MB

To illuminate the differences between high- and low-mass contact-binary evolution, in this section we show the evolution of a `MESA` model with $M_{1,\mathrm{init}} = 1.0\,M_{\odot}$, $q_{\mathrm{init}} = 0.75$ and $p_{\mathrm{init}} = 1.47$ d. We use the same `MESA` setup as used in Chap. 5, with two exceptions. The first is that we include MB, using the saturated scheme from Sills et al. (2000) of Eq. (6.3). Secondly, instead of using the prescription of Eqs. (4.8) for ET, which applies to radiative envelopes of massive stars, we adapt it to convective envelopes (see also, Shu et al. 1976):

$$L_1' = (L_1 + L_2)\frac{S_1}{S_1 + S_2}, \tag{6.5a}$$

$$L_2' = (L_1 + L_2)\frac{S_2}{S_1 + S_2}. \tag{6.5b}$$

The dependence on g drops since convective energy transport is independent on the local effective gravity (Anderson and Shu 1977). We note however an alternative model of gravity darkening of Lucy (1967a), which has a mild dependence of $T_{\mathrm{eff}} \propto g^{0.08}$, so $L \propto g^{0.32}$.

Figure 6.2 shows the late-time evolution of the radii, the mass ratio, and the period derivative. First we note that the model without ET evolves qualitatively the same as a high-mass model. It goes through fast case A MT, reverses mass ratio, and with continuing angular momentum loss, becomes a contact system. At around 7.3 Myr, the system merges as a binary with $q = 1.25$ ($\approx$0.8 when inverted). However, compared to Figs. 4.3 and 5.6, the model with ET evolves differently. Our models with convective envelopes are thermally unstable under the combination of mass and energy transfer, as predicted by Flannery (1976). The steps in the cycle occur like this:

1. The system is in a semi-detached state with the high-mass component initiating thermal-timescale MT.
2. The accretor falls out of thermal equilibrium too and swells as it accepts mass from the donor.
3. Contact is engaged between the components, and ET kicks in.
4. The energy gainer (the low-mass component) has its envelope severely inflated, and the MT direction reverses in order to keep the common surface on the same equipotential.
5. The binary separation increases due to MT from a low-mass to a high-mass component, until the low-mass component detaches from the RL.
6. ET shuts off and the low-mass component shrinks well under the RL as it thermally relaxes.
7. While the secondary relaxes, the high-mass component becomes the donor again and a new cycle starts.

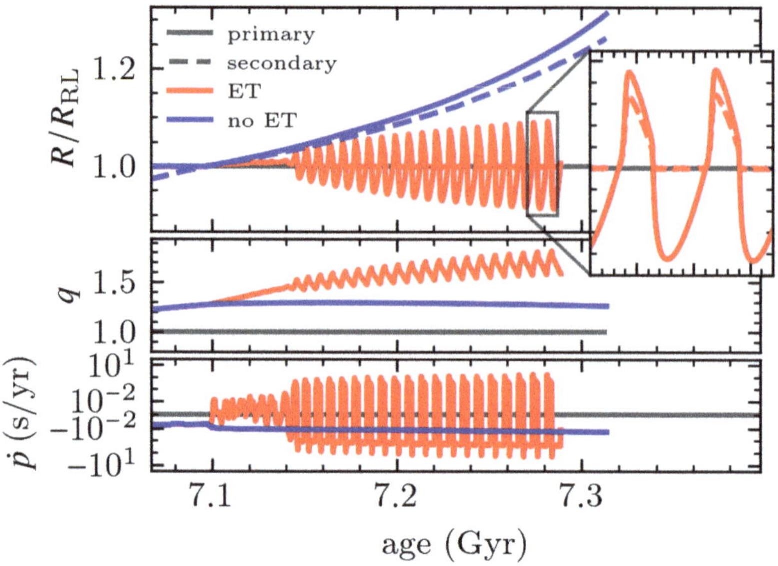

Fig. 6.2 Late-time evolution of a low-mass binary that evolves into contact. When contact is engaged, ET causes TROs, which further increase the mass ratio. The top panel shows the radii, middle panel the mass ratio, and the bottom panel the period derivative. The y-axis scale in the bottom panel is logarithmic, except between -10^{-2} and 10^{-2}, where it is linear to smoothly cross zero

In the example given here, a mass-ratio reversal happened first, making the initial primary the low-mass component and the initial secondary the high-mass component during the TROs. The inset in Fig. 6.2 shows how the high-mass (secondary) component stays attached to the RL, while it is the low-mass (primary) that switches between contact and detached states.

In the middle panel of Fig. 6.2, we see that the net result of one cycle is to slightly increase the mass ratio (i.e., farther away from unity). In this particular model, the time of the contact phase of the TRO is similar to the semi-detached phase, but there is a wide variation of possible behaviors under ET. In the top panel of Fig. 6.3, we show the model with initial parameters $M_{\rm init} = 0.9\,M_\odot$, $q_{\rm init} = 0.85$ and $p_{\rm init} = 1.91$ d. This models has a longer contact than semi-detached phase during its TROs, after having a long stable contact phase initially. The bottom panel of Fig. 6.3 shows the inverse case (here $M_{\rm init} = 0.9\,M_\odot$, $q_{\rm init} = 0.65$ and $p_{\rm init} = 1.13$ d). This model even seems to find a marginally stable configuration, where only very small radius variations remain, after having several TROs at the beginning of its contact phase. It remains to be investigated if the small radius variations are physical, or numerical in origin (e.g., due to the finite tolerances in the MT solver). If they are numerical, and (near-)constant-radius solutions exist, the properties of such low-mass contact-binary models would be of great interest as they will be evolving on the nuclear

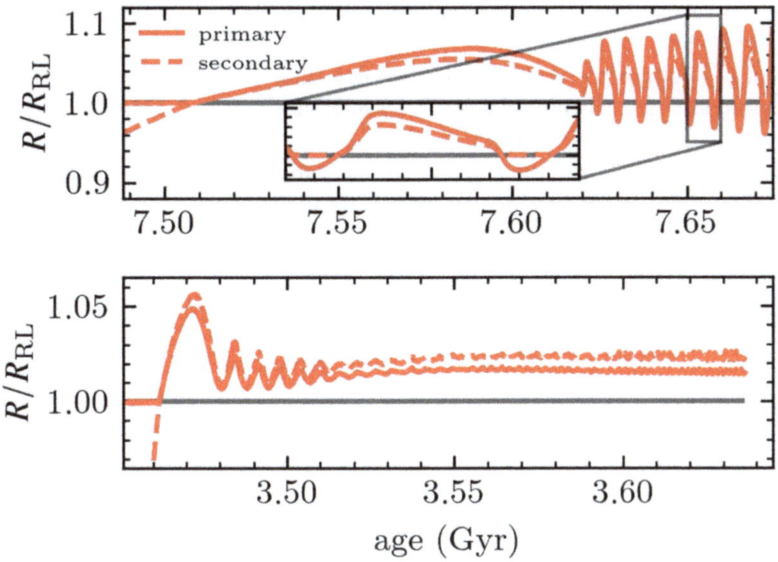

Fig. 6.3 Late-time evolution of two other W UMa models. See text for their initial parameters. The top model has its TROs develop after an initially stable contact phase, and the TRO has a longer contact vs. semi-detached phase. The bottom model shows the inverse, where, after several TROs, the model tends toward a constant-radius solution

timescale. They will therefore contribute much more to the predicted distributions of a synthetic population.

Discussing again Fig. 6.2, we see that, averaged over the oscillations, the mass ratio is increasing, meaning extreme-mass-ratio contact binaries can be formed through this type of interaction. It is expected that many Luminous Red Novae are formed through the Darwin instability (Darwin 1879), when the mass ratio of the contact binary reaches a critical mass ratio (e.g., Pešta and Pejcha 2023). However, from this model we cannot say what the ultimate fate would be, since, to save computational resources, the model was terminated after 20 TROs occurred.[1] If the amplitude of the oscillations keep growing, mass loss from the outer Lagrangian point can trigger the merger as well (see, Pejcha 2014). It is not established whether the Darwin instability is the only one causing mergers of low-mass contact binaries, or that runaway mass loss from L_2 in a contact binary also contributes. While Henneco et al. (2024b) identified that some low-mass binaries enter contact while Darwin unstable, they did not calculate the contact evolution of any low-mass ($M_{1,\mathrm{init}} < 1.3\,M_\odot$) system. Detailed evolution models (calculated through contact

[1] Since the thermal timescale of the stars needs to be well resolved for a significant fraction of the stars' nuclear timescale, such models take on the order of $\tau_{\mathrm{nuc}}/\tau_{\mathrm{KH}} \approx 100$ times longer to compute.

phases) will shed light on which instability operates in which parameter space of W UMa stars.

From the bottom panel of Fig. 6.2, we note that since TROs are characterized by continuous thermal-timescale MT, a rather large period derivative should be observed for such binaries, which is positive (for this model in particular, around 1–5 s yr^{-1}). Recall that, in the contact phases of the TROs, the period is increasing as mass flows from the low- to the high-mass component. This is in contrast for the no-ET case, where MT is on the nuclear timescale and thus $\dot{p}$ is much lower and negative (around -10^{-2} s yr^{-1} for this model). However, the presence of third bodies also affects the angular-momentum-loss rate, and thus the period derivative. Because quite some low-mass contact binaries are detected with a triple companion (Tokovinin et al. 2006), this needs to be taken into account when interpreting these predictions.

6.5 Population Properties

In this section, we apply the processes of ET and MB to a population of low-mass binaries with the initial primary mass in the range $M_{1,\mathrm{init}} = 0.5\text{–}1.4 M_{\odot}$, mass ratios of $q_{\mathrm{init}} = 0.35\text{–}0.95$ d and initial periods of 0.3–5 d. We produce four grids of models, one each with the combination of ET included and ignored, and using saturated or classical MB.

For the analysis, we use the same population-synthesis methods as in Chap. 5, but we adapt the weighting functions, $\mathcal{W}$, to be appropriate for the properties of observed populations of low-mass binaries (Chabrier 2003; Raghavan et al. 2010; Moe and Di Stefano 2017). In particular, Moe and Di Stefano (2017) report a high excess twin binary fraction (binaries with $q \geqslant 0.95$), which is absent for massive binaries. The weights are:

$$W_{M_1} dM_1 = \begin{cases} M_1^{-2.3} \mathrm{d}M_1 & M_1 > 1\,M_{\odot}, \\ A \exp\left(-\frac{(\log M_1 + 0.65)^2}{2(0.57^2)}\right) \mathrm{d}\log M_1 & M_1 < 1\,M_{\odot}. \end{cases} \tag{6.6a}$$

$$W_p \mathrm{d}\log p = \exp\left(-\frac{\log p - 5.03}{2.28^2}\right) \mathrm{d}\log p, \tag{6.6b}$$

$$W_q \mathrm{d}q = \begin{cases} a q^{-0.5} \mathrm{d}q & q < 0.95, \\ (a q^{-0.5} + 6) \mathrm{d}q & q \geqslant 0.95. \end{cases} \tag{6.6c}$$

Figure 6.4 shows the 2D probability density distributions of the model grids. We see a significant difference between the ET, saturated MB grid and the other three. First, because the saturated MB law of Eq. (6.3) is much weaker than the classical law (Eq. 6.2), the models with component masses under around 1 $M_{\odot}$ are able to survive long enough to contribute a significant fraction of the probability

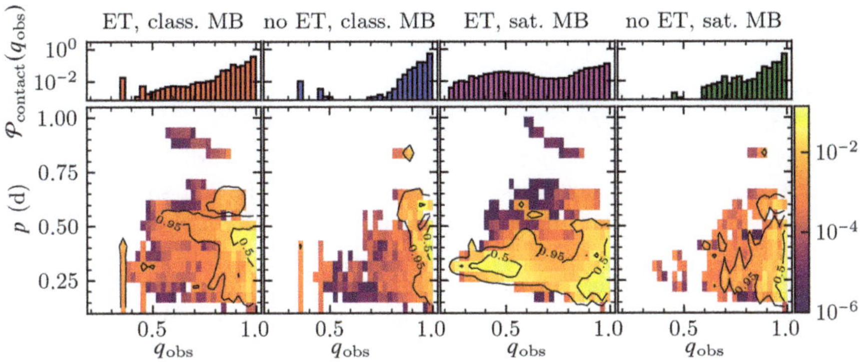

Fig. 6.4 Period-mass-ratio distribution of our simulated W UMa populations. The marginalized mass-ratio distribution is shown in the upper panels. The ET, saturated MB grid is significantly different than the other three, with a large population of W UMa binaries with mass ratios away from unity

density. Note how, in the classical MB grids, the most likely period is larger than that of the saturated grids. This is because in the classical MB grids, the highest contribution comes from the $M_{1,\rm init} \approx 1.1$–$1.4 M_\odot$ models, who are less sensitive to MB because of their shallow convective envelopes (see also, Pylyser and Savonije 1988; Podsiadlowski et al. 2002). Second, the distributions show that the models with ET allow for a good fraction of contact binaries at mass ratios away from unity. In the classical MB grid, there is a small strip in the 0.95 percentile HDI at larger periods ($p \approx 0.5$ d) that extends down to $q_{\rm obs} \approx 0.6$. In the saturated MB grid however, a large fraction of the models sit at $p \approx 0.3$ d and $q_{\rm obs} \approx 0.45$. The fact that the angular-momentum-loss timescale is longer than the thermal timescale enables the envelope of the low-mass component to inflate in response to ET from the high-mass component. This reverses the MT direction as discussed above and drives the mass ratio away from unity.

It should be noted however that the TRO models were halted prematurely for computational reasons. Therefore, if the effects of ET lowering the mass ratio persist, that population should evolve to even lower mass ratio and larger period. Currently, several extreme-mass-ratio contact binaries have been characterized (e.g., Broens 2013; Poro et al. 2025). Their reported mass ratios are lower than $q_{\rm obs} < 0.2$, which are currently not predicted by our grid. Widening the initial conditions to include $q_{\rm init} \lesssim 0.3$ could account for some systems, but a deeper understanding of the ET process is required as well.

6.6 Conclusions

From the models presented in this chapter, we can make the following initial conclusions.

First, MB must saturate in order to explain the period distribution of contact binaries. The lifetime of such systems would otherwise be too short, and only the systems with larger total mass would remain as those are, comparatively, less affected by MB. This conclusion is strengthened by the work of El-Badry et al. (2022), who studied the longer-period, detached counterparts of the population modeled here. If saturation is reached at rotational periods associated with (still tidally locked) detached binaries, then it will likely do so too at the even shorted periods in contact systems.

Second, some form of ET must be operating in the interiors of contact-binary components. If the thermal contact between the components is very poor, mimicking the no-ET case explored here, few low-mass-ratio contact binaries would be produced (compared to $q \sim 1$ ones). A mechanism is required to drive the mass ratio toward extreme values. ET is able to do it by inflating the envelope of the energy gainer, which, combined with the contact condition, causes MT from the low- to the high-mass component. s Unfortunately, the final fate of the contact-binary models cannot be assessed from the models in this chapter. The were cut short to save on computational resources, and the long-term evolution is expected to be heavily dependent on the precise implementation of the ET process. Further observational constraints and hydrodynamical models of MT and ET can provide insight into the process, and should dictate how it is modeled in evolutionary models.

Chapter 7
Conclusions and Outlook

ἓν οἶδα ὅτι οὐδὲν οἶδα.

—Socrates

In this thesis, we have investigated the structure and evolution of massive contact binaries. In Chaps. 3 and 4, we have developed the methodology to represent tidal deformation and energy transfer in close binary evolution, under the working assumption that the stars are shellular. Chapter 5 presented a detailed study of the population of massive binaries that engages in contact phases, and investigated the effect of energy transfer on the predicted mass-ratio distribution. We ended with Chap. 6, where we focused on the evolution of the low-mass W UMa-type contact binaries.

Before diving into the modeling of contact binaries, Chap. 2 first presented the analysis of the long-period massive binary 9 Sagittarii. With a period of around nine years, we do not expect this system to interact strongly, let alone form a contact binary, but nevertheless, it is an instrumental system to calibrate massive-star evolution with. Using highly accurate astrometric measurements from the VLT interferometer, we first determined the geometric orbit of the binary, putting firm constraints on the inclination, i, the period, p, and the eccentricity, e. Next, using a novel approach of grid-based spectral disentangling, we obtained new values for the RV semi-amplitudes K_1 and K_2. These values were significantly different from those from the literature, obtained with classical line-profile fitting. Our semi-amplitudes imply dynamical masses of $53\,M_\odot$ and $39\,M_\odot$, making 9 Sgr one of the most massive, galactic binary stars ever resolved interferometrically. Finally, modeling the atmospheres of both stars gave an estimate of the effective temperature and surface gravity, revealing a large mass discrepancy. This is most likely due to repeated normalization of the spectra, which is a weakness of our method. Evolutionary modeling suggested that either the stars have misaligned spin axes with respect of the orbit, or that internal mixing is stronger the more massive a star is.

Chapter 3 dealt with tidal deformation, and we used Roche equipotentials, with their corresponding volume equivalent radii, r_Ψ, as the new independent variable

M. Fabry, *Detailed Structure and Evolution Modeling of the Tightest Massive Binary Stars*, Springer Theses, https://doi.org/10.1007/978-3-032-28788-5_7

of stellar structure. In this way, the equations of stellar structure and evolution, usually cast in spherically symmetric form, Eqs. (1.5), are corrected with values f_P (Eq. 3.9) and f_T (Eq. 3.14), which depend on the mass ratio. Crucially, both f_P and f_P/f_T are smaller than unity for stars not overflowing their outer Lagrangian point (cf. Figs. 3.5 and 3.8), reducing the pressure gradient and increasing the (radiative) temperature gradient of the outer envelopes of tidally distorted stars. This has the effect of increasing the radius of the star of the same mass with respect to one that is spherically symmetric, by up to around 5%. This is a significant effect, as precision astrophysics, using measurements of eclipsing binaries, is reaching 1% accuracy or better. Inferring properties by comparing to non-rotating, or, equivalently, non-deformed stellar models, can thus lead to systematic discrepancies between observation and theory.

In Chap. 4, we considered the process of energy transfer (ET) in massive contact binaries. We argued that, if the common layers of the contact binary are shellular, there must be an energy flow occurring near the RL of both stars. We quantified the amount of energy that needs to transferred in Eq. (4.9), so that both components radiate an amount of energy proportional to $S_\Psi \langle g \rangle$, see Eqs. (4.8). As a result, the properties of the common layers of contact binaries are different compared to those that do not transfer energy. Figure 4.5 compared the density and temperature profiles of the ET models versus the no-ET models, and showed our ET implementation correctly makes the common layers of the binary shellular. In the evolutionary context, our $25M_\odot + 20M_\odot$ model with ET spent more time in contact at mass ratios far from unity than without ET, meaning some systems are measurably affected by the ET process. Furthermore, we identified that the luminosity ratio, L_2/L_1, is a good indicator of whether a binary might be in contact or not (as long as the mass ratio is not too close to unity).

The work of Chaps. 3 and 4 culminated in Chap. 5, were we investigated the effect of ET across the whole population of massive contact binaries. Our goal was to assess whether ET could alleviate the discrepancy between the predicted and observed mass-ratio distribution of the massive contact systems. From our model grids and population synthesis, we determined that ET alone cannot explain the difference. Despite changing the outer layers, thus affecting the observable properties and producing a different luminosity-ratio distribution for long-lived contact binaries, the nuclear evolution of the components drives the mass ratio of our models to unity reasonably quickly. We found that at most around 10% of all contact binaries will be significantly affected by ET in terms of their mass-ratio evolution (among which the system studied in Chap. 4). We performed an experiment by varying how far overshoot regions can penetrate molecular weight barriers, and found it has a major impact on the evolution of a contact binary. We concluded that it is high time to closely study the convective-core boundary mixing processes, such as overshooting (and its penetration through molecular weight barriers) and semiconvection, and determine their effects on the rejuvenation of mass accretors. Using asteroseismologically determined constraints can help this endeavor. Alongside these theoretical aspects, we discussed also that the birth distributions of binary stars might have to be reassessed. If we assumed our models accurately describe contact binary evolution, and the sample of observed binaries is representative of the whole population, we uncovered that the period distribution is likely not the commonly adopted Öpik's law down to periods of 1 d or less. Instead, we suggested that the close binary population we observe today would be born at longer periods, and migrate inwards over the nuclear timescale of massive stars, which is supported by other observational and theoretical studies. Of course, expanding the sample of observed contact binaries needs to happen in parallel so as to make more meaningful comparisons between the theoretical models and observed systems.

Finally, in Chap. 6 we computed a grid of low-mass contact binary models, which exhibited thermal relaxation oscillations. Energy transfer has the effect of inflating the convective envelope of the secondary, causing the mass-transfer direction to reverse. This creates the conditions necessary to form low-mass-ratio contact binaries, several of which are observed in through photometric analysis. We also considered the process of magnetic braking, which is the dominant angular-momentum sink in close binaries that have magnetic dynamos. Only under a saturated model of magnetic braking, having a shallower period dependence, $\dot{J} \propto p_{\rm orb}^{-1}$, instead of the classical $\dot{J} \propto p_{\rm orb}^{-3}$ relation, could we explain the existence of contact binaries at periods shorter than $p \approx 0.5\,$d. Magnetic braking in contact binaries is thus likely weaker than is typically assumed in other studies of low-mass binary evolution.

The work presented in this thesis opens up various new and exciting research avenues. The precise evolution of (massive) contact binaries is far from a solved question. Not so very long ago, Peter Eggleton called contact-binary evolution (Eggleton 2012) "one of the two great unsolved problems in stellar astrophysics, the other being common-envelope evolution." Below, we outline various follow-up studies that build on the conclusions of this thesis, and that use the methods developed in it. The first one draws from the work on 9 Sgr in Chap. 2, while the others from the work on contact binaries in the later chapters.

7.1 Characterizing Wide or Single-Degenerate Binaries

In Chap. 2, we showed that grid-based spectral disentangling is a powerful tool to characterize wide binaries, where RV measurements through classical spectral-line fitting is challenging. With new astrometric measurements of massive, long-period binaries, e.g., with the CHARA array (ten Brummelaar et al. 2005; Lanthermann et al. 2023), complemented with spectroscopic data, more such binaries can be characterized. This way we gain insight in the multiplicity of massive stars at all period ranges, and add more calibrators to the list of massive stars, because stars in long-period orbits do not interact strongly.

We mentioned in the introduction (Sect. 1.3) that spectral disentangling can be used to discover—or disprove—the existence of compact objects in single-lined spectroscopic binaries. Currently, a European Southern Observatory Large

Program (PIs: Shenar, Bodensteiner) has just completed the data-acquisition stage of obtaining 25 spectra for each of 929 massive stars in the Small Magellanic Cloud. With such datasets, spectral disentangling can be used not only to obtain high signal-to-noise ratio separated spectra, but also to discover black holes (such as in Shenar et al. 2022) or stripped stars (such as in Shenar et al. 2020; Bodensteiner et al. 2020). Thanks to this large survey, we will gain considerable knowledge on the properties of massive binaries in a lower-metallicity environment, and the methods developed and used in Chap. 2 are a crucial step in this process.

7.2 Stability of Mass Transfer

As introduced in Sect. 1.5.1, stable mass transfer is a promising channel to produce GW progenitors in low-metallicity environments. As the name suggests, it however depends critically on the stability of mass transfer, which depends on the response of the stars and the orbit under changes in mass. We quantify such changes by computing mass-radius exponents, ζ, for both the RL and the adiabatic response of the donor star. We thus define:

$$\zeta_{\mathrm{RL}} = \frac{\partial \ln R_{\mathrm{RL}}}{\partial \ln M_{\mathrm{d}}}, \tag{7.1}$$

$$\zeta_{\mathrm{d}} = \frac{\partial \ln R_{\mathrm{d}}}{\partial \ln M_{\mathrm{d}}}. \tag{7.2}$$

Under the assumption of conservative mass transfer, ζ_{RL} is a function only of the mass ratio and separation of the binary, while ζ_{d} depends on the precise structure of the donor. Mass transfer will become dynamically unstable if $\zeta_{\mathrm{d}} < \zeta_{\mathrm{RL}}$, i.e., when the RL shrinks faster than the radius of the donor under mass loss. However, this is a rather simplified condition. Generally, the response of the donor under mass loss is not constant at the adiabatic response, because the MT rate is not constant throughout a MT event. Furthermore, an initially stable situation, in the sense that $\zeta_{\mathrm{d}} > \zeta_{\mathrm{RL}}$ at the onset of thermal timescale MT, can become unstable over the course of a whole MT phase (which can take several thermal timescales), a situation which is called the delayed-dynamical instability (Ge et al. 2010). Still, from the condition $\zeta_{\mathrm{RL}} = \zeta_{\mathrm{d}}$, a critical mass ratio can be computed for each donor star for mass transfer to become unstable. In a systematic study by Ge et al. (2010, 2015, 2020), criteria for the stability have been derived, but only for spherically symmetric models. Those are then applied to specific systems to determine semi-analytically what parameter space can form double compact-object mergers (Picco et al. 2024). However, we have seen in Chap. 3 that rotation and/or tidal deformation changes the structure of (near-)Roche-filling stars. The mass-radius responses could thus be different than is taken into account in previous investigations. Fortunately, since we expect Roche-filling stars to be synchronized (at least near the surface), a wide family of stability curves for different rotation rates of the donor need not be calculated. Instead the structure corrections computed in Sect. 3.2.5 can be directly used to obtain new ζ_d coefficients. The intricate dependency of the mass-radius exponent on the structure of the star prevents us from making even naive predictions on whether the tidal deformation corrections stabilizes or destabilizes MT. A careful study of a wide range of donor stars is required to determine whether, and in what parameter range, ζ_d is higher, favoring stability, or lower, making donors more unstable.

7.3 Interaction on the Pre-main Sequence and with Third Bodies

In this thesis, we have discussed massive stars in binary systems with orbits of less than ten days. As mentioned in Sect. 5.3.1, a good amount of close binaries will therefore interact even before they have reached their formal ZAMS position. This is not even considering that a massive star might still be accreting while it is already on the main sequence (Bernasconi and Maeder 1996). Putting this last aspect aside, the methods in this thesis allow for the study of contact binaries under the influence of the binary formation processes. In particular, we could adopt a simple timescale for the hardening of binaries during their formation, and run simulations starting from, e.g., a fully radiative pre-main-sequence model. This will then also take into account the fact that a more massive star can already pass a significant fraction of its main-sequence lifetime before its companion even reaches the ZAMS.

A similar effect caused by third bodies can be taken into account. It is known that tight binaries have a higher chance of hosting a third star in a wide orbit (Tokovinin et al. 2006). Therefore, contact binaries or their progenitors, which have the shortest periods, are likely to have a triple star affecting it. The net effect of the von Zeipel-Kozai-Lidov cycles on the inner binary, i.e., its secular evolution, can be modeled as an extra torque within binary-evolution instruments, such as `MESA`. Currently, triple-star interactions have been considered only with rapid, parameterized, stellar-evolution modeling (Toonen et al. 2016, 2020; Kummer et al. 2023), but no detailed models that include triple-star effects have been computed. Combining the methods of this thesis, which concerns the detailed physics of close binaries, with a prescription for triple interactions is the next step in triple-star population synthesis.

7.4 Initial Conditions of Hydrodynamical Simulations

In Fig. 1.28, we showed the results from a 3D magneto-hydrodynamical simulation of merging stars from Schneider et al. (2019). In the setup of these simulations, the authors mapped 1D `MESA` models onto spherical shells using the distribution from Górski et al. (2005). Afterwards, the models need to be relaxed, because of the used numerical discretization (as discussed in Ohlmann et al. 2017). With current computational resources, it is unfeasible to follow the merger from the onset of mass loss through L_2. Therefore, in the next step, the authors artificially shrink the orbit by adding a sink term to the momentum of all cells so that the initially spherical models gradually come into contact. Once sufficiently deep contact is reached, the sink term is removed, and the simulation proceeds as normal.

While Schneider et al. (2019) indicate that the initial separation (before the shrinking procedure) has only a marginal effect on the outcome, the methods in this thesis can be used to skip the shrinking procedure altogether. Since the models computed in Chaps. 3 and 4 use the fundamental assumption of shellularity, instead of mapping this 1D structure to spherical shells, such profiles can be directly mapped to Roche-deformed shells. In essence, you can map any property $\mathcal{X}(r_\Psi)$ of the 1D shellular profile to $\mathcal{X}(x, y, z)$ of 3D space, because we can invert $\Psi = \Psi(r_\Psi)$ from the numerical integrations of Sect. 3.2.5, and the Roche potential defines $\Psi = \Psi(x, y, z)$ (Eq. 1.38). Therefore, a contact binary of arbitrary mass ratio and degree of overflow that is hydrostatic can be initialized in 3D starting from 1D models computed through the methods in this thesis. However, we recognize that initial conditions defined on Roche-deformed shells might still be numerically unstable as a result of discretization. It should be investigated if damping techniques are needed (as done in Ohlmann et al. 2017) to bring 3D contact-binary models back to hydrostatic equilibrium.

Additionally, care needs to be taken in the region beyond the photospheres of both stars (where detailed structure modeling like `MESA` does not work). In a 3D finite-volume simulation, there cannot be a hard boundary between a dense region (in the sense that $\rho \neq 0$) and a vacuum, as that would slow down the calculations tremendously. With massive stars, we are lucky, since they launch a strong stellar wind, so that $\rho \neq 0$ in the whole simulation domain. However, an initial wind profile would need be assumed.

7.5 Low-Mass Contact Binaries

The results on a population of low-mass W UMa binaries, presented in Chap. 6 are, to our knowledge, based on the first wide-range grids of low-mass contact binary models. However, from these models, only a couple conclusions could be drawn. First, the effect of magnetic braking on the orbital evolution of contact binaries must be tempered with respect to the classical understanding of the process as described by the Skumanich law. Second, some form of energy transfer must operate to form extreme-mass-ratio contact binaries.

Because the contact binaries in this mass-regime are, generally, not in thermal equilibrium (unlike the massive ones), the precise prescription of energy transfer, i.e., its location and efficiency, matters for the long-term evolution of these systems. Therefore, constraints on how the energy transfer process operates are dearly needed. Some insight can be gained by studying the surface properties of

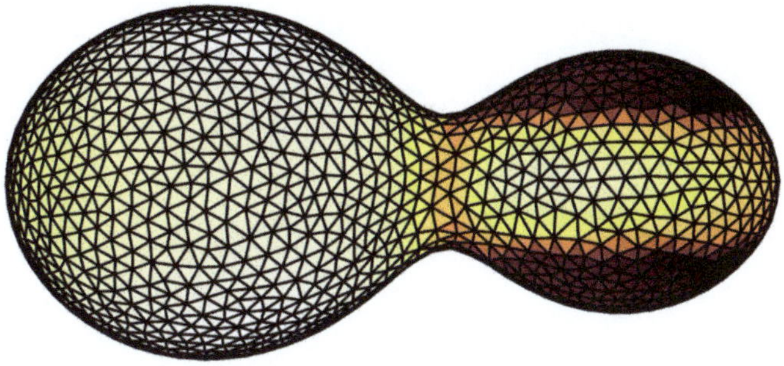

Fig. 7.1 PHOEBE mesh of a contact binary with a parameterized prescription of surface energy transfer. The color scale indicated temperature (with arbitrary units), and shows an energy transfer stream

the observed contact-binary sample. In photometric analyses with PHOEBE, the underlying mesh that is used to compute the predicted flux can be adapted to include, for example, a hot stream from the hotter to the cooler component, see Fig. 7.1. Fitting such meshes to the $\sim$300 contact binaries in the *KEPLER* sample (Kirk et al. 2016) could reveal patterns in certain parameters. For example, if contact binaries require a particularly wide energy-transfer stream, much wider than the bridge itself, this would suggest that energy transfer operates mostly internally to drive their effective temperatures to equal values. Instead, if strong, equatorially confined streams are preferred, this would indicate energy transfer occurs more toward the surface of the stars.

A mesh like that of Fig. 7.1 can also be used to explain the O'Connell effect, which is the observational phenomenon where the two maxima of a light curve are unequal. Such asymmetries are commonly explained with large star spots (created by magnetic fields), sometimes taking up more than half of the star's disk (e.g., Austin et al. 2007). The choice to fit star spots instead of another model is usually because it offers a lot of freedom to optimize the light-curve model. More than one spot can be added, and for each the location, size, and temperature can be adjusted. Instead, the energy-transfer stream shown in Fig. 7.1 has (only) 4 parameters.

According to the model of Stępień (2009), the latitudinal extent of the stream relates to the thermal structure of the stream, which in turn can inform how energy transfer should be implemented in evolutionary calculations. In this way, we steadily make progress in unraveling the life of contact binaries from the first time they touch, until they finally merge.

Appendix A
Additional Figures and Tables to Chapter 2

This appendix is taken from:

Resolving the dynamical mass tension of the massive binary 9 Sgr
M. Fabry, C. Hawcroft, A. J. Frost, L. Mahy, P. Marchant, J-B. Le Bouquin, H. Sana
Astronomy & Astrophysics, 651, A119 (2021)

This appendix contains supplementary figures (Figs. A.1, A.2, A.3, and A.4) and tables (Tables A.1, A.2, and A.3) to Chap. 2.

M. Fabry, *Detailed Structure and Evolution Modeling of the Tightest Massive Binary Stars*, Springer Theses, https://doi.org/10.1007/978-3-032-28788-5

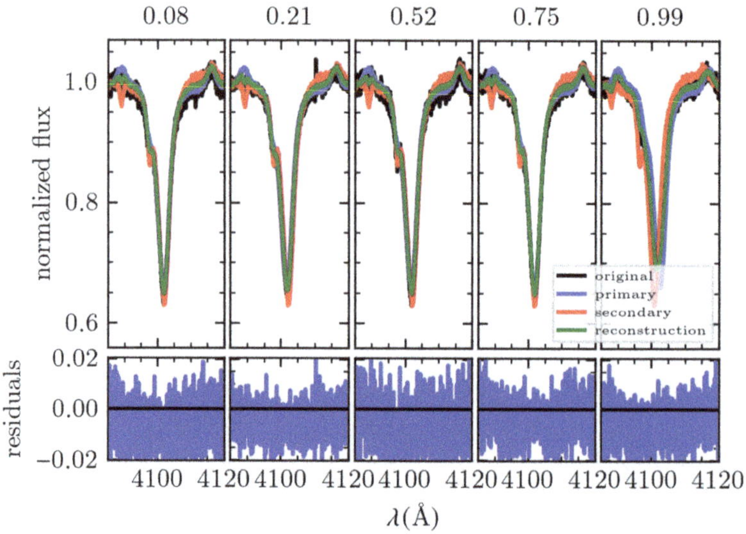

Fig. A.1 Reconstruction of the Hδ line at orbital phases indicated above each column. At phase 0.99 the reconstruction is slightly too high (∼1%) near the core, hinting at a possible inconsistent normalization across the spectra. The well resolved N III feature in the left wing of Hδ is noted

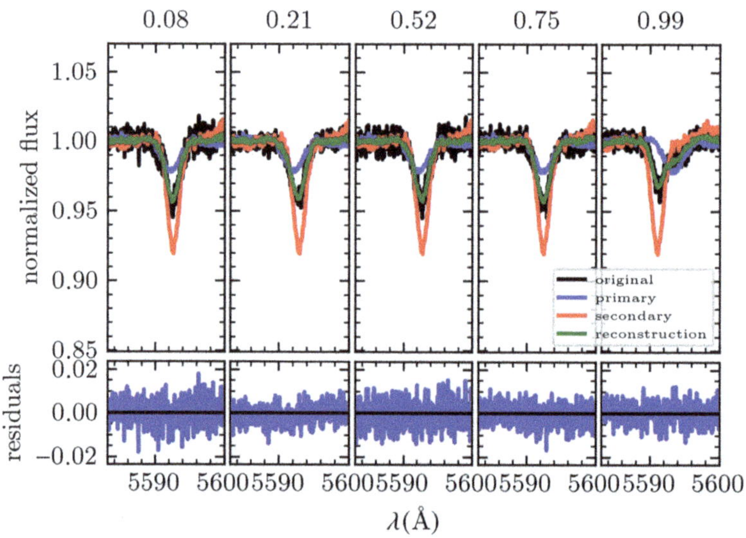

Fig. A.2 Reconstruction of the O III λ 5592 line at orbital phases indicated above each column. A good correspondence is observed, even for this relatively weak metal line, although the reconstruction is slightly too shallow at phase 0.99

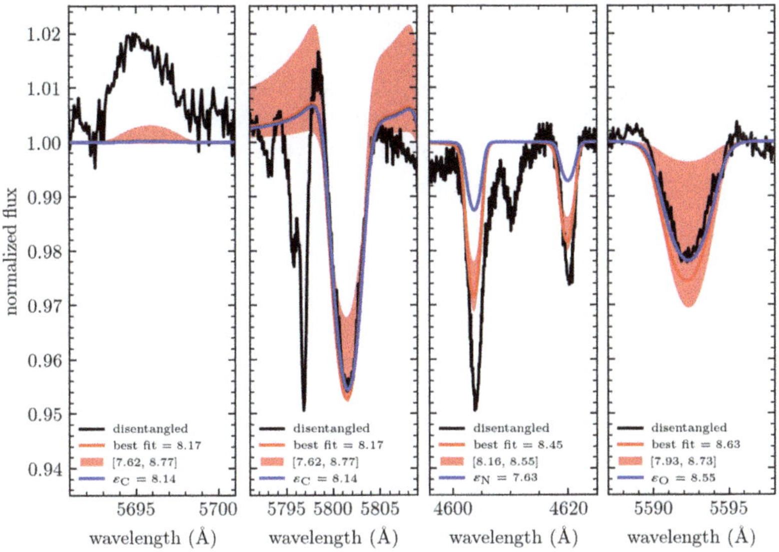

Fig. A.3 Disentangled (black) and best-fit model (red) spectra of the primary star with error ranges, along with a model having the Brott et al. (2011) CNO baseline abundances (blue). From left to right, the lines C III λ 5696, C IV λ 5801, N V λλ 4604, 4620, and O III λ 5592 are shown, and the legend gives CNO abundance values [X/H] + 12 and the baseline values ε_X. The third panel provides evidence for significant nitrogen enrichment

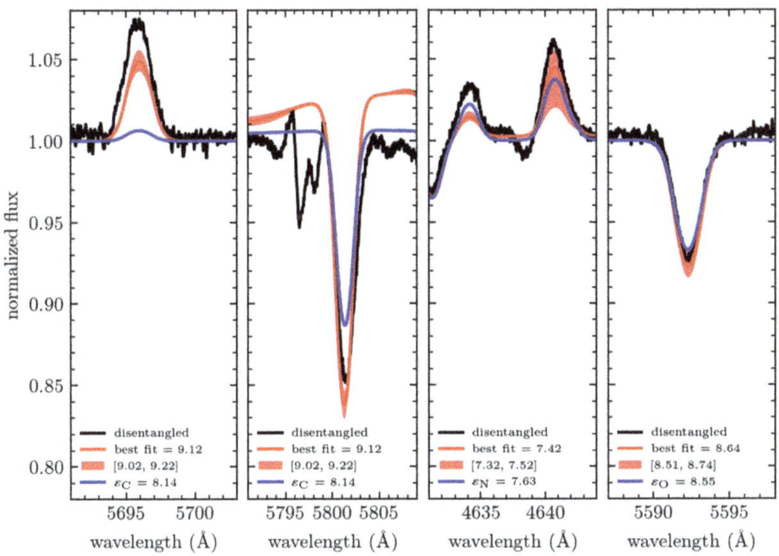

Fig. A.4 Comparison of the disentangled (black) and best-fit model (red) spectra of the secondary star with error ranges, along with a model having the Brott et al. (2011) CNO baseline abundances (blue). From left to right, the lines C III λ 5696, C IV λ 5801, N III λλ 4634, 4641, and O III λ 5592 are shown, with a similar legend as Fig. A.3. The two leftmost panels argue for an elevated carbon abundance with respect to baseline, although as discussed in Sect. 2.4.2, the issues of C III λ 5696 prohibits making quantitative conclusions

Table A.1 Previously unanalyzed spectra of 9 Sgr. The columns signify the modified Julian date (MJD) of observation, spectrograph, exposure time and estimated S/N at 4500 Å

MJD	Instrument	Exposure time (s)	S/N
57,171.1874	HERMES	571	90
57,172.1426	HERMES	325	96
57,173.1646	HERMES	478	75
57,201.0247	HERMES	360	138
57,236.9486	HERMES	800	161
57,634.0119	HARPS	180	143

Table A.2 Journal of the interferometric measurements of 9 Sgr. Given are the VLTI instrument (Instr.), the MJD of observation, the relative angular separation (Sep.), the position angle (PA) east of north, and the major (σ_{maj}) and minor (σ_{min}) axes of the 1σ error ellipses along with its PA east of north (PA_σ)

Instr.	MJD	Sep. (mas)	PA (deg)	σ_{maj} (mas)	σ_{min} (mas)	PA_σ (deg)
AMBER	54,995.306	20.74	69.87	1.12	0.75	154
AMBER	55,644.304	11.85	71.58	1.20	0.50	116
AMBER	55,648.389	11.76	72.19	0.83	0.63	113
PIONIER	56,154.134	0.74	149.10	1.85	0.82	16
PIONIER	56,189.027	1.09	207.39	1.03	0.48	153
PIONIER	56,191.087	1.05	201.56	0.57	0.21	138
PIONIER	56,376.318	4.87	242.59	0.50	0.23	44
PIONIER	56,383.324	4.94	241.83	1.82	0.31	112
PIONIER	56,438.244	5.57	244.91	0.13	0.06	151
PIONIER	56,549.143	4.40	247.37	2.12	0.36	141
PIONIER	56,714.359	3.23	59.89	0.21	0.18	150
PIONIER	56,751.389	4.92	62.51	0.24	0.10	135
PIONIER	56,783.350	6.23	63.05	0.22	0.09	121
PIONIER	56,787.340	6.43	63.33	0.12	0.09	127
PIONIER	56,818.282	7.56	64.46	0.50	0.23	137
PIONIER	56,903.045	10.63	65.49	0.25	0.22	6
PIONIER	57,558.303	21.25	67.89	0.43	0.24	176
GRAVITY	57,558.305	21.194	68.028	0.101	0.102	90
GRAVITY	57,560.312	21.45	67.90	0.24	0.22	90
GRAVITY	57,647.081	21.664	68.246	0.101	0.102	90
PIONIER	57,601.215	21.69	68.08	0.42	0.33	62
PIONIER	57,900.386	22.17	68.53	0.93	0.27	144
PIONIER	57,995.105	21.86	69.04	0.43	0.25	130

Table A.3 Spectral lines and their respective wavelength ranges, which were fitted against the FASTWIND atmosphere modeling in Sect. 2.4, for the primary and secondary star

Spectral line	λ_{min}[Å]	λ_{max}[Å]	Spectral line	λ_{min}[Å]	λ_{max}[Å]
He I 4026	4022.0	4029.8	He II 4686	4680.6	4690.4
N IV 4058	4056.5	4059.9	$H\beta$	4850.8	4870.9
$H\delta$	4093.6	4107.9	He II 5411	5405.4	5417.1
He II 4200	4195.5	4204.3	O III 5592	5589.3	5595.4
$H\gamma$	4332.5	4348.4	C III 5696	5692.4	5698.8
He I 4471	4468.6	4474.0	C IV 5801	5798.8	5805.0
He II 4541	4535.5	4547.3	C IV 5812	5809.3	5815.0
N V 4604	4600.3	4608.5	He I 5875	5872.0	5878.9
N V 4619	4617.4	4622.8	$H\alpha$	6551.7	6571.2

Appendix B
Distortion in the Single Rotating Star Potential

This appendix is from:

Modeling Overcontact Binaries, I. The Effect of Tidal Distortion
M. Fabry, P. Marchant and H. Sana
Astronomy & Astrophysics, 661, A123 (2022)

B.1 Polynomial Fits in the Rotating Shell Potential

In the case of the rotating star potential, Eq. (3.15), integral expressions can be written down to compute all relevant quantities required to use the modified stellar structure equations presented in Sect. 3.2.1. Paxton et al. (2019) numerically integrated these and provided polynomial fits in the variable $\omega = \Omega/\Omega_{\mathrm{crit}}$, where $\Omega_{\mathrm{crit}} = \sqrt{GM/r_{\mathrm{e}}^3}$ is the classical critical velocity of equatorial radius r_{e}.

To improve on the errors introduced by such fits, we recalculated the integrals and used the Levenberg-Marquardt non-linear least-squares algorithm from `scipy` (Virtanen et al. 2020) to find more accurate polynomial fits.

First, for the equipotential volumes V_Ψ, we have the fit:

$$V_\Psi(\omega) \simeq \frac{4\pi}{3} r_{\mathrm{e}}^3 \left(1 - \frac{\omega^2}{2} + 0.1180\omega^4 - 0.07688\omega^6\right), \tag{B.1}$$

with a maximal error of 0.20% in $\omega \in [0, 1]$. We note that since $V_\Psi(1)$ is an analytical result (Kopal 1959) and the leading order term in the expansion is set by requiring that a slow rotating shell is ellipsoidal, this fit has only one free parameter.

M. Fabry, *Detailed Structure and Evolution Modeling of the Tightest Massive Binary Stars*, Springer Theses, https://doi.org/10.1007/978-3-032-28788-5

For the volume equivalent radius, r_Ψ, and equatorial radius, r_e, have the following fits:

$$r_\Psi(\omega) \simeq r_\mathrm{e}\left(1 - \frac{\omega^2}{6} + 0.02025\omega^4 - 0.03870\omega^6\right), \tag{B.2}$$

with a maximal error of 0.11% in $\omega \in [0, 1]$, and,

$$r_\mathrm{e} \simeq r_\Psi(\omega)\left(1 + \frac{\omega^2}{6} - 0.005124\omega^4 + 0.06562\omega^6\right), \tag{B.3}$$

with a maximal error of 0.15% in $\omega \in [0, 1]$. Next, the surface area S_Ψ and average effective gravity $\langle g\rangle$ are approximated by:

$$S_\Psi \simeq 4\pi r_\mathrm{e}^2\left(1 - \frac{\omega^2}{3} + 0.08696\omega^4 - 0.05079\omega^6\right), \tag{B.4}$$

$$S_\Psi\,\langle g\rangle \simeq 4\pi G M_\Psi\left(1 - \frac{2}{3}\omega^2 + 0.2989\omega^4 + 0.006978\omega^6\right), \tag{B.5}$$

with a maximal error of 0.07% and 0.25% in $\omega \in [0, 1]$, respectively. Since the inverse gravity diverges when $\omega \to 1$, we require a non-polynomial term to represent it appropriately in that limit. Therefore, constructing the auxiliary functions:

$$A(\omega) = 1 + 0.3293\omega^4 - 0.4926\omega^6 - 0.5560\ln(1 - \omega^{5.626}), \tag{B.6}$$

$$B(\omega) = 1 + \frac{\omega^2}{5} + 0.4140\omega^4 - 0.8650\omega^6 - \frac{3}{2} \times 0.5580\ln(1 - \omega^{5.626}), \tag{B.7}$$

we have then for the average inverse effective gravity $\langle g^{-1}\rangle$, and the specific moment of inertia i_rot, the following fits:

$$S_\Psi\langle g^{-1}\rangle \simeq \frac{4\pi r_\mathrm{e}^4}{G M_\Psi}A(\omega), \tag{B.8}$$

$$i_\mathrm{rot}(\omega) \simeq \frac{2}{3}r_\mathrm{e}^2\frac{B(\omega)}{A(\omega)}, \tag{B.9}$$

with a maximal error of 0.85% and 0.68% in $\omega \in [0, 0.9999]$, respectively.

For the correction factors, f_P and f_T, we construct the following fits:

$$f_P(\omega) \simeq \frac{1 - \frac{2}{3}\omega^2 + 0.2133\omega^4 - 0.1068\omega^6}{A(\omega)}, \tag{B.10}$$

$$f_T(\omega) \simeq \frac{1 - 0.07955\omega^4 - 0.2322\omega^6}{A(\omega)}, \tag{B.11}$$

with a maximal error of 0.56% and 0.58% in $\omega \in [0, 0.9999]$, respectively.

Finally, with the auxiliary function:

$$C(\omega) = 1 + \frac{17}{60}\omega^2 + 0.4010\omega^4 - 0.8606\omega^6 - 0.9305\ln(1 - \omega^{5.626}), \tag{B.12}$$

the implicit equation from which ω can be obtained from j_{rot} is:

$$\frac{j_{\text{rot}}}{\sqrt{GM_\Psi r_\Psi}} \simeq \frac{2}{3}\frac{\omega C(\omega)}{A(\omega)}, \tag{B.13}$$

with a maximal error of 0.54% in $\omega \in [0, 0.9999]$.

With these fits, we improve all the fits of Paxton et al. (2019), especially the maximal error of f_T, taking it from 1.6% to 0.58%.

B.2 Matching the Tidal and Single Rotating Star Distortion Models

In Fig. 3.5, we make the comparison of the structure correction factor, f_P, of the tidal model as well as the single rotating star model in one plot. This is meant to illustrate the corrections that would be used in a simulation with the MESA code if rotation is included in a model for a binary system without including tidal deformation. In practice, a tidally distorted star does not have a well defined equatorial radius with which to define ω, but nevertheless the current implementation of MESA makes use of the equipotential radius r_Ψ, the mass contained within the shell M_Ψ and the angular frequency of the shell Ω_Ψ to derive the corrections as if these properties corresponded to a single rotating star with cylindrical symmetry.

For a tidally synchronized binary, the angular velocity is derived from Kepler's third law:

$$\Omega = \sqrt{\frac{GM_1(1+q)}{a^3}}, \tag{B.14}$$

with $q = M_2/M_1$ as the mass ratio and a as the separation of the binary. Next, we consider the equivalent single rotating star. Given the mass M_1 and equatorial radius r_e, it has a critical rotation rate:

$$\Omega_{\text{crit}} = \sqrt{\frac{GM_1}{r_e^3}}. \tag{B.15}$$

Dividing these equations, and setting $\omega = \Omega/\Omega_{\rm crit}$ as the dimensionless rotation rate, we arrive at an implicit equation for ω:

$$\omega^2 = \left(\frac{r_e(\omega, r_\Psi)}{a}\right)^3 (1+q) = \left(\frac{r_\Psi}{a} f_{r_{\rm e}}(\omega)\right)^3 (1+q). \tag{B.16}$$

where $f_{r_{\rm e}}$ relates the volume equivalent radius to the equatorial radius as a function of the dimensionless rotation rate in a single rotating star. For a given value of r_Ψ/a and q, this implicit equation gives the value of ω that a MESA simulation would determine to compute the corrections to the stellar-structure equations, despite there not being a well-defined equatorial radius. The values shown in Fig. 3.5 for the single star equivalent correspond to evaluating the results of Appendix B.1 for $f_P(\omega)$, with ω derived from r_Ψ/a and q with the implicit Eq. (B.16).

Appendix C
Additional Considerations of Energy Transfer

This appendix is from:

Modeling contact binaries, II. The effect of energy transfer
M. Fabry, P. Marchant, N. Langer and H. Sana
Astronomy & Astrophysics, 672, A175 (2023)

C.1 Energy Transfer Location

Because our ET formalism is implemented explicitly and the radius of a stellar model can change during the solver iterations, which is also coupled to mass added or subtracted due to MT, a non-negligible mismatch between the location of energy input and the RL can occur. In particular, when mass is removed due to MT, the ET location is then evaluated on the remaining cells, without those cells being adjusted since the previous step. Therefore, in phases with a high MT rate, it is possible that all mass above the RL is removed in one step (although after the solver performs the integration, the star will expand due to thermal relaxation). This means that the location of ET in terms of radius is rather poorly defined. To mitigate this, instead of choosing directly to put energy in cells with radii:

$$r \in [1.00, 1.01] R_{\mathrm{RL}}, \tag{C.1}$$

we opt to do following. At the beginning of the evolutionary step, we evaluate:

$$x_{\mathrm{depth}} = 1 - m(r = R_{\mathrm{RL}})/M, \tag{C.2}$$

M. Fabry, *Detailed Structure and Evolution Modeling of the Tightest Massive Binary Stars*, Springer Theses, https://doi.org/10.1007/978-3-032-28788-5

as the depth of the current RL in fractional mass coordinates. Then during the next integration, the location of energy input is in the cells with radii:

$$r \in [1.00, 1.01]r(x = x_{\mathrm{depth}}). \tag{C.3}$$

This treatment mitigates mismatches in ET locations introduced by MT during evolutionary steps to the degree that the depth fraction changes from step to step.

C.2 Convergence Study

This appendix details the convergence study we used to validate our stellar model with and without energy transfer. We have implemented the ET method in both MESA version r15140, as well as version r22.11.1.

C.2.1 Explicit Evaluation

In MESA r15140, both the tidal deformation corrections of Sect. 3.2.5 and the ET of Sect. 4.3.4 are evaluated explicitly in the differential equation solver, meaning that only information on the previous evolutionary step was used.

We performed a time resolution study where we took a base template and increased the time resolution by successive factors of two. Figure C.1 shows the

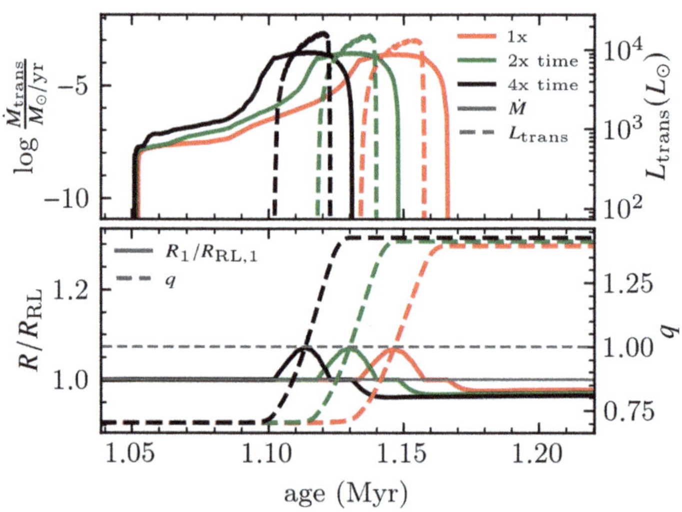

Fig. C.1 Evolution of the MT rate and ET rate (top panel) and the fractional RL radius and mass ratio (bottom panel) during fast case A MT for a $20\,M_\odot + 14\,M_\odot$ binary, computed using the explicit evaluation of the tidal deformation corrections in MESA r15140. The color indicates what time resolution scale has been used

evolution of the MT rate, ET rate, the mass ratio, and the donor radius during the fast case A MT phase. While the time of the onset of MT, which coincides with the donor reaching the RL, remains roughly constant for the different resolutions, the MT rate evolves significantly differently. For higher time resolutions, the MT rate is higher, resulting in slightly shorter contact phase and a higher final mass ratio. This effect is independent of our implementation of ET since the changes occur before the contact phase (signified by the bump in the donor radius). Furthermore, these models do not yet converge to a certain value for the quantities of interest, which are the contact phase duration and the mass ratio after the MT phase. This means we would need to increase the time resolution further to get to a convergent solution. This is prohibitive however keeping in mind the aim is to run a full population of models in later work. We therefore elected to implement the tidal deformation corrections in `MESA` version 22.11.1 to allow for implicit evaluation during integrations.

C.2.2 Implicit Evaluation

In `MESA` r22.11.1, infrastructure is provided to supply numerically approximated derivatives necessary for the solver to implicitly evaluate the tidal deformation corrections during an integration step. Generally, implicit evaluation is beneficial for numerical stability, and allows for longer timesteps. We note that the amount of energy to be transferred and its location in the stellar structure is still evaluated explicitly, as detailed in Sect. 4.3.4.

We carried out a similar resolution study as the one described in Sect. C.2.1 of a $20M_\odot + 14M_\odot$ binary, were we progressively increase the time resolution between simulations. Compared to version r15140, we observe a much better convergence of the MT rate evolution, as pictured in Fig. C.2.

Finally, we performed a spacial resolution convergence study for the same initial conditions as before. Figure C.3 shows that our models are well converged in the spacial coordinate already at the 1x level. Even during the contact phase where ET occurs, and a sharp increase in luminosity is introduced, the models depend very weakly on the spacial resolution.

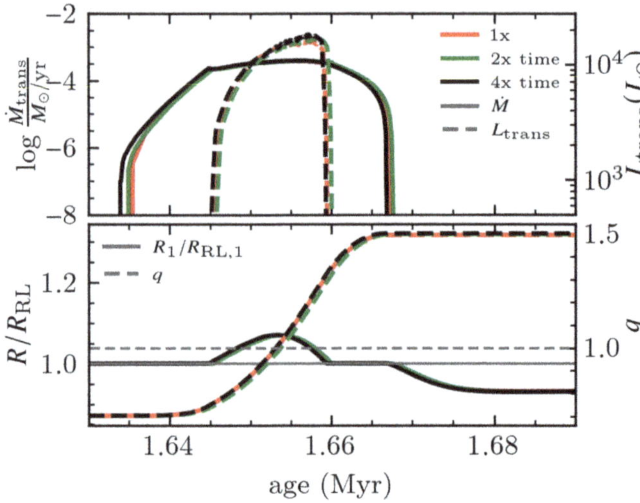

Fig. C.2 Evolution of the MT rate and ET rate (top panel) and the fractional RL radius and mass ratio (bottom panel) during fast case A MT for a $20\,M_\odot + 14\,M_\odot$ binary, but, as opposed to Fig. C.1, now the models were computed using implicit evaluation of the tidal deformation corrections in MESA r22.11.1. The color indicates what time resolution scale has been used

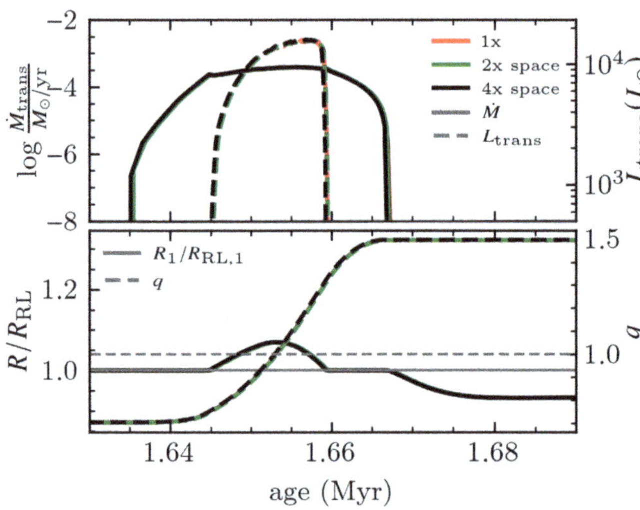

Fig. C.3 Evolution of the MT rate and ET rate (top panel) and the fractional RL radius and mass ratio (bottom panel) during fast case A MT for a $20\,M_\odot + 14\,M_\odot$ binary, using implicit evaluation of the tidal deformation corrections in MESA r22.11.1. Here the color indicates the increase in spatial resolution

Appendix D
Additional Figures and Tables to Chapter 5

This appendix is from:

Modeling contact binaries, III. Properties of a population of close, massive binaries
M. Fabry, P. Marchant, N. Langer and H. Sana
Astronomy & Astrophysics, 695, A109 (2025)

This appendix contains supplementary figures (Figs. D.1 and D.2) and tables (Tables D.1 and D.2) to Chap. 5.

M. Fabry, *Detailed Structure and Evolution Modeling of the Tightest Massive Binary Stars*, Springer Theses, https://doi.org/10.1007/978-3-032-28788-5

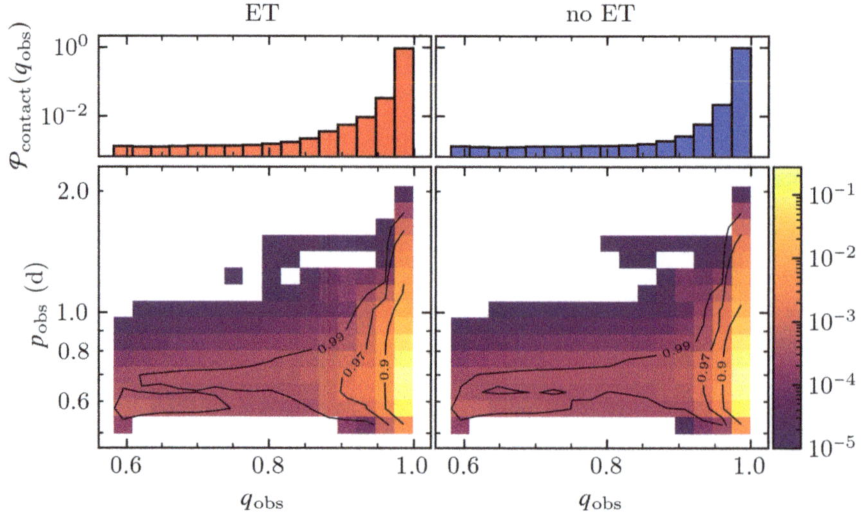

Fig. D.1 Probability density in ($q_{\rm obs}$, $p_{\rm obs}$) (lower panels) and $q_{\rm obs}$ space (upper panels) for the models that experience RLOF at the ZAMS. The thin black lines are highest density intervals of the indicated percentiles, i.e., we expect to observe at most 3% of systems with a mass ratio significantly away from unity

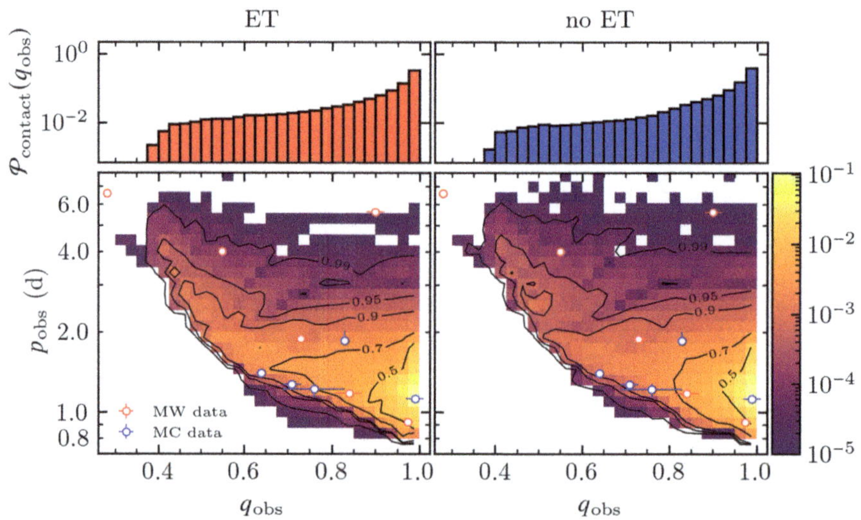

Fig. D.2 Probability density in ($q_{\rm obs}$, $p_{\rm obs}$) (lower panels) and $q_{\rm obs}$ space (upper panels) for the models with initial primary mass higher than 20 $M_{\odot}$. The thin black lines are HDI's of the indicated percentiles. ET has a measurable effect on a larger part of the parameter space than in Fig. 5.8

Table D.1 Parameters of observed, massive (near-)contact systems in the Milky Way and M31. For each system, the first (second) line gives the mass, radius, effective temperature and relative overflow for the primary (secondary)

Name	p(d)	$q = M_2/M_1$	$M(M_\odot)$	$R(R_\odot)$	$T_{\rm eff}$(kK)	$R/R_{\rm RL}$	Reference
V382 Cyg	1.886	0.726 ± 0.017	26.1 ± 0.4	9.4 ± 0.2	$37.20 \pm 0.69 : 0.72$	1.01	A21
			19.0 ± 0.3	8.7 ± 0.2	$38.25 \pm 0.73 : 0.75$	1.08	
TU Mus	1.387	0.625 ± 0.009	16.7 ± 0.4	7.2 ± 0.5	38.7	1.09	P08
			10.4 ± 0.4	5.7 ± 0.5	33.2	1.06	
LY Aur	4.002	0.550 ± 0.007	25.5	16.1	31.0	1.03	Mr13
			14.0	12.6	31.2	1.03	
V701 Sco	0.762	0.995 ± 0.002	9.78 ± 0.22	4.137 ± 0.316	23.5 ± 1.0	1.14	Y19
			9.74 ± 0.22	4.132 ± 0.312	23.44 ± 0.05	1.15	
CT Tau	0.667	0.983 ± 0.003	14.25 ± 3.30	4.89 ± 0.41	25.45 ± 2.25	1.31	Y19
			14.01 ± 3.40	4.89 ± 0.42	25.64 ± 0.12	1.33	
GU Mon	0.897	0.976 ± 0.003	8.79 ± 0.13	4.636 ± 0.252	28.0 ± 2.0	1.19	Y19
			8.58 ± 0.12	4.596 ± 0.246	27.82 ± 0.07	1.21	
XZ Cep	5.097	0.50 ± 0.02	18.7 ± 1.3	14.2 ± 0.1	28.0 ± 1.0	0.85	Ms17
			9.3 ± 0.5	14.2 ± 0.1	24.0 ± 3.0	1.17	
LSS 3074	2.184	0.86 ± 0.04	17.2 ± 1.4	8.2 ± 0.7	39.9 ± 1.5	0.91	R17
			14.8 ± 1.1	7.5 ± 0.6	34.1 ± 1.5	0.93	
MY Cam	1.175	0.839 ± 0.027	37.7 ± 1.6	7.60 ± 0.10	42.0 ± 1.5	1.00	A21
			31.6 ± 1.4	7.01 ± 0.09	39.0 ± 1.5	1.01	
V348 Car	5.562	0.90 ± 0.02	32 ± 4	18.8 ± 1.4	29.7 ± 1.3	0.93	HE85
			29 ± 4	19.3 ± 1.4	26.2	1.00	
V729 Cyg	6.598	0.290 ± 0.012	31.6 ± 2.9	25.6 ± 1.1	28.0	1.03	YY14
			8.8 ± 0.3	14.5 ± 1.0	21.26 ± 0.37	1.03	
BH Cen	0.792	0.844 ± 0.003	9.4 ± 5.4	4.0 ± 0.7	17.9	1.09	Lg84
			7.9 ± 5.4	3.7 ± 0.7	17.43 ± 0.02	1.09	
SV Cen	1.659	0.800 ± 0.078	9.6	7.8	16	1.01	LS81
			7.7	7.3	24	1.08	
V606 Cen	1.495	0.548 ± 0.001	14.7 ± 0.4	6.83 ± 0.06	29.2	1.01	Lz99
			7.96 ± 0.22	5.19 ± 0.05	21.77 ± 0.02	1.08	
HD 64315 B	1.019	1.00 ± 0.09	14.6 ± 2.3	5.52 ± 0.55	32	1.10	Lo17
			14.6 ± 2.3	5.33 ± 0.52	32	1.07	
V4741 M31A	1.604	0.924 ± 0.015	18[a]	7.2[a]	31.6	0.99	Li22
			16.8[a]	6.7[a]	27.3	0.97	
V1555 M31A	0.917	0.974 ± 0.012	23[a]	6.7[a]	35.1	1.24	Li22
			22.4[a]	6.4[a]	34.4	1.24	

References A21: Abdul-Masih et al. (2021); P08: Penny et al. (2008); Mr13: Mayer et al. (2013); Y19: Yang et al. (2019); Ms17: Martins et al. (2017); R17: Raucq et al. (2017); HE85: Hilditch and Evans (1985); YY14: Yaşarsoy and Yakut (2014); Lg84: Leung et al. (1984); LS81: Linnell and Scheick (1991); Lz99: Lorenz et al. (1999); Lo17: Lorenzo et al. (2017); Li22: Li et al. (2022)

[a] M_1 is adopted, from which M_2 follows via q. The radii are computed from Kepler's third law and the fractional radii from the photometric solution

Table D.2 Parameters of observed, massive (near-)contact systems in the Magellanic Clouds. For each system, the first (second) line gives the mass, radius, effective temperature and relative overflow for the primary (secondary)

Name	p(d)	$q = M_2/M_1$	$M(M_\odot)$	$R(R_\odot)$	$T_{\rm eff}$(kK)	$R/R_{\rm RL}$	Reference
VFTS 352	1.124	0.99 ± 0.01	28.85 ± 0.30	7.25 ± 0.02	$41.45^{+1.17}_{-0.80}$	1.01	A21
			28.63 ± 0.30	7.22 ± 0.02	$44.15^{+0.11}_{-0.12}$	1.08	
VFTS 066	1.141	0.52 ± 0.05	$13.0^{+7.0}_{-5.0}$	$5.8^{+0.5}_{-0.8}$	$32.8^{+1.7}_{-1.0}$	1.07	M20
			$6.6^{+3.5}_{-2.8}$	$4.4^{+0.4}_{-0.8}$	$29.0^{+1.0}_{-1.2}$	1.09	
VFTS 661	1.266	0.71 ± 0.02	$27.3^{+0.9}_{-1.0}$	$6.80^{+0.04}_{-0.01}$	$38.4^{+0.9}_{-0.4}$	0.94	M20
			$19.4^{+0.6}_{-0.7}$	$5.70^{+0.03}_{-0.01}$	$31.8^{+1.4}_{-0.6}$	0.92	
VFTS 217	1.855	0.83 ± 0.01	46.8 ± 11.7	$10.1^{+1.5}_{-1.2}$	$45.0^{+0.6}_{-0.4}$	0.91	M20
			38.9 ± 9.7	$9.4^{+1.4}_{-1.0}$	$41.8^{+1.7}_{-0.6}$	0.92	
VFTS 563	1.217	0.76 ± 0.01	$26.2^{+11.9}_{-5.2}$	$6.6^{+0.4}_{-0.7}$	$32.4^{+1.0}_{-0.9}$	0.95	M20
			$20.0^{+9.1}_{-3.9}$	$5.8^{+0.4}_{-0.6}$	$32.4^{+1.2}_{-0.9}$	0.95	
MACHO CB[a]	1.217	0.64 ± 0.01	41.2 ± 1.2	9.56 ± 0.02	50	1.06	O01
			27.0 ± 1.2	7.99 ± 0.05	49.5	1.09	
SMC 108086[b]	0.883	0.85 ± 0.11	16.9 ± 1.2	5.7 ± 0.2	$36.00^{+0.69}_{-0.66}$	1.19	A21
			14.3 ± 1.7	5.3 ± 0.2	$35.20^{+0.64}_{-0.72}$	1.19	

References A21: Abdul-Masih et al. (2021); M20: Mahy et al. (2020a); O01: Ostrov (2001)

[a] MACHO* J053441.3-693139

[b] OGLE SMC-SC10 108086

References

Abbasi R, Ackermann M, Adams J, et al (2009) The IceCube data acquisition system: Signal capture, digitization, and timestamping. Nucl Instrum Methods Phys Res A 601:294–316, https://doi.org/10.1016/j.nima.2009.01.001, https://ui.adsabs.harvard.edu/abs/2009NIMPA.601..294A

Abbott DC, Bieging JH, Churchwell E (1984) The detection of variable, nonthermal radio emission from two O type stars. Astrophys J 280:671–678, https://doi.org/10.1086/162040, http://adsabs.harvard.edu/abs/1984ApJ...280..671A

Abbott BP, Abbott R, Abbott TD et al (2016) Observation of gravitational waves from a binary black hole merger. Phys Rev Lett 116(6):1102. https://doi.org/10.1103/PhysRevLett.116.061102, https://journals.aps.org/prl/abstract/10.1103/PhysRevLett.116.061102

Abbott BP, Abbott R, Abbott TD et al (2018) Prospects for observing and localizing gravitational-wave transients with Advanced LIGO, Advanced Virgo and KAGRA. Living Rev Relat 21:3, https://doi.org/10.1007/s41114-018-0012-9, https://ui.adsabs.harvard.edu/abs/2018LRR....21....3A

Abdul-Masih M, Sana H, Sundqvist J, Mahy L, Menon A, Almeida LA, De Koter A, de Mink SE, Justham S, Langer N, Puls J, Shenar T, Tramper F (2019) Clues on the origin and evolution of massive contact binaries: atmosphere analysis of VFTS 352. Astrophys J 880(2):115, https://doi.org/10.3847/1538-4357/ab24d4, http://arxiv.org/abs/1906.01066, 1906.01066

Abdul-Masih M, Sana H, Conroy KE, Sundqvist J, Prša A, Kochoska A, Puls J (2020) Spectroscopic patch model for massive stars using PHOEBE II and FASTWIND. Astron Astrophys 636:A59, https://doi.org/10.1051/0004-6361/201937341, https://ui.adsabs.harvard.edu/abs/2020A&A...636A..59A

Abdul-Masih M, Sana H, Hawcroft C, Almeida LA, Brands SA, de Mink SE, Justham S, Langer N, Mahy L, Marchant P, Menon A, Puls J, Sundqvist J (2021) Constraining the overcontact phase in massive binary evolution. I. Mixing in V382 Cyg, VFTS 352, and OGLE SMC-SC10 108086. Astron Astrophys 651:A96, https://doi.org/10.1051/0004-6361/202040195, https://ui.adsabs.harvard.edu/abs/2021A&A...651A..96A/abstract

Abdul-Masih M, Escorza A, Menon A, Mahy L, Marchant P (2022) Constraining the overcontact phase in massive binary evolution. II. Period stability of known O+O overcontact systems. Astron Astrophys 666:A18, https://doi.org/10.1051/0004-6361/202244148, https://ui.adsabs.harvard.edu/abs/2022A%26A...666A..18A/abstract

Abel T, Bryan GL, Norman ML (2002) The formation of the first star in the universe. Science 295:93–98, https://doi.org/10.1126/science.295.5552.93, https://ui.adsabs.harvard.edu/abs/2002Sci...295...93A

M. Fabry, *Detailed Structure and Evolution Modeling of the Tightest Massive Binary Stars*, Springer Theses, https://doi.org/10.1007/978-3-032-28788-5

Adams FC, Laughlin G (1997) A dying universe: the long term fate and evolution of astrophysical objects. Rev Mod Phys 69(2):337–372, https://doi.org/10.1103/RevModPhys.69.337, http://arxiv.org/abs/astro-ph/9701131, astro-ph/9701131

Aerts C, Tkachenko A (2023) Asteroseismic modelling of fast rotators and its opportunities for astrophysics. https://doi.org/10.48550/arXiv.2311.08453, https://ui.adsabs.harvard.edu/abs/2023arXiv231108453A

Aerts C, Thoul A, Daszyńska J, Scuflaire R, Waelkens C, Dupret MA, Niemczura E, Noels A (2003) Asteroseismology of HD 129929: core overshooting and nonrigid rotation. Science 300:1926–1928, https://doi.org/10.1126/science.1084993, https://ui.adsabs.harvard.edu/abs/2003Sci...300.1926A

Aerts C, Christensen-Dalsgaard J, Kurtz DW (2010) Asteroseismology. Springer, https://doi.org/10.1007/978-1-4020-5803-5, https://ui.adsabs.harvard.edu/abs/2010aste.book.....A

Aitken RG (1935) The binary stars. Dover, New York, https://ui.adsabs.harvard.edu/abs/1935bist.book.....A

Almeida LA, Sana H, de Mink SE, Tramper F, Soszyński I, Langer N, Barbá RH, Cantiello M, Damineli A, de Koter A, Garcia M, Gräfener G, Herrero A, Howarth I, Maíz Apellániz J, Norman C, Ramírez-Agudelo OH, Vink JS (2015) Discovery of the massive overcontact binary VFTS352: evidence for enhanced internal mixing. Astrophys J 812:102, https://doi.org/10.1088/0004-637X/812/2/102, https://ui.adsabs.harvard.edu/abs/2015ApJ...812..102A

Almeida LA, Sana H, Taylor W, Barbá R, Bonanos AZ, Crowther P, Damineli A, de Koter A, de Mink SE, Evans CJ, Gieles M, Grin NJ, Hénault-Brunet V, Langer N, Lennon D, Lockwood S, Maíz Apellániz J, Moffat AFJ, Neijssel C, Norman C, Ramírez-Agudelo OH, Richardson ND, Schootemeijer A, Shenar T, Soszyński I, Tramper F, Vink JS (2017) The Tarantula massive binary monitoring. I. Observational campaign and OB-type spectroscopic binaries. Astron Astrophys 598:A84, https://doi.org/10.1051/0004-6361/201629844

Anderson L, Shu FH (1977) On the light curves of W Ursae Majoris stars. Astrophys J 214:798–808, https://doi.org/10.1086/155309, https://ui.adsabs.harvard.edu/abs/1977ApJ...214..798A

Andrassy R, Leidi G, Higl J, Edelmann PVF, Schneider FRN, Röpke FK (2024) Towards a self-consistent model of the convective core boundary in upper main sequence stars. I. 2.5D and 3D simulations. Astron Astrophys 683:A97, https://doi.org/10.1051/0004-6361/202347407, https://ui.adsabs.harvard.edu/abs/2024A&A...683A..97A

Ashfaque SM (1977) Astronomy in the Indus valley civilization: a survey of the problems and possibilities of the ancient Indian astronomy and cosmology in the light of Indus script decipherment by the Finnish scholars. Centaurus 21(2):149–193, https://doi.org/10.1111/j.1600-0498.1977.tb00351.x, https://onlinelibrary.wiley.com/doi/abs/10.1111/j.1600-0498.1977.tb00351.x

Asplund M, Grevesse N, Sauval AJ, Scott P (2009) The chemical composition of the Sun. Annu Rev Astron Astrophys 47:481–522, https://doi.org/10.1146/annurev.astro.46.060407.145222, https://ui.adsabs.harvard.edu/abs/2009ARA&A..47..481A

Aston FW (1925) Atoms and X-rays. Nature 116(2929):902–904, https://doi.org/10.1038/116902a0, https://www.nature.com/articles/116902a0

Atkinson RDE (1936) Atomic synthesis and stellar energy. III. Astrophys J 84:73, https://doi.org/10.1086/143750, https://ui.adsabs.harvard.edu/abs/1936ApJ....84...73A

Atkinson RDE, Houtermans FG (1929) Zur Frage der Aufbaumöglichkeit der Elemente in Sternen. Z Phys 54:656–665, https://doi.org/10.1007/BF01341595, https://ui.adsabs.harvard.edu/abs/1929ZPhy...54..656A

Austin SJ, Robertson JW, Tycner C, Campbell T, Honeycutt RK (2007) Late-type near-contact eclipsing binary [HH97] FS Aur-79. Astron J 133:1934–1946, https://doi.org/10.1086/512614, https://ui.adsabs.harvard.edu/abs/2007AJ....133.1934A

Bailer-Jones CAL, Rybizki J, Fouesneau M, Demleitner M, Andrae R (2021) Estimating distances from parallaxes. V. Geometric and photogeometric distances to 1.47 billion stars in Gaia early data release 3. Astron J 161:147, https://doi.org/10.3847/1538-3881/abd806, https://ui.adsabs.harvard.edu/abs/2021AJ....161..147B, 2012.05220

Balbus SA, Hawley JF (1991) A powerful local shear instability in weakly magnetized disks. I. Linear analysis. Astrophys J 376:214, https://doi.org/10.1086/170270, https://ui.adsabs.harvard.edu/abs/1991ApJ...376..214B

Banyard G, Sana H, Mahy L, Bodensteiner J, Villaseñor JI, Evans CJ (2022) The observed multiplicity properties of B-type stars in the Galactic young open cluster NGC 6231. Astron Astrophys 658:A69, https://doi.org/10.1051/0004-6361/202141037, https://ui.adsabs.harvard.edu/abs/2022A&A...658A..69B/abstract

Baron F, Monnier JD, Pedretti E, Zhao M, Schaefer G, Parks R, Che X, Thureau N, ten Brummelaar TA, McAlister HA, Ridgway ST, Farrington C, Sturmann J, Sturmann L, Turner N (2012) Imaging the Algol triple system in the H band with the CHARA interferometer. Astrophys J 752(1):20, https://doi.org/10.1088/0004-637X/752/1/20, https://dx.doi.org/10.1088/0004-637X/752/1/20

Bastian N, Covey KR, Meyer MR (2010) A Universal stellar initial mass function? A critical look at variations. Annu Rev Astron Astrophys 48:339–389, https://doi.org/10.1146/annurev-astro-082708-101642, https://ui.adsabs.harvard.edu/abs/2010ARA&A..48..339B

Bateson FM (1959) Polynesian navigation. Publ Astron Soc Pac 71(420):187, https://doi.org/10.1086/127363

Bernasconi PA, Maeder A (1996) About the absence of a proper zero age main sequence for massive stars. Astron Astrophys 307:829–839

Bethe HA (1939) Energy production in stars. Phys Rev 55:434–456, https://doi.org/10.1103/PhysRev.55.434, https://ui.adsabs.harvard.edu/abs/1939PhRv...55..434B

Bethe HA, Critchfield CL (1938) The formation of deuterons by proton combination. Phys Rev 54:248–254, https://doi.org/10.1103/PhysRev.54.248, https://ui.adsabs.harvard.edu/abs/1938PhRv...54..248B

Biermann P, Thomas HC (1972) Models for contact binaries. Astron Astrophys 16:60, https://ui.adsabs.harvard.edu/abs/1972A&A....16...60B

Binnendijk L (1970) The orbital elements of W Ursae Majoris systems. Vistas Astron 12:217–256, https://doi.org/10.1016/0083-6656(70)90041-3, https://www.sciencedirect.com/science/article/pii/0083665670900413

Bionta RM, Blewitt G, Bratton CB, Casper D, Ciocio A, Claus R, Cortez B, Crouch M, Dye ST, Errede S, Foster GW, Gajewski W, Ganezer KS, Goldhaber M, Haines TJ, Jones TW, Kielczewska D, Kropp WR, Learned JG, LoSecco JM, Matthews J, Miller R, Mudan MS, Park HS, Price LR, Reines F, Schultz J, Seidel S, Shumard E, Sinclair D, Sobel HW, Stone JL, Sulak LR, Svoboda R, Thornton G, van der Velde JC, Wuest C (1987) Observation of a neutrino burst in coincidence with supernova 1987A in the Large Magellanic Cloud. Phys Rev Lett 58(14):1494–1496, https://doi.org/10.1103/PhysRevLett.58.1494, https://link.aps.org/doi/10.1103/PhysRevLett.58.1494

Björklund R, Sundqvist JO, Puls J, Najarro F (2021) New predictions for radiation-driven, steady-state mass-loss and wind-momentum from hot, massive stars II. A grid of O-type stars in the Galaxy and the Magellanic Clouds. Astron Astrophys 648:A36, https://doi.org/10.1051/0004-6361/202038384, http://arxiv.org/abs/2008.06066, 2008.06066

Björklund R, Sundqvist JO, Singh SM, Puls J, Najarro F (2023) New predictions for radiation-driven, steady-state mass-loss and wind-momentum from hot, massive stars. III. Updated mass-loss rates for stellar evolution. Astron Astrophys 676:A109, https://doi.org/10.1051/0004-6361/202141948, https://ui.adsabs.harvard.edu/abs/2023A&A...676A.109B

Blagorodnova N, Kotak R, Polshaw J, Kasliwal MM, Cao Y, Cody AM, Doran GB, Elias-Rosa N, Fraser M, Fremling C, Gonzalez-Fernandez C, Harmanen J, Jencson J, Kankare E, Kudritzki RP, Kulkarni SR, Magnier E, Manulis I, Masci FJ, Mattila S, Nugent P, Ochner P, Pastorello A, Reynolds T, Smith K, Sollerman J, Taddia F, Terreran G, Tomasella L, Turatto M, Vreeswijk PM, Wozniak P, Zaggia S (2017) Common envelope ejection for a luminous red nova in M101. Astrophys J 834:107, https://doi.org/10.3847/1538-4357/834/2/107, https://ui.adsabs.harvard.edu/abs/2017ApJ...834..107B

Blagorodnova N, Karambelkar V, Adams SM, Kasliwal MM, Kochanek CS, Dong S, Campbell H, Hodgkin S, Jencson JE, Johansson J, Kozłowski S, Laher RR, Masci F, Nugent P, Rebbapragada

U (2020) Progenitor, precursor, and evolution of the dusty remnant of the stellar merger M31-LRN-2015. Mon Not R Astron Soc 496:5503–5517, https://doi.org/10.1093/mnras/staa1872, https://ui.adsabs.harvard.edu/abs/2020MNRAS.496.5503B

Blagorodnova N, Klencki J, Pejcha O, Vreeswijk PM, Bond HE, Burdge KB, De K, Fremling C, Gehrz RD, Jencson JE, Kasliwal MM, Kupfer T, Lau RM, Masci FJ, Rich MR (2021) The luminous red nova AT 2018bwo in NGC 45 and its binary yellow supergiant progenitor. Astron Astrophys 653:A134, https://doi.org/10.1051/0004-6361/202140525

Blouin S, Shaffer NR, Saumon D, and Starrett CE (2020) New conductive opacities for white dwarf envelopes. Astrophys J 899(1):46, https://doi.org/10.3847/1538-4357/ab9e75, 2006.16390

Bodenheimer PH (2011) Principles of star formation. Springer, Berlin Heidelberg, https://doi.org/10.1007/978-3-642-15063-0, https://ui.adsabs.harvard.edu/abs/2011psf..book.....B

Bodensteiner J, Shenar T, Mahy L, Fabry M, Marchant P, Abdul-Masih M, Banyard G, Bowman DM, Dsilva K, Frost AJ, Hawcroft C, Reggiani M, Sana H (2020) Is HR 6819 a triple system containing a black hole? An alternative explanation. Astron Astrophys 641:A43, https://doi.org/10.1051/0004-6361/202038682

Bodensteiner J, Sana H, Wang C, Langer N, Mahy L, Banyard G, de Koter A, de Mink SE, Evans CJ, Götberg Y, Patrick LR, Schneider FRN, Tramper F (2021) The young massive SMC cluster NGC 330 seen by MUSE - II. Multiplicity properties of the massive-star population. Astron Astrophys 652:A70, https://doi.org/10.1051/0004-6361/202140507, https://www.aanda.org/articles/aa/abs/2021/08/aa40507-21/aa40507-21.html, 2104.13409

Böhm-Vitense E (1958) über die Wasserstoffkonvektionszone in Sternen verschiedener Effektivtemperaturen und Leuchtkräfte. Mit 5 Textabbildungen. Z Astrophys 46:108, https://ui.adsabs.harvard.edu/abs/1958ZA.....46..108B

Bohr N (1913) On the constitutions of atoms and molecules. Pt. I. Philos Mag 26:1–25, https://doi.org/10.1080/14786441308634955, https://ui.adsabs.harvard.edu/abs/1913PMag...26....1B

Bolton CT (1972) Identification of Cygnus X-1 with HDE 226868. Nature 235(5336):271–273, https://doi.org/10.1038/235271b0, https://www.nature.com/articles/235271b0

Bondi H, Hoyle F (1944) On the mechanism of accretion by stars. Mon Not R Astron Soc 104:273, https://doi.org/10.1093/mnras/104.5.273, https://ui.adsabs.harvard.edu/abs/1944MNRAS.104..273B

Bonnell IA, Bate MR (2005) Binary systems and stellar mergers in massive star formation. Mon Not R Astron Soc 362:915–920, https://doi.org/10.1111/j.1365-2966.2005.09360.x, https://ui.adsabs.harvard.edu/abs/2005MNRAS.362..915B

Bordier E, Frost AJ, Sana H, Reggiani M, Mérand A, Rainot A, Ramírez-Tannus MC, de Wit WJ (2022) The origin of close massive binaries in the M17 star-forming region. Astron Astrophys 663:A26, https://doi.org/10.1051/0004-6361/202141849, https://ui.adsabs.harvard.edu/abs/2022A&A...663A..26B

Borucki WJ, Koch D, Basri G et al (2010) Kepler planet-detection mission: introduction and first results. Science 327:977, https://doi.org/10.1126/science.1185402, https://ui.adsabs.harvard.edu/abs/2010Sci...327..977B

Broens E (2013) Light-curve analysis and orbital period change of the extreme mass-ratio overcontact binary AW CrB. Mon Not R Astron Soc 430:3070–3077, https://doi.org/10.1093/mnras/stt113, https://ui.adsabs.harvard.edu/abs/2013MNRAS.430.3070B

Bromm V, Coppi PS, Larson RB (2002) The formation of the first stars. I. The primordial star-forming cloud. Astrophys J 564:23–51, https://doi.org/10.1086/323947, https://ui.adsabs.harvard.edu/abs/2002ApJ...564...23B

Brott I, de Mink SE, Cantiello M, Langer N, de Koter A, Evans CJ, Hunter I, Trundle C, Vink JS (2011) Rotating massive main-sequence stars - I. Grids of evolutionary models and isochrones. Astron Astrophys 530:A115, https://doi.org/10.1051/0004-6361/201016113, https://www.aanda.org/articles/aa/abs/2011/06/aa16113-10/aa16113-10.html

Brown NJ, Waagen EO, Scovil C, Nelson P, Oksanen A, Solonen J, Price A (2002) Peculiar variable in Monoceros. Int Astron Union Circ 7785:1, https://ui.adsabs.harvard.edu/abs/2002IAUC.7785....1B

Browning MK, Brun AS, Toomre J (2004) Simulations of core convection in rotating A-type stars: differential rotation and overshooting. Astrophys J 601:512–529, https://doi.org/10.1086/380198, https://ui.adsabs.harvard.edu/abs/2004ApJ...601..512B

Burbidge EM, Burbidge GR, Fowler WA, Hoyle F (1957) Synthesis of the elements in stars. Rev Mod Phys 29(4):547–650, https://doi.org/10.1103/RevModPhys.29.547, https://link.aps.org/doi/10.1103/RevModPhys.29.547

Burssens S, Bowman DM, Michielsen M, Simón-Díaz S, Aerts C, Vanlaer V, Banyard G, Nardetto N, Townsend RHD, Handler G, Mombarg JSG, Vanderspek R, Ricker G (2023) A calibration point for stellar evolution from massive star asteroseismology. Nat Astron 7:913–930, https://doi.org/10.1038/s41550-023-01978-y, https://ui.adsabs.harvard.edu/abs/2023NatAs...7..913B

Cameron AGW, Truran JW (1977) The supernova trigger for formation of the solar system. Icarus 30(3):447–461, https://doi.org/10.1016/0019-1035(77)90101-4, https://www.sciencedirect.com/science/article/pii/0019103577901014

Cannon AJ, Pickering EC (1901) Spectra of bright southern stars photographed with the 13-inch Boyden telescope as part of the Henry Draper Memorial. Ann Harvard College Observ 28:129–P.6, https://ui.adsabs.harvard.edu/abs/1901AnHar..28..129C

Canuto VM (1992) Turbulent convection with overshooting: Reynolds stress approach. Astrophys J 392:218, https://doi.org/10.1086/171420, https://ui.adsabs.harvard.edu/abs/1992ApJ...392..218C

Carneiro LP, Puls J, Sundqvist JO, Hoffmann TL (2016) Atmospheric NLTE models for the spectroscopic analysis of blue stars with winds. III. X-ray emission from wind-embedded shocks. Astron Astrophys 590:A88, https://doi.org/10.1051/0004-6361/201527718

Cassisi S, Potekhin AY, Pietrinferni A, Catelan M, Salaris M (2007) Updated electron-conduction opacities: The impact on low-mass stellar models. Astrophys J 661:1094–1104, https://doi.org/10.1086/516819, astro-ph/0703011

Castor JI, Abbott DC, Klein RI (1975) Radiation-driven winds in Of stars. Astrophys J 195:157–174, https://doi.org/10.1086/153315, http://adsabs.harvard.edu/abs/1975ApJ...195..157C

Chabrier G (2003) Galactic stellar and substellar initial mass function. Publ Astron Soc Pac 115:763–795, https://doi.org/10.1086/376392, https://ui.adsabs.harvard.edu/abs/2003PASP..115..763C

Chandrasekhar S (1931a) The highly collapsed configurations of a stellar mass. Mon Not R Astron Soc 91:456–466, https://doi.org/10.1093/mnras/91.5.456, https://ui.adsabs.harvard.edu/abs/1931MNRAS..91..456C

Chandrasekhar S (1931b) The maximum mass of ideal white dwarfs. Astrophys J 74:81, https://doi.org/10.1086/143324, https://ui.adsabs.harvard.edu/abs/1931ApJ....74...81C

Chandrasekhar S (1939) An introduction to the study of stellar structure. University of Chicago press, Chicago, https://ui.adsabs.harvard.edu/abs/1939isss.book.....C

Chandrasekhar S (1960) The stability of non-dissipative Couette flow in hydromagnetics. Proc Natl Acad Sci 46:253–257, https://doi.org/10.1073/pnas.46.2.253, https://ui.adsabs.harvard.edu/abs/1960PNAS...46..253C

Charbonneau P (1995) Genetic algorithms in astronomy and astrophysics. Astrophys J Suppl Ser 101:309, https://doi.org/10.1086/192242, http://adsabs.harvard.edu/abs/1995ApJS..101..309C

Clagett M (1989) Ancient Egyptian science: a source book. American Philosophical Society

Clayton DD (1983) Principles of stellar evolution and nucleosynthesis. Chicago University Press, https://ui.adsabs.harvard.edu/abs/1983psen.book.....C

Conroy KE, Kochoska A, Hey D, Pablo H, Hambleton KM, Jones D, Giammarco J, Abdul-Masih M, Prša A (2020) Physics of eclipsing binaries. V. General framework for solving the inverse problem. Astrophys J Suppl Ser 250:34, https://doi.org/10.3847/1538-4365/abb4e2, https://ui.adsabs.harvard.edu/abs/2020ApJS..250...34C

Conti PS, Alschuler WR (1971) Spectroscopic studies of O-type stars. I. Classification and absolute magnitudes. Astrophys J 170:325, https://doi.org/10.1086/151218, http://adsabs.harvard.edu/abs/1971ApJ...170..325C

Cook CW, Fowler WA, Lauritsen CC, Lauritsen T (1957) B12, C12, and the Red Giants. Phys Rev 107:508–515, https://doi.org/10.1103/PhysRev.107.508, https://ui.adsabs.harvard.edu/abs/1957PhRv..107..508C

Cooley JW, Tukey JW (1965) An algorithm for the machine calculation of complex Fourier series. Math Comput 19(90):297–301, https://doi.org/10.1090/S0025-5718-1965-0178586-1, https://www.ams.org/mcom/1965-19-090/S0025-5718-1965-0178586-1/

Cowan CL, Reines F, Harrison FB, Kruse HW, McGuire AD (1956) Detection of the free neutrino: a confirmation. Science 124(3212):103–104, https://www.jstor.org/stable/1751492, 1751492

Cox JP, Giuli RT (1968) Principles of stellar structure. Gordon and Breach, New York, https://ui.adsabs.harvard.edu/abs/1968pss..book.....C

Cristini A, Meakin C, Hirschi R, Arnett D, Georgy C, Viallet M, Walkington I (2017) 3D hydrodynamic simulations of carbon burning in massive stars. Mon Not R Astron Soc 471:279–300, https://doi.org/10.1093/mnras/stx1535, https://ui.adsabs.harvard.edu/abs/2017MNRAS.471..279C

Curd P (2019) Anaxagoras. In: Zalta EN (ed) The Stanford encyclopedia of philosophy, winter 2019 edn. Metaphysics Research Lab, Stanford University, https://plato.stanford.edu/archives/win2019/entries/anaxagoras/

Cyburt RH, Amthor AM, Ferguson R, Meisel Z, Smith K, Warren S, Heger A, Hoffman RD, Rauscher T, Sakharuk A, Schatz H, Thielemann FK, Wiescher M (2010) The JINA REACLIB database: its recent updates and impact on type-I X-ray bursts. Astrophys J Suppl Ser 189:240–252, https://doi.org/10.1088/0067-0049/189/1/240, https://ui.adsabs.harvard.edu/abs/2010ApJS..189..240C

Darquier de Pellepoix A, Aubert J, Vidal G (1782) Observations astronomiques faites a Toulouse. A Avignon : chez Jean Aubert, imprimeur-libraire, reu de lμÉpicerie, http://archive.org/details/A077240126127

Darwin GH (1879) The determination of the secular effects of tidal friction by a graphical method. Proc R Soc Lond 29:168–181, https://www.jstor.org/stable/113751, 113751

Davis R, Harmer DS, Hoffman KC (1968) Search for neutrinos from the Sun. Phys Rev Lett 20:1205–1209, https://doi.org/10.1103/PhysRevLett.20.1205, https://ui.adsabs.harvard.edu/abs/1968PhRvL..20.1205D

de Mink SE, Cantiello M, Langer N, Pols OR, Brott I, Yoon SC (2009) Rotational mixing in massive binaries. Detached short-period systems. Astron Astrophys 497(1):243, https://doi.org/10.1051/0004-6361/200811439, https://ui.adsabs.harvard.edu/abs/2009A%26A...497..243D/abstract

de Mink SE, Sana H, Langer N, Izzard RG, Schneider FRN (2014) The incidence of stellar mergers and mass gainers among massive stars. Astrophys J 782:7, https://doi.org/10.1088/0004-637X/782/1/7, https://ui.adsabs.harvard.edu/abs/2014ApJ...782....7D

De Rosa RJ, Patience J, Wilson PA, Schneider A, Wiktorowicz SJ, Vigan A, Marois C, Song I, Macintosh B, Graham JR, Doyon R, Bessell MS, Thomas S, Lai O (2014) The VAST survey - III. The multiplicity of A-type stars within 75 pc. Mon Not R Astron Soc 437:1216–1240, https://doi.org/10.1093/mnras/stt1932, https://ui.adsabs.harvard.edu/abs/2014MNRAS.437.1216D

deBoer RJ, Görres J, Wiescher M, Azuma RE, Best A, Brune CR, Fields CE, Jones S, Pignatari M, Sayre D, Smith K, Timmes FX, Uberseder E (2017) The 12C(α, γ)16O reaction and its implications for stellar helium burning. Rev Mod Phys 89:035007, https://doi.org/10.1103/RevModPhys.89.035007, https://ui.adsabs.harvard.edu/abs/2017RvMP...89c5007D

Dekker H, D'Odorico S, Kaufer A, Delabre B, Kotzłowski H (2000) Design, construction, and performance of UVES, the echelle spectrograph for the UT2 Kueyen Telescope at the ESO Paranal Observatory. In: Optical and IR telescope instrumentation and detectors. International Society for Optics and Photonics, vol 4008, pp 534–545, https://doi.org/10.1117/12.395512, https://www.spiedigitallibrary.org/conference-proceedings-of-spie/4008/0000/Design-construction-and-performance-of-UVES-the-echelle-spectrograph-for/10.1117/12.395512.short

Deupree RG (2001) Stellar evolution with arbitrary rotation laws. IV. Survey of zero-age main-sequence models. Astrophys J 552:268–277, https://doi.org/10.1086/320464, https://ui.adsabs.harvard.edu/abs/2001ApJ...552..268D

Driessen FA, Sundqvist JO, Kee ND (2019) Theoretical wind clumping predictions of OB supergiants from line-driven instability simulations across the bi-stability jump. Astron Astrophys 631:A172, https://doi.org/10.1051/0004-6361/201936331, https://ui.adsabs.harvard.edu/abs/2019A&A...631A.172D

Dunbar DN, Pixley RE, Wenzel WA, Whaling W (1953) The 7.68-Mev state in C12. Phys Rev 92:649–650, https://doi.org/10.1103/PhysRev.92.649, https://ui.adsabs.harvard.edu/abs/1953PhRv...92..649D

Dunstall PR, Dufton PL, Sana H, Evans CJ, Howarth ID, Simón-Díaz S, de Mink SE, Langer N, Maíz Apellániz J, Taylor WD (2015) The VLT-FLAMES Tarantula Survey. XXII. Multiplicity properties of the B-type stars. Astron Astrophys 580:A93, https://doi.org/10.1051/0004-6361/201526192

Eddington AS (1916) On the radiative equilibrium of the stars. Mon Not R Astron Soc 77:16, https://doi.org/10.1093/mnras/77.1.16, https://ui.adsabs.harvard.edu/abs/1916MNRAS..77...16E/abstract

Eddington AS (1925) Circulating currents in rotating stars. Observatory 48:73–75, https://ui.adsabs.harvard.edu/abs/1925Obs....48...73E

Eddington AS (1926) The internal constitution of the stars. Cambridge University Press, Cambridge, https://ui.adsabs.harvard.edu/abs/1926ics..book.....E

Edelmann PVF, Ratnasingam RP, Pedersen MG, Bowman DM, Prat V, Rogers TM (2019) Three-dimensional simulations of massive stars. I. Wave generation and propagation. Astrophys J 876:4, https://doi.org/10.3847/1538-4357/ab12df, https://ui.adsabs.harvard.edu/abs/2019ApJ...876....4E

Edmonson MS (1988) The book of the year: Middle American calendrical systems. University of Utah Press, Salt Lake City

Eggen OJ (1967) Contact binaries, II. Mem R Astron Soc 70:111, https://ui.adsabs.harvard.edu/abs/1967MmRAS..70..111E

Eggleton PP (1983) Approximations to the radii of Roche lobes. Astrophys J 268:368, https://doi.org/10.1086/160960, http://adsabs.harvard.edu/abs/1983ApJ...268..368E

Eggleton PP (2012) Formation and evolution of contact binaries. J Astron Space Sci 29:145–149, https://doi.org/10.5140/JASS.2012.29.2.145, https://ui.adsabs.harvard.edu/abs/2012JASS...29..145E

Einstein A (1915) Die Feldgleichungen der Gravitation. Sitzungsberichte Königlich Preuss Akad Wiss:844–847, https://ui.adsabs.harvard.edu/abs/1915SPAW.......844E

Einstein A (1918) über Gravitationswellen. Sitzungsberichte Königlich Preuss Akad Wiss:154–167, https://ui.adsabs.harvard.edu/abs/1918SPAW.......154E

El-Badry K, Conroy C, Fuller J, Kiman R, van Roestel J, Rodriguez AC, Burdge KB (2022) Magnetic braking saturates: Evidence from the orbital period distribution of low-mass detached eclipsing binaries from ZTF. Mon Not R Astron Soc 517:4916–4939, https://doi.org/10.1093/mnras/stac2945, https://ui.adsabs.harvard.edu/abs/2022MNRAS.517.4916E

El-Badry K, Fabry M, Sana H, Shenar T, Seeburger R (2025) Complex spectral variability and hints of a luminous companion in the Be star + black hole binary candidate ALS 8814. Open J Astrophys 8:128, https://doi.org/10.33232/001c.143907, https://ui.adsabs.harvard.edu/abs/2025OJAp....8E.128E

Emden R (1907) Gaskugeln. B.G. Teubner, https://ui.adsabs.harvard.edu/abs/1907gask.book.....E

Endal AS, Sofia S (1976) The evolution of rotating stars. I - Method and exploratory calculations for a 7-solar-mass star. Astrophys J 210:184–198, https://doi.org/10.1086/154817, http://adsabs.harvard.edu/abs/1976ApJ...210..184E

Epelbaum E, Krebs H, Lee D, Meißner UG (2011) Ab initio calculation of the Hoyle state. Phys Rev Lett 106:192501, https://doi.org/10.1103/PhysRevLett.106.192501, https://ui.adsabs.harvard.edu/abs/2011PhRvL.106s2501E

Espinosa Lara F, Rieutord M (2011) Gravity darkening in rotating stars. Astron Astrophys 533:A43, https://doi.org/10.1051/0004-6361/201117252, https://ui.adsabs.harvard.edu/abs/2011A&A...533A..43E

Espinosa Lara F, Rieutord M (2013) Self-consistent 2D models of fast-rotating early-type stars. Astron Astrophys 552:A35, https://doi.org/10.1051/0004-6361/201220844, https://ui.adsabs.harvard.edu/abs/2013A%26A...552A..35E/abstract

Evans CJ, Taylor WD, Hénault-Brunet V, Sana H, de Koter A, Simón-Díaz S, Carraro G, Bagnoli T, Bastian N, Bestenlehner JM, Bonanos AZ, Bressert E, Brott I, Campbell MA, Cantiello M, Clark JS, Costa E, Crowther PA, de Mink SE, Doran E, Dufton PL, Dunstall PR, Friedrich K, Garcia M, Gieles M, Gräfener G, Herrero A, Howarth ID, Izzard RG, Langer N, Lennon DJ, Maíz Apellániz J, Markova N, Najarro F, Puls J, Ramirez OH, Sabín-Sanjulián C, Smartt SJ, Stroud VE, van Loon JTh, Vink JS, Walborn NR (2011) The VLT-FLAMES Tarantula Survey. I. Introduction and observational overview. Astron Astrophys 530:A108, https://doi.org/10.1051/0004-6361/201116782, https://ui.adsabs.harvard.edu/abs/2011A&A...530A.108E

Evans NE, Bond HE, Schaefer GH, Mason BD, Karovska M, Tingle E (2013) Binary Cepheids: separations and mass ratios in 5 M ⊙ binaries. Astron J 146:93, https://doi.org/10.1088/0004-6256/146/4/93, https://ui.adsabs.harvard.edu/abs/2013AJ....146...93E

Event Horizon Telescope Collaboration, Akiyama K, Alberdi A et al (2019) First M87 event horizon telescope results. I. The shadow of the supermassive black hole. Astrophys J 875:L1, https://doi.org/10.3847/2041-8213/ab0ec7, https://ui.adsabs.harvard.edu/abs/2019ApJ...875L...1E

Event Horizon Telescope Collaboration, Akiyama K, Alberdi A et al (2022) First Sagittarius A* event horizon telescope results. I. The shadow of the supermassive black hole in the center of the Milky Way. Astrophys J 930:L12, https://doi.org/10.3847/2041-8213/ac6674, https://ui.adsabs.harvard.edu/abs/2022ApJ...930L..12E

Farag E, Renzo M, Farmer R, Chidester MT, Timmes FX (2022) Resolving the peak of the black hole mass spectrum. Astrophys J 937:112, https://doi.org/10.3847/1538-4357/ac8b83, https://ui.adsabs.harvard.edu/abs/2022ApJ...937..112F

Ferguson JW, Alexander DR, Allard F, Barman T, Bodnarik JG, Hauschildt PH, Heffner-Wong A, Tamanai A (2005) Low-temperature opacities. Astrophys J 623:585–596, https://doi.org/10.1086/428642, https://ui.adsabs.harvard.edu/abs/2005ApJ...623..585F, astro-ph/0502045

Ferrario L, Pringle JE, Tout CA, Wickramasinghe DT (2009) The origin of magnetism on the upper main sequence. Mon Not R Astron Soc 400:L71–L74, https://doi.org/10.1111/j.1745-3933.2009.00765.x, https://ui.adsabs.harvard.edu/abs/2009MNRAS.400L..71F

Fick A (1855) Ueber diffusion. Ann Phys 170(1):59–86, https://doi.org/10.1002/andp.18551700105, https://onlinelibrary.wiley.com/doi/abs/10.1002/andp.18551700105

Fitzpatrick EL (1999) Correcting for the effects of interstellar extinction. Publ Astron Soc Pac 111:63–75, https://doi.org/10.1086/316293, http://adsabs.harvard.edu/abs/1999PASP..111...63F

Flannery BP (1976) A cyclic thermal instability in contact binary stars. Astrophys J 205:217–225, https://doi.org/10.1086/154266, https://ui.adsabs.harvard.edu/abs/1976ApJ...205..217F

Fleming WP (1917) Spectra and photographic magnitudes of stars in standard regions. Ann Harvard College Observ 71:27–45, https://ui.adsabs.harvard.edu/abs/1917AnHar..71...27F

Förster KU (2023) Evolution of massive contact binaries. https://astro.uni-bonn.de/~nlanger/thesis/kai_bachelor.pdf

Fowler RH (1926) On dense matter. Mon Not R Astron Soc 87:114–122, https://doi.org/10.1093/mnras/87.2.114, https://ui.adsabs.harvard.edu/abs/1926MNRAS..87..114F

Fraunhofer J (1817) Bestimmung des Brechungs- und des Farbenzerstreungs-Vermögens verschiedener Glasarten, in Bezug auf die Vervollkommnung achromatischer Fernröhre. Ann Phys 56(7):264–313, https://doi.org/10.1002/andp.18170560706, https://ui.adsabs.harvard.edu/abs/1817AnP....56..264F/abstract

Frost AJ, Sana H, Mahy L, Wade G, Barron J, Le Bouquin JB, Mérand A, Schneider FRN, Shenar T, Barbá RH, Bowman DM, Fabry M, Farhang A, Marchant P, Morrell NI, Smoker JV (2024)

A magnetic massive star has experienced a stellar merger. Science 384(6692):214–217, https://doi.org/10.1126/science.adg7700, https://www.science.org/doi/10.1126/science.adg7700

Fukuda S, Fukuda Y, Ishitsuka M et al (2001) Solar 8B and hep neutrino measurements from 1258 days of super-Kamiokande data. Phys Rev Lett 86:5651–5655, https://doi.org/10.1103/PhysRevLett.86.5651, https://ui.adsabs.harvard.edu/abs/2001PhRvL..86.5651F

Fukuda S, Fukuda Y, Hayakawa T et al (2003) The super-Kamiokande detector. Nucl Instrum Methods Phys Res Sect A Accelerators Spectrom Detectors Assoc Equip 501(2):418–462, https://doi.org/10.1016/S0168-9002(03)00425-X, https://www.sciencedirect.com/science/article/pii/S016890020300425X

Gaia Collaboration, Prusti T, de Bruijne JHJ et al (2016) The Gaia mission. Astron Astrophys 595:A1, https://doi.org/10.1051/0004-6361/201629272, https://ui.adsabs.harvard.edu/abs/2016A&A...595A...1G/abstract

Gaia Collaboration, Vallenari A, Brown AGA et al (2023) Gaia data release 3. Summary of the content and survey properties. Astron Astrophys 674:A1, https://doi.org/10.1051/0004-6361/202243940, https://ui.adsabs.harvard.edu/abs/2023A&A...674A...1G

Gallegos-Garcia M, Berry CPL, Marchant P, Kalogera V (2021) Binary black hole formation with detailed modeling: stable mass transfer leads to lower merger rates. Astrophys J 922:110, https://doi.org/10.3847/1538-4357/ac2610, https://ui.adsabs.harvard.edu/abs/2021ApJ...922..110G

Gando A, Gando Y, Ichimura K et al (2011) Partial radiogenic heat model for Earth revealed by geoneutrino measurements. Nat Geosci 4(9):647–651, https://doi.org/10.1038/ngeo1205, https://www.nature.com/articles/ngeo1205

Ge H, Hjellming MS, Webbink RF, Chen X, Han Z (2010) Adiabatic mass loss in binary stars. I. Computational Method. Astrophys J 717:724–738, https://doi.org/10.1088/0004-637X/717/2/724, http://adsabs.harvard.edu/abs/2010ApJ...717..724G

Ge H, Webbink RF, Chen X, Han Z (2015) Adiabatic mass loss in binary stars. II. From zero-age main sequence to the base of the giant branch. Astrophys J 812:40, https://doi.org/10.1088/0004-637X/812/1/40, http://adsabs.harvard.edu/abs/2015ApJ...812...40G

Ge H, Webbink RF, Chen X, Han Z (2020) Adiabatic mass loss in binary stars. III. From the base of the red giant branch to the tip of the asymptotic giant branch. Astrophys J 899:132, https://doi.org/10.3847/1538-4357/aba7b7, http://adsabs.harvard.edu/abs/2020ApJ...899..132G

Giacobbo N, Mapelli M (2018) The progenitors of compact-object binaries: Impact of metallicity, common envelope and natal kicks. Mon Not R Astron Soc 480:2011–2030, https://doi.org/10.1093/mnras/sty1999, https://ui.adsabs.harvard.edu/abs/2018MNRAS.480.2011G

Gingerich O (1978) Interview of Rudolf Kippenhahn. https://www.aip.org/history-programs/niels-bohr-library/oral-histories/5091

Gingerich O (1985) Did Copernicus owe a debt to Aristarchus. J History Astron 16:37, https://doi.org/10.1177/002182868501600102, https://ui.adsabs.harvard.edu/abs/1985JHA.....16...37G

González JF, Levato H (2006) Separation of composite spectra: The spectroscopic detection of an eclipsing binary star. Astron Astrophys 448:283–292, https://doi.org/10.1051/0004-6361:20053177, https://ui.adsabs.harvard.edu/abs/2006A&A...448..283G

Goodricke J (1783) A series of observations on, and a discovery of, the period of the variation of the light of the bright star in the head of medusa, called algol. In a letter from John Goodricke, Esq. to the Rev. Anthony Shepherd, D. D. F. R. S. and Plumian Professor at Cambridge. Philos Trans R Soc Lond Ser I 73:474–482, https://ui.adsabs.harvard.edu/abs/1783RSPT...73..474G

Górski KM, Hivon E, Banday AJ, Wandelt BD, Hansen FK, Reinecke M, Bartelmann M (2005) HEALPix: a framework for high-resolution discretization and fast analysis of data distributed on the sphere. Astrophys J 622:759–771, https://doi.org/10.1086/427976, https://ui.adsabs.harvard.edu/abs/2005ApJ...622..759G

Graczyk D, Soszyński I, Poleski R, Pietrzyński G, Udalski A, Szymański MK, Kubiak M, Wyrzykowski Ł, Ulaczyk K (2011) The optical gravitational lensing experiment. The OGLE-III catalog of variable stars. XII. Eclipsing binary stars in the large magellanic cloud. Acta Astron 61:103–122, https://ui.adsabs.harvard.edu/abs/2011AcA....61..103G

Graczyk D, Pietrzyński G, Thompson IB, Gieren W, Pilecki B, Konorski P, Villanova S, Górski M, Suchomska K, Karczmarek P, Stepień K, Storm J, Taormina M, Kołaczkowski Z, Wielgórski P, Narloch W, Zgirski B, Gallenne A, Ostrowski J, Smolec R, Udalski A, Soszyński I, Kervella P, Nardetto N, Szymański MK, Wyrzykowski Ł, Ulaczyk K, Poleski R, Pietrukowicz P, Kozłowski S, Skowron J, Mróz P (2018) The late-type eclipsing binaries in the large magellanic cloud: catalog of fundamental physical parameters. Astrophys J 860:1, https://doi.org/10.3847/1538-4357/aac2bf, https://ui.adsabs.harvard.edu/abs/2018ApJ...860....1G

Gräfener G, Vink JS, de Koter A, Langer N (2011) The Eddington factor as the key to understand the winds of the most massive stars. Evidence for a Γ-dependence of Wolf-Rayet type mass loss. Astron Astrophys 535:A56, https://doi.org/10.1051/0004-6361/201116701, https://ui.adsabs.harvard.edu/abs/2011A%26A...535A..56G/abstract

GRAVITY Collaboration, Abuter R, Accardo M et al (2017) First light for GRAVITY: Phase referencing optical interferometry for the very large telescope interferometer. Astron Astrophys 602:A94, https://doi.org/10.1051/0004-6361/201730838

Greenspan T (2014) Stability of the Lagrange points, L4 and L5. https://web.archive.org/web/20180418011225/http://www.math.cornell.edu/~templier/junior/final_paper/Thomas_Greenspan-Stability_of_Lagrange_points.pdf

Grunhut JH, Wade GA, Neiner C, Oksala ME, Petit V, Alecian E, Bohlender DA, Bouret JC, Henrichs HF, Hussain GAJ, Kochukhov O, MiMeS Collaboration (2017) The MiMeS survey of magnetism in massive stars: magnetic analysis of the O-type stars. Mon Not R Astron Soc 465:2432–2470, https://doi.org/10.1093/mnras/stw2743, https://ui.adsabs.harvard.edu/abs/2017MNRAS.465.2432G

Guo Z (2021) Asteroseismology of close binary stars: tides and mass transfer. Front Astron Space Sci 8:67, https://doi.org/10.3389/fspas.2021.663026, https://ui.adsabs.harvard.edu/abs/2021FrASS...8...67G

Hadrava P (1995) Orbital elements of multiple spectroscopic stars. Astron Astrophys Suppl Ser 114:393

Hamann WR, Koesterke L, Wessolowski U (1995) Spectral analyses of the Galactic Wolf-Rayet stars: Hydrogen-helium abundances and improved stellar parameters for the WN class. Astron Astrophys 299:151, https://ui.adsabs.harvard.edu/abs/1995A%26A...299..151H/abstract

Hastings B, Langer N, Koenigsberger G (2020) Internal circulation in tidally locked massive binary stars: Consequences for double black hole formation. Astron Astrophys 641:A86, https://doi.org/10.1051/0004-6361/202038499, https://www.aanda.org/10.1051/0004-6361/202038499

Hazlehurst J (1985) The dissipation factor in contact binaries. Astron Astrophys 145:25–40, https://ui.adsabs.harvard.edu/abs/1985A&A...145...25H

Hazlehurst J (1993) The equilibrium of a contact binary. Astron Astrophys 271:209, https://ui.adsabs.harvard.edu/abs/1993A%26A...271..209H/abstract

Hazlehurst J, Refsdal S (1980) On the stability of age-zero contact binaries. I. Astron Astrophys 84:200–211, https://ui.adsabs.harvard.edu/abs/1980A&A....84..200H

Heger A, Langer N, Woosley SE (2000) Presupernova evolution of rotating massive stars. I. Numerical method and evolution of the internal stellar structure. Astrophys J 528:368–396, https://doi.org/10.1086/308158, https://ui.adsabs.harvard.edu/abs/2000ApJ...528..368H

Heger A, Fryer CL, Woosley SE, Langer N, Hartmann DH (2003) How massive single stars end their life. Astrophys J 591:288–300, https://doi.org/10.1086/375341, https://ui.adsabs.harvard.edu/abs/2003ApJ...591..288H

Heger A, Müller B, Mandel I (2023) Black holes as the end state of stellar evolution: Theory and simulations. In: Haiman Z (ed) The encyclopedia of cosmology (set 2): black holes. World Scientific, New Jersey, p 53, https://ui.adsabs.harvard.edu/abs/2023arXiv230409350H

Henderson (1839) II. On the parallax of α centauri. Mon Not R Astron Soc 4(19):168–169, https://doi.org/10.1093/mnras/4.19.168, https://doi.org/10.1093/mnras/4.19.168

Henneco J, Schneider FRN, Hekker S, Aerts C (2024a) Merger seismology: Distinguishing massive merger products from genuine single stars using asteroseismology. Astron Astrophys 690:A65, https://doi.org/10.1051/0004-6361/202450508, https://ui.adsabs.harvard.edu/abs/2024A&A...690A..65H

Henneco J, Schneider FRN, Laplace E (2024b) Contact tracing of binary stars: Pathways to stellar mergers. Astron Astrophys 682:A169, https://doi.org/10.1051/0004-6361/202347893, https://ui.adsabs.harvard.edu/abs/2024A&A...682A.169H

Henyey LG, Wilets L, Böhm KH, Lelevier R, Levee RD (1959) A method for automatic computation of stellar evolution. Astrophys J 129:628, https://doi.org/10.1086/146661, https://ui.adsabs.harvard.edu/abs/1959ApJ...129..628H

Henyey LG, Forbes JE, Gould NL (1964) A new method of automatic computation of stellar evolution. Astrophys J 139:306, https://doi.org/10.1086/147754, https://ui.adsabs.harvard.edu/abs/1964ApJ...139..306H

Herrero A, Kudritzki RP, Vilchez JM, Kunze D, Butler K, Haser S (1992) Intrinsic parameters of galactic luminous OB stars. Astron Astrophys 261:209–234, https://ui.adsabs.harvard.edu/abs/1992A&A...261..209H

Herschel W (1802) Catalogue of 500 new Nebulae, nebulous stars, planetary Nebulae, and clusters of stars; with remarks on the construction of the heavens. Philos Trans R Soc Lond Ser I 92:477–528, https://ui.adsabs.harvard.edu/abs/1802RSPT...92..477H

Herschel W (1803) Account of the changes that have happened, during the last twenty-five years, in the relative situation of double-stars; with an investigation of the cause to which they are owing. Philos Trans R Soc Lond Ser I 93:339–382, https://ui.adsabs.harvard.edu/abs/1803RSPT...93..339H

Herschel Mr, Watson Dr (1782) Catalogue of double stars. By Mr. Herschel, F. R. S. communicated by Dr. Watson. Philos Trans R Soc Lond Ser I 72:112–162, https://ui.adsabs.harvard.edu/abs/1782RSPT...72..112H

Hertzsprung E (1911) über die Verwendung photographischer effektiver Wellenlängen zur Bestimmung von Farbenäquivalenten. Publ Astrophys Observ Potsdam 63, https://ui.adsabs.harvard.edu/abs/1911POPot..63.....H

Hewish A, Bell SJ, Pilkington JDH, Scott PF, Collins RA (1968) Observation of a rapidly pulsating radio source. Nature 217(5130):709–713, https://doi.org/10.1038/217709a0, https://www.nature.com/articles/217709a0

Hilditch RW (2001) An introduction to close binary stars. Cambridge University Press

Hilditch RW, Evans TL (1985) The massive near-contact binary system V348 Carinae (HD 90707) in IC 2581. Mon Not R Astron Soc 213(1):75–83, https://doi.org/10.1093/mnras/213.1.75

Hills JG, Day CA (1976) Stellar collisions in globular clusters. Astrophys Lett 17:87, https://ui.adsabs.harvard.edu/abs/1976ApL.....17...87H

Hirata K, Kajita T, Koshiba M, Nakahata M, Oyama Y, Sato N, Suzuki A, Takita M, Totsuka Y, Kifune T, Suda T, Takahashi K, Tanimori T, Miyano K, Yamada M, Beier EW, Feldscher LR, Kim SB, Mann AK, Newcomer FM, Van R, Zhang W, Cortez BG (1987) Observation of a neutrino burst from the supernova SN1987A. Phys Rev Lett 58(14):1490–1493, https://doi.org/10.1103/PhysRevLett.58.1490, https://link.aps.org/doi/10.1103/PhysRevLett.58.1490

Holgado G, Simón-Díaz S, Haemmerlé L, Lennon DJ, Barbá RH, Cerviño M, Castro N, Herrero A, Meynet G, Arias JI (2020) The IACOB project. VI. On the elusive detection of massive O-type stars close to the ZAMS. Astron Astrophys 638:A157, https://doi.org/10.1051/0004-6361/202037699, https://ui.adsabs.harvard.edu/abs/2020A&A...638A.157H

Horvat M, Conroy KE, Pablo H, Hambleton KM, Kochoska A, Giammarco J, Prša A (2018) Physics of eclipsing binaries. III. Spin-orbit misalignment. Astrophys J Suppl Ser 237:26, https://doi.org/10.3847/1538-4365/aacd0f, https://ui.adsabs.harvard.edu/abs/2018ApJS..237...26H

Hoyle F (1954) On nuclear reactions occuring in very hot STARS.I. The synthesis of elements from carbon to nickel. Astrophys J Suppl Ser 1:121, https://doi.org/10.1086/190005, https://ui.adsabs.harvard.edu/abs/1954ApJS....1..121H

Huang WJ, Wang M, Kondev FG, Audi G, Naimi S (2021) The AME 2020 atomic mass evaluation (I). Evaluation of input data, and adjustment procedures. Chin Phys C 45:030002, https://doi.org/10.1088/1674-1137/abddb0, https://ui.adsabs.harvard.edu/abs/2021ChPhC..45c0002H

Hulse RA, Taylor JH (1975) Discovery of a pulsar in a binary system. Astrophys J 195:L51–L53, https://doi.org/10.1086/181708, https://ui.adsabs.harvard.edu/abs/1975ApJ...195L..51H

Hunger H, Pingree D (1999) Astral sciences in mesopotamia. Brill, https://brill.com/display/title/824

Hurley JR, Tout CA, Pols OR (2002) Evolution of binary stars and the effect of tides on binary populations. Mon Not R Astron Soc 329:897–928, https://doi.org/10.1046/j.1365-8711.2002.05038.x, https://ui.adsabs.harvard.edu/abs/2002MNRAS.329..897H

Hut P (1980) Stability of tidal equilibrium. Astron Astrophys 92:167–170, https://ui.adsabs.harvard.edu/abs/1980A&A....92..167H

IceCube Collaboration, Aartsen MG, Ackermann M et al (2018) Multimessenger observations of a flaring blazar coincident with high-energy neutrino IceCube-170922A. Science 361:eaat1378, https://doi.org/10.1126/science.aat1378, https://ui.adsabs.harvard.edu/abs/2018Sci...361.1378I

Iglesias CA, Rogers FJ (1996) Updated opal opacities. Astrophys J 464:943, https://doi.org/10.1086/177381, https://ui.adsabs.harvard.edu/abs/1996ApJ...464..943I

Ilijić S (2017) Fd3: Spectral disentangling of double-lined spectroscopic binary stars. Astrophys Source Code Library p ascl:1705.012, https://ui.adsabs.harvard.edu/abs/2017ascl.soft05012I

Ilijić S, Hensberge H, Pavlovski K, Freyhammer LM (2004) Obtaining normalised component spectra with FDBinary. In: Astronomical Society of the Pacific conference series, San Francisco, California, USA, vol 318, pp 111–113, http://adsabs.harvard.edu/abs/2004ASPC..318..111I

Ivanova N, Justham S, Chen X, De Marco O, Fryer CL, Gaburov E, Ge H, Glebbeek E, Han Z, Li XD, Lu G, Marsh T, Podsiadlowski P, Potter A, Soker N, Taam R, Tauris TM, van den Heuvel EPJ, Webbink RF (2013) Common envelope evolution: Where we stand and how we can move forward. Astron Astrophys Rev 21:59, https://doi.org/10.1007/s00159-013-0059-2

Ivezić Ž, Kahn SM, Tyson JA et al (2019) LSST: From science drivers to reference design and anticipated data products. Astrophys J 873:111, https://doi.org/10.3847/1538-4357/ab042c, https://ui.adsabs.harvard.edu/abs/2019ApJ...873..111I

Janssens S, Shenar T, Degenaar N, Bodensteiner J, Sana H, Audenaert J, Frost AJ (2023) MWC 656 is unlikely to contain a black hole. Astron Astrophys 677:L9, https://doi.org/10.1051/0004-6361/202347318

Jermyn AS, Schwab J, Bauer E, Timmes FX, Potekhin AY (2021) Skye: A differentiable equation of state. Astrophys J 913(1):72, https://doi.org/10.3847/1538-4357/abf48e, 2104.00691

Jermyn AS, Bauer EB, Schwab J, Farmer R, Ball WH, Bellinger EP, Dotter A, Joyce M, Marchant P, Mombarg JSG, Wolf WM, Sunny Wong TL, Cinquegrana GC, Farrell E, Smolec R, Thoul A, Cantiello M, Herwig F, Toloza O, Bildsten L, Townsend RHD, Timmes FX (2023) Modules for experiments in stellar astrophysics (MESA): Time-dependent convection, energy conservation, automatic differentiation, and infrastructure. Astrophys J Suppl Ser 265:arXiv:2208.03651, https://doi.org/10.3847/1538-4365/acae8d, https://ui.adsabs.harvard.edu/abs/2023ApJS..265...15J, 2208.03651

Johnston C (2021) One size does not fit all: Evidence for a range of mixing efficiencies in stellar evolution calculations. Astron Astrophys 655:A29, https://doi.org/10.1051/0004-6361/202141080, https://ui.adsabs.harvard.edu/abs/2021A&A...655A..29J

Johnstone CP, Bartel M, Güdel M (2021) The active lives of stars: A complete description of the rotation and XUV evolution of F, G, K, and M dwarfs. Astron Astrophys 649:A96, https://doi.org/10.1051/0004-6361/202038407, https://ui.adsabs.harvard.edu/abs/2021A&A...649A..96J

Jones D, Conroy KE, Horvat M, Giammarco J, Kochoska A, Pablo H, Brown AJ, Sowicka P, Prša A (2020) Physics of eclipsing binaries. IV. The impact of interstellar extinction on the light curves of eclipsing binaries. Astrophys J Suppl Ser 247:63, https://doi.org/10.3847/1538-4365/ab7927, https://ui.adsabs.harvard.edu/abs/2020ApJS..247...63J

Joss PC, Salpeter EE, Ostriker JP (1973) On the "critical luminosity" in stellar interiors and stellar surface boundary conditions. Astrophys J 181:429–438, https://doi.org/10.1086/152060, https://ui.adsabs.harvard.edu/abs/1973ApJ...181..429J

Kähler H (1989) The structure equations of contact binaries and the light curve paradox. Astron Astrophys 209:67–84, https://ui.adsabs.harvard.edu/abs/1989A&A...209...67K

Kähler H (2004) The structure of contact binaries. Astron Astrophys 414(1):317–333, https://doi.org/10.1051/0004-6361:20031629, http://www.aanda.org/10.1051/0004-6361:20031629

Kasliwal MM, Bally J, Masci F, Cody AM, Bond HE, Jencson JE, Tinyanont S, Cao Y, Contreras C, Dykhoff DA, Amodeo S, Armus L, Boyer M, Cantiello M, Carlon RL, Cass AC, Cook D, Corgan DT, Faella J, Fox OD, Green W, Gehrz RD, Helou G, Hsiao E, Johansson J, Khan RM, Lau RM, Langer N, Levesque E, Milne P, Mohamed S, Morrell N, Monson A, Moore A, Ofek EO, O' Sullivan D, Parthasarathy M, Perez A, Perley DA, Phillips M, Prince TA, Shenoy D, Smith N, Surace J, Van Dyk SD, Whitelock PA, Williams R (2017) SPIRITS: uncovering unusual infrared transients with Spitzer. Astrophys J 839:88, https://doi.org/10.3847/1538-4357/aa6978, https://ui.adsabs.harvard.edu/abs/2017ApJ...839...88K

Kaufer A, Stahl O, Tubbesing S, Nørregaard P, Avila G, Francois P, Pasquini L, Pizzella A (1999) Commissioning FEROS, the new high-resolution spectrograph at La-Silla. Messenger 95:8–12, http://adsabs.harvard.edu/abs/1999Msngr..95....8K

Keppens R, MacGregor KB, Charbonneau P (1995) On the evolution of rotational velocity distributions for solar-type stars. Astron Astrophys 294:469–487, https://ui.adsabs.harvard.edu/abs/1995A&A...294..469K

Kharchenko NV, Piskunov AE, Röser S, Schilbach E, Scholz RD (2005) Astrophysical parameters of Galactic open clusters. Astron Astrophys 438:1163–1173, https://doi.org/10.1051/0004-6361:20042523

Kinugawa T, Inayoshi K, Hotokezaka K, Nakauchi D, and Nakamura T (2014) Possible indirect confirmation of the existence of Pop III massive stars by gravitational wave. Mon Not R Astron Soc 442:2963–2992, https://doi.org/10.1093/mnras/stu1022, https://ui.adsabs.harvard.edu/abs/2014MNRAS.442.2963K

Kippenhahn R, Thomas HC (1970) A simple method for the solution of the stellar structure equations including rotation and tidal forces. In: Stellar rotation. Gordon and Breach Science Publishers, Columbus, https://ui.adsabs.harvard.edu/abs/1970stro.coll...20K

Kippenhahn R, Kohl K, Weigert A (1967) Entwicklung in engen Doppelsternsystemen II. Z Astrophys 66:58, https://ui.adsabs.harvard.edu/abs/1967ZA.....66...58K

Kippenhahn R, Ruschenplatt G, Thomas HC (1980) The time scale of thermohaline mixing in stars. Astron Astrophys 91:175–180, https://ui.adsabs.harvard.edu/abs/1980A&A....91..175K

Kippenhahn R, Weigert A, Weiss A (2012) Stellar structure and evolution. Springer Berlin Heidelberg, https://doi.org/10.1007/978-3-642-30304-3, https://ui.adsabs.harvard.edu/abs/2013sse..book.....K

Kirk B, Conroy K, Prša A, Abdul-Masih M, Kochoska A, Matijevič G, Hambleton K, Barclay T, Bloemen S, Boyajian T, Doyle LR, Fulton BJ, Hoekstra AJ, Jek K, Kane SR, Kostov V, Latham D, Mazeh T, Orosz JA, Pepper J, Quarles B, Ragozzine D, Shporer A, Southworth J, Stassun K, Thompson SE, Welsh WF, Agol E, Derekas A, Devor J, Fischer D, Green G, Gropp J, Jacobs T, Johnston C, LaCourse DM, Saetre K, Schwengeler H, Toczyski J, Werner G, Garrett M, Gore J, Martinez AO, Spitzer I, Stevick J, Thomadis PC, Vrijmoet EH, Yenawine M, Batalha N, Borucki W (2016) Kepler eclipsing binary stars. VII. The catalog of eclipsing binaries found in the entire Kepler data set. Astron J 151:68, https://doi.org/10.3847/0004-6256/151/3/68, https://ui.adsabs.harvard.edu/abs/2016AJ....151...68K

Klencki J, Nelemans G, Istrate AG, Chruslinska M (2021) It has to be cool: Supergiant progenitors of binary black hole mergers from common-envelope evolution. Astron Astrophys 645:A54, https://doi.org/10.1051/0004-6361/202038707

Knigge C, Baraffe I, Patterson J (2011) The evolution of cataclysmic variables as revealed by their donor stars. Astrophys J Suppl Ser 194:28, https://doi.org/10.1088/0067-0049/194/2/28, https://ui.adsabs.harvard.edu/abs/2011ApJS..194...28K

Kobulnicky HA, Kiminki DC, Lundquist MJ, Burke J, Chapman J, Keller E, Lester K, Rolen EK, Topel E, Bhattacharjee A, Smullen RA, Álvarez CAV, Runnoe JC, Dale DA, Brotherton MM (2014) Toward complete statistics of massive binary stars: penultimate results from the Cygnus OB2 radial velocity survey. Astrophys J Suppl Ser 213(2):34, https://doi.org/10.1088/0067-0049/213/2/34, https://dx.doi.org/10.1088/0067-0049/213/2/34

Kobulnicky HA, Molnar LA, Cook EM, Henderson LE (2022) A Bayesian analysis of physical parameters for 783 Kepler close binaries: extreme-mass-ratio systems and a new mass ratio versus period lower limit. Astrophys J Suppl Ser 262:12, https://doi.org/10.3847/1538-4365/ac75bd, https://ui.adsabs.harvard.edu/abs/2022ApJS..262...12K

Köhler K, Langer N, de Koter A, de Mink SE, Crowther PA, Evans CJ, Gräfener G, Sana H, Sanyal D, Schneider FRN, Vink JS (2015) The evolution of rotating very massive stars with LMC composition. Astron Astrophys 573:A71, https://doi.org/10.1051/0004-6361/201424356, https://ui.adsabs.harvard.edu/abs/2015A&A...573A..71K

Kolb U, Ritter H (1990) A comparative study of the evolution of a close binary using a standard and an improved technique for computing mass transfer. Astron Astrophys 236:385–392

Kopal Z (1959) Close binary systems. Wiley, https://ui.adsabs.harvard.edu/abs/1959cbs..book.....K

Kozai Y (1962) Secular perturbations of asteroids with high inclination and eccentricity. Astron J 67:591–598, https://doi.org/10.1086/108790, http://adsabs.harvard.edu/abs/1962AJ.....67..591K

Kragh H (2010) An anthropic myth: Fred Hoyle's carbon-12 resonance level. Arch History Exact Sci 64(6):721–751, https://doi.org/10.1007/s00407-010-0068-8, https://doi.org/10.1007/s00407-010-0068-8

Kramers HA (1923) On the theory of X-ray absorption and of the continuous X-ray spectrum. Lond Edinb Dublin Philos Mag J Sci 46(275):836–871, https://doi.org/10.1080/14786442308565244, https://www.tandfonline.com/doi/full/10.1080/14786442308565244

Kratter KM, Matzner CD (2006) Fragmentation of massive protostellar discs. Mon Not R Astron Soc 373:1563–1576, https://doi.org/10.1111/j.1365-2966.2006.11103.x, https://ui.adsabs.harvard.edu/abs/2006MNRAS.373.1563K

Kroupa P (2001) On the variation of the initial mass function. Mon Not R Astron Soc 322:231–246, https://doi.org/10.1046/j.1365-8711.2001.04022.x, https://ui.adsabs.harvard.edu/abs/2001MNRAS.322..231K

Krumholz MR, Bonnell IA (2007) Models for the formation of massive stars, https://doi.org/10.48550/arXiv.0712.0828, http://arxiv.org/abs/0712.0828, 0712.0828

Kuiper GP (1941) On the interpretation of β Lyrae and other close binaries. Astrophys J 93:133, https://doi.org/10.1086/144252, https://ui.adsabs.harvard.edu/abs/1941ApJ....93..133K

Kummer F, Toonen S, de Koter A (2023) The main evolutionary pathways of massive hierarchical triple stars. Astron Astrophys 678:A60, https://doi.org/10.1051/0004-6361/202347179, https://ui.adsabs.harvard.edu/abs/2023A&A...678A..60K

Kurahashi N, Murase K, Santander M (2022) High-energy extragalactic neutrino astrophysics. Annu Rev Nucl Particle Sci 72:365–387, https://doi.org/10.1146/annurev-nucl-011122-061547, https://ui.adsabs.harvard.edu/abs/2022ARNPS..72..365K

Landau LD, Lifshitz EM (1975) The classical theory of fields. Pergamon Press, Oxford, https://ui.adsabs.harvard.edu/abs/1975ctf..book.....L

Lane HJ (1870) On the theoretical temperature of the Sun, under the hypothesis of a gaseous mass maintaining its volume by its internal heat, and depending on the laws of gases as known to terrestrial experiment. Am J Sci 50:57–74, https://doi.org/10.2475/ajs.s2-50.148.57, https://ui.adsabs.harvard.edu/abs/1870AmJS...50...57L

Langer N (1997) The Eddington limit in rotating massive stars. In: ASP conference series, vol 120, p 83, https://ui.adsabs.harvard.edu/abs/1997ASPC..120...83L

Langer N (2022) Open questions in massive star research. arXiv e-prints 2209.04165, https://doi.org/10.48550/arXiv.2209.04165, https://ui.adsabs.harvard.edu/abs/2022arXiv220904165L

Langer N, Fricke KJ, Sugimoto D (1983) Semiconvective diffusion and energy transport. Astron Astrophys 126:207, https://ui.adsabs.harvard.edu/abs/1983A&A...126..207L

Langer N, Hamann WR, Lennon M, Najarro F, Pauldrach AWA, Puls J (1994) Towards an understanding of very massive stars. A new evolutionary scenario relating O stars, LBVs and Wolf-Rayet stars. Astron Astrophys 290:819–833

Lanthermann C, Le Bouquin JB, Sana H, Mérand A, Monnier JD, Perraut K, Frost AJ, Mahy L, Gosset E, De Becker M, Kraus S, Anugu N, Davies CL, Ennis J, Gardner T, Labdon A, Setterholm B, ten Brummelaar T, Schaefer GH (2023) Multiplicity of northern bright O-type stars with optical long baseline interferometry. Results of the pilot survey. Astron Astrophys 672:A6, https://doi.org/10.1051/0004-6361/202245364

Lapeyrere V, Kervella P, Lacour S, Azouaoui N, Garcia-Dabo CE, Perrin G, Eisenhauer F, Perraut K, Straubmeier C, Amorim A, Brandner W (2014) GRAVITY data reduction software. In: Proc. SPIE, Society of Photo-Optical Instrumentation Engineers (SPIE) conference series, vol 9146, p 91462D, https://doi.org/10.1117/12.2056850, http://adsabs.harvard.edu/abs/2014SPIE.9146E..2DL

Larson J (2007) Ancient Greek cults: a guide. Routledge, New York

Laughlin G, Bodenheimer P, Adams FC (1997) The end of the main sequence. Astrophys J 482:420–432, https://doi.org/10.1086/304125, https://ui.adsabs.harvard.edu/abs/1997ApJ...482..420L

Lauterborn D (1970) Evolution with mass exchange of case C for a binary system of total mass 7 M sun. Astron Astrophys 7:150, https://ui.adsabs.harvard.edu/abs/1970A&A.....7..150L

Le Bouquin JB, Berger JP, Lazareff B, Zins G, Haguenauer P, Jocou L, Kern P, Millan-Gabet R, Traub W, Absil O, Augereau JC, Benisty M, Blind N, Bonfils X, Bourget P, Delboulbe A, Feautrier P, Germain M, Gitton P, Gillier D, Kiekebusch M, Kluska J, Knudstrup J, Labeye P, Lizon JL, Monin JL, Magnard Y, Malbet F, Maurel D, Ménard F, Micallef M, Michaud L, Montagnier G, Morel S, Moulin T, Perraut K, Popovic D, Rabou P, Rochat S, Rojas C, Roussel F, Roux A, Stadler E, Stefl S, Tatulli E, Ventura N (2011) PIONIER: A 4-telescope visitor instrument at VLTI. Astron Astrophys 535:A67, https://doi.org/10.1051/0004-6361/201117586

Le Bouquin JB, Sana H, Gosset E, Becker MD, Duvert G, Absil O, Anthonioz F, Berger JP, Ertel S, Grellmann R, Guieu S, Kervella P, Rabus M, Willson M (2017) Resolved astrometric orbits of ten O-type binaries. Astron Astrophys 601:A34, https://doi.org/10.1051/0004-6361/201629260, https://www.aanda.org/articles/aa/abs/2017/05/aa29260-16/aa29260-16.html

Leavitt HS, Pickering EC (1912) Periods of 25 variable stars in the small magellanic cloud. Harv Coll Observ Circ 173:1–3, https://ui.adsabs.harvard.edu/abs/1912HarCi.173....1L

Ledoux P (1947) Stellar models with convection and with discontinuity of the mean molecular weight. Astrophys J 105:305, https://doi.org/10.1086/144905, https://ui.adsabs.harvard.edu/abs/1947ApJ...105..305L

Lehmann-Filhés R (1894) über die Bestimmung einer Doppelsternbahn aus spectroskopichen Messungen der im Visionsradius liegenden Geschwindigkeitscomponente. Astron Nachr 136:17, https://doi.org/10.1002/asna.18941360202, https://ui.adsabs.harvard.edu/abs/1894AN....136...17L

Leonard PJT (1989) Stellar collisions in globular clusters and the blue straggler problem. Astron J 98:217, https://doi.org/10.1086/115138, https://ui.adsabs.harvard.edu/abs/1989AJ.....98..217L

Leung KC, Sistero RF, Zhai DS, Grieco A, Candellero B (1984) Revised UBV photometric solution of the early-type contact system BH Centauri. Astron J 89:872–875, https://doi.org/10.1086/113582, https://ui.adsabs.harvard.edu/abs/1984AJ.....89..872L

Li L, Han Z, Zhang F (2005) Structure and evolution of low-mass W Ursae Majoris type systems - III. The effects of the spins of the stars. Mon Not R Astron Soc 360:272–281, https://doi.org/10.1111/j.1365-2966.2005.09024.x, https://ui.adsabs.harvard.edu/abs/2005MNRAS.360..272L

Li FX, Qian SB, Jiao CL, Ma WW (2022) Two massive close binaries with twin components in the nearby galaxy M31. Astrophys J 932:14, https://doi.org/10.3847/1538-4357/ac6c81, https://ui.adsabs.harvard.edu/abs/2022ApJ...932...14L

Lidov ML (1962) The evolution of orbits of artificial satellites of planets under the action of gravitational perturbations of external bodies. Planet Space Sci 9:719–759, https://doi.org/10.1016/0032-0633(62)90129-0, http://adsabs.harvard.edu/abs/1962P%26SS....9..719L

LIGO Scientific Collaboration, Virgo Collaboration, KAGRA Collaboration et al (2023) GWTC-3: Compact binary coalescences observed by LIGO and Virgo during the second part of the third observing run. Phys Rev X 13(4):041039, https://doi.org/10.1103/PhysRevX.13.041039, https://link.aps.org/doi/10.1103/PhysRevX.13.041039

LIGO Scientific Collaboration, Virgo Collaboration, KAGRA Collaboration (2025) GWTC-4.0: updating the gravitational-wave transient catalog with observations from the first part of the fourth LIGO-Virgo-KAGRA observing run, https://doi.org/10.48550/arXiv.2508.18082, https://ui.adsabs.harvard.edu/abs/2025arXiv250818082T

Linnell AP, Scheick X (1991) Does SV centauri harbor an accretion disk? Astrophys J 379:721, https://doi.org/10.1086/170547, https://ui.adsabs.harvard.edu/abs/1991ApJ...379..721L

Longair MS (2013) The cosmic century. Cambridge University Press, https://ui.adsabs.harvard.edu/abs/2013coce.book.....L

Lorenz R, Mayer P, Drechsel H (1999) V606 centauri: an early-type eclipsing contact binary. Astron Astrophys 345:531–546

Lorenzo J, Negueruela I, Baker AKFV, García M, Simón-Díaz S, Pastor P, Méndez Majuelos M (2014) MY Camelopardalis, a very massive merger progenitor. Astron Astrophys 572:A110, https://doi.org/10.1051/0004-6361/201424345, https://ui.adsabs.harvard.edu/abs/2014A&A...572A.110L/abstract

Lorenzo J, Simón-Díaz S, Negueruela I, Vilardell F, Garcia M, Evans CJ, Montes D (2017) The massive multiple system HD 64315. Astron Astrophys 606:A54, https://doi.org/10.1051/0004-6361/201731352, https://ui.adsabs.harvard.edu/abs/2017A&A...606A..54L/abstract

Lubow SH, Shu FH (1977) On the structure of contact binaries. II. Zero-age models. Astrophys J 216:517–525, https://doi.org/10.1086/155493, https://ui.adsabs.harvard.edu/abs/1977ApJ...216..517L

Lubow SH, Shu FH (1979) On the structure of contact binaries. IV. High-mass models. Astrophys J 229:657–660, https://doi.org/10.1086/156999, https://ui.adsabs.harvard.edu/abs/1979ApJ...229..657L

Lucy LB (1967a) Gravity-darkening for stars with convective envelopes. Z Astrophys 65:89, https://ui.adsabs.harvard.edu/abs/1967ZA.....65...89L

Lucy LB (1967b) Structure of W Ursae Majoris stars. Astron J 72:309–310, https://ui.adsabs.harvard.edu/abs/1967AJ.....72S.309L

Lucy LB (1968) The structure of contact binaries. Astrophys J 151:1123, https://doi.org/10.1086/149510, https://ui.adsabs.harvard.edu/abs/1968ApJ...151.1123L

Lucy LB (1976) W Ursae Majoris systems with marginal contact. Astrophys J 205:208–216, https://doi.org/10.1086/154265, https://ui.adsabs.harvard.edu/abs/1976ApJ...205..208L

Lucy LB, Solomon PM (1970) Mass loss by hot stars. Astrophys J 159:879, https://doi.org/10.1086/150365, https://ui.adsabs.harvard.edu/abs/1970ApJ...159..879L

Maeder A (1987) Evidences for a bifurcation in massive star evolution. The ON-blue stragglers. Astron Astrophys 178:159, https://ui.adsabs.harvard.edu/abs/1987A%26A...178..159M/abstract

Maeder A, Meynet G (2000) Stellar evolution with rotation. VI. The Eddington and Omega -limits, the rotational mass loss for OB and LBV stars. Astron Astrophys 361:159–166, https://doi.org/10.48550/arXiv.astro-ph/0006405, https://ui.adsabs.harvard.edu/abs/2000A&A...361..159M

Maggiore M, Van Den Broeck C, Bartolo N, Belgacem E, Bertacca D, Bizouard MA, Branchesi M, Clesse S, Foffa S, García-Bellido J, Grimm S, Harms J, Hinderer T, Matarrese S, Palomba C, Peloso M, Ricciardone A, Sakellariadou M (2020) Science case for the Einstein telescope. J Cosmol Astropart Phys 2020:050, https://doi.org/10.1088/1475-7516/2020/03/050, https://ui.adsabs.harvard.edu/abs/2020JCAP...03..050M

Mahy L, Gosset E, Sana H, Damerdji Y, De Becker M, Rauw G, Nitschelm C (2012) Evidence for a physically bound third component in HD 150136. Astron Astrophys 540:A97, https://doi.org/10.1051/0004-6361/201118199, https://ui.adsabs.harvard.edu/abs/2012A&A...540A..97M

Mahy L, Gosset E, Manfroid J, Nitschelm C, Hervé A, Semaan T, Sana H, Le Bouquin JB, Toonen S (2018) The triple system HD 150136: from periastron passage to actual masses. Astron Astrophys 616:A75, https://doi.org/10.1051/0004-6361/201832810

Mahy L, Almeida LA, Sana H, Clark JS, de Koter A, de Mink SE, Evans CJ, Grin NJ, Langer N, Moffat AFJ, Schneider FRN, Shenar T, Tramper F (2020a) The Tarantula massive binary monitoring. IV. Double-lined photometric binaries. Astron Astrophys 634:A119, https://doi.org/10.1051/0004-6361/201936152

Mahy L, Sana H, Abdul-Masih M, Almeida LA, Langer N, Shenar T, de Koter A, de Mink SE, de Wit S, Grin NJ, Evans CJ, Moffat AFJ, Schneider FRN, Barbá R, Clark JS, Crowther

P, Gräfener G, Lennon DJ, Tramper F, Vink JS (2020b) The Tarantula massive binary monitoring. III. Atmosphere analysis of double-lined spectroscopic systems. Astron Astrophys 634:A118, https://doi.org/10.1051/0004-6361/201936151, https://ui.adsabs.harvard.edu/abs/2020A&A...634A.118M

Mahy L, Sana H, Shenar T, Sen K, Langer N, Marchant P, Abdul-Masih M, Banyard G, Bodensteiner J, Bowman DM, Dsilva K, Fabry M, Hawcroft C, Janssens S, Van Reeth T, Eldridge C (2022) Identifying quiescent compact objects in massive Galactic single-lined spectroscopic binaries. Astron Astrophys 664:A159, https://doi.org/10.1051/0004-6361/202243147, https://ui.adsabs.harvard.edu/abs/2022A&A...664A.159M

Maíz Apellániz J, Sana H, Barbá RH, Le Bouquin JB, Gamen RC (2017) A close encounter of the massive kind. Mon Not R Astron Soc 464:3561–3567, https://doi.org/10.1093/mnras/stw2618, http://adsabs.harvard.edu/abs/2017MNRAS.464.3561M

Maíz Apellániz J, Barbá RH (2018) Optical-NIR dust extinction towards Galactic O stars. Astron Astrophys 613:A9, https://doi.org/10.1051/0004-6361/201732050

Mandel I, Broekgaarden FS (2022) Rates of compact object coalescences. Living Rev Relat 25:1, https://doi.org/10.1007/s41114-021-00034-3, https://ui.adsabs.harvard.edu/abs/2022LRR....25....1M

Mandel I, de Mink SE (2016) Merging binary black holes formed through chemically homogeneous evolution in short-period stellar binaries. Mon Not R Astron Soc 458:2634–2647, https://doi.org/10.1093/mnras/stw379, https://ui.adsabs.harvard.edu/abs/2016MNRAS.458.2634M

Mandel I, Farmer A (2022) Merging stellar-mass binary black holes. Phys Rep 955:1–24, https://doi.org/10.1016/j.physrep.2022.01.003, https://ui.adsabs.harvard.edu/abs/2022PhR...955....1M

Marchant P (2018) The impact of tides and mass transfer on the evolution of metal-poor massive binary stars. Thesis, Universitäts- und Landesbibliothek Bonn, https://bonndoc.ulb.uni-bonn.de/xmlui/handle/20.500.11811/7507

Marchant P, Langer N, Podsiadlowski P, Tauris TM, Moriya TJ (2016) A new route towards merging massive black holes. Astron Astrophys 588:A50, https://doi.org/10.1051/0004-6361/201628133, https://ui.adsabs.harvard.edu/abs/2016A&A...588A..50M

Marchant P, Pappas KMW, Gallegos-Garcia M, Berry CPL, Taam RE, Kalogera V, Podsiadlowski P (2021) The role of mass transfer and common envelope evolution in the formation of merging binary black holes. Astron Astrophys 650:A107, https://doi.org/10.1051/0004-6361/202039992, https://www.aanda.org/10.1051/0004-6361/202039992

Marchenko SV, Moffat AFJ, Eenens PRJ (1998) The Wolf-Rayet binary WR 141 (WN5O + O5 V-III) revisited. Publ Astron Soc Pac 110:1416–1422, https://doi.org/10.1086/316280, https://ui.adsabs.harvard.edu/abs/1998PASP..110.1416M

Mardling RA, Aarseth SJ (2001) Tidal interactions in star cluster simulations. Mon Not R Astron Soc 321:398–420, https://doi.org/10.1046/j.1365-8711.2001.03974.x, https://ui.adsabs.harvard.edu/abs/2001MNRAS.321..398M

Martins F, Hillier DJ (2012) On the formation of C III 4647-50-51 and C III 5696 in O star atmospheres. Astron Astrophys 545:A95, https://doi.org/10.1051/0004-6361/201219788

Martins F, Plez B (2006) UBVJHK synthetic photometry of Galactic O stars. Astron Astrophys 457:637–644, https://doi.org/10.1051/0004-6361:20065753

Martins F, Mahy L, Hillier DJ, Rauw G (2012) A quantitative study of O stars in NGC 2244 and the Monoceros OB2 association. Astron Astrophys 538:A39, https://doi.org/10.1051/0004-6361/201117458, https://www.aanda.org/articles/aa/abs/2012/02/aa17458-11/aa17458-11.html

Martins F, Mahy L, Hervé A (2017) Properties of six short-period massive binaries: A study of the effects of binarity on surface chemical abundances. Astron Astrophys 607:A82, https://doi.org/10.1051/0004-6361/201731593

Mason E, Diaz M, Williams RE, Preston G, Bensby T (2010) The peculiar nova V1309 Scorpii/nova Scorpii 2008. A candidate twin of V838 Monocerotis. Astron Astrophys 516:A108, https://doi.org/10.1051/0004-6361/200913610, https://ui.adsabs.harvard.edu/abs/2010A&A...516A.108M

Maury AC, Pickering EC (1897) Spectra of bright stars photographed with the 11-inch Draper Telescope as part of the Henry Draper Memorial. Ann Harv Coll Observ 28:1–128, https://ui.adsabs.harvard.edu/abs/1897AnHar..28....1M

Mayer P, Drechsel H, Harmanec P, Yang S, Šlechta M (2013) The O-type eclipsing contact binary LY Aurigae – member of a quadruple system. Astron Astrophys 559:A22, https://doi.org/10.1051/0004-6361/201322153, https://www.aanda.org/articles/aa/abs/2013/11/aa22153-13/aa22153-13.html

Mayer P, Harmanec P, Sana H, Le Bouquin JB (2014) The three-body system δ Circini. Astron J 148:114, https://doi.org/10.1088/0004-6256/148/6/114, http://adsabs.harvard.edu/abs/2014AJ....148..114M

Mayor M, Pepe F, Queloz D, Bouchy F, Rupprecht G, Lo Curto G, Avila G, Benz W, Bertaux JL, Bonfils X, Dall Th, Dekker H, Delabre B, Eckert W, Fleury M, Gilliotte A, Gojak D, Guzman JC, Kohler D, Lizon JL, Longinotti A, Lovis C, Megevand D, Pasquini L, Reyes J, Sivan JP, Sosnowska D, Soto R, Udry S, van Kesteren A, Weber L, Weilenmann U (2003) Setting new standards with HARPS. Messenger 114:20–24, http://adsabs.harvard.edu/abs/2003Msngr.114...20M

McCrea WH (1964) Extended main-sequence of some stellar clusters. Mon Not R Astron Soc 128:147, https://doi.org/10.1093/mnras/128.2.147, https://ui.adsabs.harvard.edu/abs/1964MNRAS.128..147M

McKee CF, Cowie LL (1975) The interaction between the blast wave of a supernova remnant and interstellar clouds. Astrophys J 195:715–725, https://doi.org/10.1086/153373, https://ui.adsabs.harvard.edu/abs/1975ApJ...195..715M

Mena JG, Peres TSC (2018) The Chandrasekhar-Eddington dispute. https://mathshistory.st-andrews.ac.uk/HistTopics/Chandrasekhar_Eddington/

Menon A, Langer N, de Mink SE, Justham S, Sen K, Szécsi D, de Koter A, Abdul-Masih M, Sana H, Mahy L, Marchant P (2021) Detailed evolutionary models of massive contact binaries - I. Model grids and synthetic populations for the Magellanic Clouds. Mon Not R Astron Soc 507:5013–5033, https://doi.org/10.1093/mnras/stab2276, https://ui.adsabs.harvard.edu/abs/2021MNRAS.507.5013M

Menon A, Pawlak M, Lennon DJ, Sen K, Langer N (2024) A study of sixty-one massive over-contact binary candidates from the OGLE survey of the Magellanic Clouds. ArXiv e-prints https://doi.org/10.48550/arXiv.2410.16427, http://arxiv.org/abs/2410.16427, 2410.16427

Mestel L (1968) Magnetic braking by a stellar wind—I. Mon Not R Astron Soc 138(3):359–391, https://doi.org/10.1093/mnras/138.3.359

Meynet G, Maeder A (2000) Stellar evolution with rotation. V. Changes in all the outputs of massive star models. Astron Astrophys 361:101–120, https://doi.org/10.48550/arXiv.astro-ph/0006404, https://ui.adsabs.harvard.edu/abs/2000A&A...361..101M

Meynet G, Maeder A (2005) Stellar evolution with rotation. XI. Wolf-Rayet star populations at different metallicities. Astron Astrophys 429:581–598, https://doi.org/10.1051/0004-6361:20047106

Michell J (1767) An inquiry into the probable parallax, and magnitude of the fixed stars, from the quantity of light which they afford us, and the particular circumstances of their situation, by the Rev. John Michell, B. D. F. R. S. Philos Trans (1683–1775) 57:234–264, https://www.jstor.org/stable/105952, 105952

Michielsen M, Aerts C, Bowman DM (2021) Probing the temperature gradient in the core boundary layer of stars with gravito-inertial modes. The case of KIC 7760680. Astron Astrophys 650:A175, https://doi.org/10.1051/0004-6361/202039926, https://ui.adsabs.harvard.edu/abs/2021A&A...650A.175M

Mirabel IF, Rodrigues I (2003) Formation of a Black Hole in the Dark. Science 300(5622):1119–1120, https://doi.org/10.1126/science.1083451, https://www.science.org/doi/10.1126/science.1083451

Mochnacki SW (1984) Accurate integrations of the Roche model. Astrophys J Suppl Ser 55:551–561, https://doi.org/10.1086/190967, http://adsabs.harvard.edu/abs/1984ApJS...55..551M

Moe M (2019) Multiplicity statistics and properties across the HR diagram. Mem Soc Astron Ital 90:347, https://ui.adsabs.harvard.edu/abs/2019MmSAI..90..347M

Moe M, Di Stefano R (2017) Mind your Ps and Qs: the interrelation between period (P) and mass-ratio (Q) distributions of binary stars. Astrophys J Suppl Ser 230:15, https://doi.org/10.3847/1538-4365/aa6fb6, http://adsabs.harvard.edu/abs/2017ApJS..230...15M

Moens N, Poniatowski LG, Hennicker L, Sundqvist JO, El Mellah I, Kee ND (2022) First 3D radiation-hydrodynamic simulations of Wolf-Rayet winds. Astron Astrophys 665:A42, https://doi.org/10.1051/0004-6361/202243451, https://ui.adsabs.harvard.edu/abs/2022A&A...665A..42M

Mokiem MR, de Koter A, Puls J, Herrero A, Najarro F, Villamariz MR (2005) Spectral analysis of early-type stars using a genetic algorithm based fitting method. Astron Astrophys 441:711–733, https://doi.org/10.1051/0004-6361:20053522

Mokiem MR, de Koter A, Evans CJ, Puls J, Smartt SJ, Crowther PA, Herrero A, Langer N, Lennon DJ, Najarro F, Villamariz MR, Vink JS (2007) The VLT-FLAMES survey of massive stars: Wind properties and evolution of hot massive stars in the Large Magellanic Cloud. Astron Astrophys 465(3):1003–1019, https://doi.org/10.1051/0004-6361:20066489, https://www.aanda.org/articles/aa/abs/2007/15/aa6489-06/aa6489-06.html

Mombarg JSG, Rieutord M, Espinosa Lara F (2023) The first two-dimensional stellar structure and evolution models of rotating stars. Calibration to β Cephei pulsator HD 192575. Astron Astrophys 677:L5, https://doi.org/10.1051/0004-6361/202347454

Moravveji E, Townsend RHD, Aerts C, Mathis S (2016) Sub-inertial gravity modes in the B8V Star KIC 7760680 reveal moderate core overshooting and low vertical diffusive mixing. Astrophys J 823:130, https://doi.org/10.3847/0004-637X/823/2/130, https://ui.adsabs.harvard.edu/abs/2016ApJ...823..130M

Morton DC (1967) Mass loss from three OB supergiants in Orion. Astrophys J 150:535, https://doi.org/10.1086/149356, https://ui.adsabs.harvard.edu/abs/1967ApJ...150..535M

Moss D (2003) The survival of fossil magnetic fields during pre-main sequence evolution. Astron Astrophys 403:693–697, https://doi.org/10.1051/0004-6361:20030431

Muijres L, Vink JS, de Koter A, Hirschi R, Langer N, Yoon SC (2012) Mass-loss predictions for evolved very metal-poor massive stars. Astron Astrophys 546:A42, https://doi.org/10.1051/0004-6361/201118666

Muñoz DJ, Miranda R, Lai D (2019) Hydrodynamics of circumbinary accretion: angular momentum transfer and binary orbital evolution. Astrophys J 871:84, https://doi.org/10.3847/1538-4357/aaf867, https://ui.adsabs.harvard.edu/abs/2019ApJ...871...84M

Nakano S, Nishiyama K, Kabashima F, Sakurai Y, Jacques C, Pimentel E, Chekhovich D, Korotkiy S, Kryachko T, Samus NN (2008) V1309 Scorpii = Nova Scorpii 2008. Int Astron Union Circ 8972:1, https://ui.adsabs.harvard.edu/abs/2008IAUC.8972....1N

Naoz S (2016) The eccentric Kozai-Lidov effect and its applications. Annu Rev Astron Astrophys 54:441–489, https://doi.org/10.1146/annurev-astro-081915-023315, https://ui.adsabs.harvard.edu/abs/2016ARA&A..54..441N

Nazé Y, Becker MD, Rauw G, Barbieri C (2008) A binary signature in the non-thermal radio-emitter Cyg OB2 #9. Astron Astrophys 483(2):543–546, https://doi.org/10.1051/0004-6361:200809491, https://www.aanda.org/articles/aa/abs/2008/20/aa09491-08/aa09491-08.html

Needham J (1959) Science and civilisation in China, Vol. 3, Mathematics and the sciences of the heavens and the Earth. Cambridge University Press, http://archive.org/details/science-and-civilisation-in-china-vol.-3-mathematics-and-the-sciences-of-the-hea

Nelson CA, Eggleton PP (2001) A complete survey of case a binary evolution with comparison to observed Algol-type systems. Astrophys J 552:664–678, https://doi.org/10.1086/320560, https://ui.adsabs.harvard.edu/abs/2001ApJ...552..664N

Newton I (1687) Philosophiae naturalis principia mathematica. Royal Society, London, https://doi.org/10.3931/e-rara-440, https://ui.adsabs.harvard.edu/abs/1687pnpm.book.....N

Newton ER, Irwin J, Charbonneau D, Berlind P, Calkins ML, Mink J (2017) The Hα emission of nearby M dwarfs and its relation to stellar rotation. Astrophys J 834:85, https://doi.org/10.3847/1538-4357/834/1/85, https://ui.adsabs.harvard.edu/abs/2017ApJ...834...85N

Nieuwenhuijzen H, de Jager C (1995) Atmospheric accelerations and the stability of dynamic supergiant atmospheres. Astron Astrophys 302:811, https://ui.adsabs.harvard.edu/abs/1995A&A...302..811N

Offner SSR, Moe M, Kratter KM, Sadavoy SI, Jensen ELN, Tobin JJ (2023) The Origin and evolution of multiple star systems. In: Protostars and planets VII, Astronomical Society of the Pacific, San Francisco, Kyoto, Japan, Astronomical Society of the Pacific conference series, vol 534, p 275, https://doi.org/10.26624/LSUL6799, https://ui.adsabs.harvard.edu/abs/2023ASPC..534..275O

Ohlmann ST, Röpke FK, Pakmor R, Springel V (2017) Constructing stable 3D hydrodynamical models of giant stars. Astron Astrophys 599:A5, https://doi.org/10.1051/0004-6361/201629692, https://ui.adsabs.harvard.edu/abs/2017A&A...599A...5O

Olejak A, Belczynski K, Ivanova N (2021) Impact of common envelope development criteria on the formation of LIGO/Virgo sources. Astron Astrophys 651:A100, https://doi.org/10.1051/0004-6361/202140520

Ondra L (1999) A new view of Mizar. https://leosondra.cz/en/mizar

Öpik E (1924) Statistical studies of double stars: on the distribution of relative luminosities and distances of double stars in the Harvard revised photometry north of declination −31°. Publ Tartu Astrofiz Observ 25(6):1, https://ui.adsabs.harvard.edu/abs/1924PTarO..25f...1O

Öpik E (1938) Stellar structure, source of energy, and evolution. Publ Tartu Astrofiz Observ 30:C1–C115, https://ui.adsabs.harvard.edu/abs/1938PTarO..30C...1O

Ostrov PG (2001) Orbital solution for the MACHO*05:34:41.3-69:31:39 O3If*+O6:V eclipsing binary system in the Large Magellanic Cloud. Mon Not R Astron Soc 321:L25–L28, https://doi.org/10.1046/j.1365-8711.2001.04267.x, https://ui.adsabs.harvard.edu/abs/2001MNRAS.321L..25O

Owocki SP, Rybicki GB (1984) Instabilities in line-driven stellar winds. I. Dependence on perturbation wavelength. Astrophys J 284:337–350, https://doi.org/10.1086/162412, https://ui.adsabs.harvard.edu/abs/1984ApJ...284..337O

Paczynski B, Szczygiel D, Pilecki B, Pojmanski G (2006) Eclipsing binaries in ASAS catalog. Mon Not R Astron Soc 368(3):1311–1318, https://doi.org/10.1111/j.1365-2966.2006.10223.x, http://arxiv.org/abs/astro-ph/0601026, astro-ph/0601026

Palate M, Rauw G, Koenigsberger G, Moreno E (2013) Spectral modelling of massive binary systems. Astron Astrophys 552:A39, https://doi.org/10.1051/0004-6361/201219754, https://ui.adsabs.harvard.edu/abs/2013A%26A...552A..39P/abstract

Papaloizou J, Pringle JE (1979) On the maintenance of a temperature discontinuity in contact binary stars. Mon Not R Astron Soc 189:5P–8P, https://doi.org/10.1093/mnras/189.1.5P, https://ui.adsabs.harvard.edu/abs/1979MNRAS.189P...5P

Parker EN (1955) The formation of sunspots from the solar toroidal field. Astrophys J 121:491, https://doi.org/10.1086/146010, https://ui.adsabs.harvard.edu/abs/1955ApJ...121..491P

Pastorello A, Chen TW, Cai YZ, Morales-Garoffolo A, Cano Z, Mason E, Barsukova EA, Benetti S, Berton M, Bose S, Bufano F, Callis E, Cannizzaro G, Cartier R, Chen P, Dong S, Dyrbye S, Elias-Rosa N, Flörs A, Fraser M, Geier S, Goranskij VP, Kann DA, Kuncarayakti H, Onori F, Reguitti A, Reynolds T, Losada IR, Carracedo AS, Schweyer T, Smartt SJ, Tatarnikov AM, Valeev AF, Vogl C, Wevers T, Postigo AdU, Izzo L, Inserra C, Kankare E, Maguire K, Smith KW, Stalder B, Tartaglia L, Thöne CC, Valerin G, Young DR (2019) The evolution of luminous red nova AT 2017jfs in NGC 4470. Astron Astrophys 625:L8, https://doi.org/10.1051/0004-6361/201935511, https://www.aanda.org/articles/aa/abs/2019/05/aa35511-19/aa35511-19.html

Patterson J (1984) The evolution of cataclysmic and low-mass X-ray binaries. Astrophys J Suppl Ser 54:443–493, https://doi.org/10.1086/190940, https://ui.adsabs.harvard.edu/abs/1984ApJS...54..443P

Pauli W (1925) Über den Zusammenhang des Abschlusses der Elektronengruppen im Atom mit der Komplexstruktur der Spektren. Z Phys 31(1):765–783, https://doi.org/10.1007/BF02980631, http://link.springer.com/10.1007/BF02980631

Pauli D, Langer N, Aguilera-Dena DR, Wang C, Marchant P (2022) A synthetic population of Wolf-Rayet stars in the LMC based on detailed single and binary star evolution models. Astron Astrophys 667:A58, https://doi.org/10.1051/0004-6361/202243965

Pavlovski K, Southworth J, Tamajo E (2018) Physical properties and CNO abundances for high-mass stars in four main-sequence detached eclipsing binaries: V478 Cyg, AH Cep, V453 Cyg, and V578 Mon. Mon Not R Astron Soc 481:3129–3147, https://doi.org/10.1093/mnras/sty2516, https://ui.adsabs.harvard.edu/abs/2018MNRAS.481.3129P

Paxton B, Bildsten L, Dotter A, Herwig F, Lesaffre P, Timmes F (2011) Modules for experiments in stellar astrophysics (MESA). Astrophys J Suppl Ser 192:3, https://doi.org/10.1088/0067-0049/192/1/3, http://adsabs.harvard.edu/abs/2011ApJS..192....3P, 1009.1622

Paxton B, Cantiello M, Arras P, Bildsten L, Brown EF, Dotter A, Mankovich C, Montgomery MH, Stello D, Timmes FX, Townsend R (2013) Modules for experiments in stellar astrophysics (MESA): planets, oscillations, rotation, and massive stars. Astrophys J Suppl Ser 208:4, https://doi.org/10.1088/0067-0049/208/1/4, http://adsabs.harvard.edu/abs/2013ApJS..208....4P, 1301.0319

Paxton B, Marchant P, Schwab J, Bauer EB, Bildsten L, Cantiello M, Dessart L, Farmer R, Hu H, Langer N, Townsend RHD, Townsley DM, Timmes FX (2015) Modules for experiments in stellar astrophysics (MESA): binaries, pulsations, and explosions. Astrophys J Suppl Ser 220:15, https://doi.org/10.1088/0067-0049/220/1/15, http://adsabs.harvard.edu/abs/2015ApJS..220...15P, 1506.03146

Paxton B, Schwab J, Bauer EB, Bildsten L, Blinnikov S, Duffell P, Farmer R, Goldberg JA, Marchant P, Sorokina E, Thoul A, Townsend RHD, Timmes FX (2018) Modules for experiments in stellar astrophysics (MESA): convective boundaries, element diffusion, and massive star explosions. Astrophys J Suppl Ser 234:34, https://doi.org/10.3847/1538-4365/aaa5a8, http://adsabs.harvard.edu/abs/2018ApJS..234...34P, 1710.08424

Paxton B, Smolec R, Schwab J, Gautschy A, Bildsten L, Cantiello M, Dotter A, Farmer R, Goldberg JA, Jermyn AS, Kanbur SM, Marchant P, Thoul A, Townsend RHD, Wolf WM, Zhang M, Timmes FX (2019) Modules for experiments in stellar astrophysics (MESA): pulsating variable stars, rotation, convective boundaries, and energy conservation. Astrophys J Suppl Ser 243:10, https://doi.org/10.3847/1538-4365/ab2241, http://adsabs.harvard.edu/abs/2019ApJS..243...10P, 1903.01426

Payne CH (1925) Stellar atmospheres; a contribution to the observational study of high temperature in the reversing layers of stars. PhD thesis, Harvard University, https://ui.adsabs.harvard.edu/abs/1925PhDT.........1P

Pedersen MG, Aerts C, Pápics PI, Rogers TM (2018) The shape of convective core overshooting from gravity-mode period spacings. Astron Astrophys 614:A128, https://doi.org/10.1051/0004-6361/201732317, https://ui.adsabs.harvard.edu/abs/2018A&A...614A.128P

Pedersen MG, Aerts C, Pápics PI, Michielsen M, Gebruers S, Rogers TM, Molenberghs G, Burssens S, Garcia S, Bowman DM (2021) Internal mixing of rotating stars inferred from dipole gravity modes. Nat Astron 5:715–722, https://doi.org/10.1038/s41550-021-01351-x, https://ui.adsabs.harvard.edu/abs/2021NatAs...5..715P

Pejcha O (2014) Burying a binary: dynamical mass loss and a continuous optically thick outflow explain the candidate stellar merger V1309 Scorpii. Astrophys J 788:22, https://doi.org/10.1088/0004-637X/788/1/22, https://ui.adsabs.harvard.edu/abs/2014ApJ...788...22P

Pejcha O, Metzger BD, Tomida K (2016) Cool and luminous transients from mass-losing binary stars. Mon Not R Astron Soc 455:4351–4372, https://doi.org/10.1093/mnras/stv2592, https://ui.adsabs.harvard.edu/abs/2016MNRAS.455.4351P

Penny LR, Ouzts C, Gies DR (2008) Tomographic separation of composite spectra. XI. The physical properties of the massive close binary HD 100213 (TU Muscae). Astrophys J 681:554–561, https://doi.org/10.1086/587509, https://ui.adsabs.harvard.edu/abs/2008ApJ...681..554P

Pešta M, Pejcha O (2023) Mass-ratio distribution of contact binary stars. Astron Astrophys 672:A176, https://doi.org/10.1051/0004-6361/202245613

Picco A, Marchant P, Sana H, Nelemans G (2024) Forming merging double compact objects with stable mass transfer. Astron Astrophys 681:A31, https://doi.org/10.1051/0004-6361/202347090, https://ui.adsabs.harvard.edu/abs/2024A&A...681A..31P

Pickering EC (1890) On the spectrum of zeta Ursae Majoris. Observatory 13:80–81, https://ui.adsabs.harvard.edu/abs/1890Obs....13...80P

Pietrzyński G, Graczyk D, Gallenne A, Gieren W, Thompson IB, Pilecki B, Karczmarek P, Górski M, Suchomska K, Taormina M, Zgirski B, Wielgórski P, Kołaczkowski Z, Konorski P, Villanova S, Nardetto N, Kervella P, Bresolin F, Kudritzki RP, Storm J, Smolec R, Narloch W (2019) A distance to the Large Magellanic Cloud that is precise to one percent. Nature 567:200–203, https://doi.org/10.1038/s41586-019-0999-4, https://ui.adsabs.harvard.edu/abs/2019Natur.567..200P

Pittard JM, Dougherty SM (2006) Radio, X-ray, and γ-ray emission models of the colliding-wind binary WR140. Mon Not R Astron Soc 372:801–826, https://doi.org/10.1111/j.1365-2966.2006.10888.x, http://adsabs.harvard.edu/abs/2006MNRAS.372..801P

Planck M (1901) Ueber das Gesetz der Energieverteilung im Normalspectrum. Ann Phys 309:553–563, https://doi.org/10.1002/andp.19013090310, https://ui.adsabs.harvard.edu/abs/1901AnP...309..553P

Planck Collaboration, Aghanim N, Akrami Y et al (2020) Planck 2018 results. VI. Cosmological parameters. Astron Astrophys 641:A6, https://doi.org/10.1051/0004-6361/201833910

Podsiadlowski Ph, Rappaport S, Pfahl ED (2002) Evolutionary sequences for low- and intermediate-mass X-ray binaries. Astrophys J 565:1107–1133, https://doi.org/10.1086/324686, https://ui.adsabs.harvard.edu/abs/2002ApJ...565.1107P

Pols OR (1994) Case A evolution of massive close binaries: Formation of contact systems and possible reversal of the supernova order. Astron Astrophys 290:119–128

Poro A, Li K, Paki E, Baudart S, Michel R, Wang L, Fernández Lajús E, Alicavus F, Foschino S, Aceves H, Tamayo FJ, Chavez CE (2025) BSN-I: The first in-depth photometric study of seven total-eclipse contact binary systems. Mon Not R Astron Soc 537:3160–3174, https://doi.org/10.1093/mnras/staf222, https://ui.adsabs.harvard.edu/abs/2025MNRAS.537.3160P

Potekhin AY, Chabrier G (2010) Thermodynamic functions of dense plasmas: analytic approximations for astrophysical applications. Contrib Plasma Phys 50:82–87, https://doi.org/10.1002/ctpp.201010017, https://ui.adsabs.harvard.edu/abs/2010CoPP...50...82P, 1001.0690

Poutanen J (2017) Rosseland and flux mean opacities for compton scattering. Astrophys J 835(2):119, https://doi.org/10.3847/1538-4357/835/2/119, 1606.09466

Prisinzano L, Damiani F, Micela G, Sciortino S (2005) The star formation region NGC 6530: Distance, ages and initial mass function. Astron Astrophys 430:941–957, https://doi.org/10.1051/0004-6361:20040432

Prša A, Zwitter T (2005) A computational guide to physics of eclipsing binaries. I. Demonstrations and perspectives. Astrophys J 628:426–438, https://doi.org/10.1086/430591, https://ui.adsabs.harvard.edu/abs/2005ApJ...628..426P

Prša A, Batalha N, Slawson RW, Doyle LR, Welsh WF, Orosz JA, Seager S, Rucker M, Mjaseth K, Engle SG, Conroy K, Jenkins J, Caldwell D, Koch D, Borucki W (2011a) Kepler eclipsing binary stars. I. Catalog and principal characterization of 1879 eclipsing binaries in the first data release. Astron J 141:83, https://doi.org/10.1088/0004-6256/141/3/83, https://ui.adsabs.harvard.edu/abs/2011AJ....141...83P

Prša A, Pepper J, Stassun KG (2011b) Expected large synoptic survey telescope (LSST) yield of eclipsing binary stars. Astron J 142:52, https://doi.org/10.1088/0004-6256/142/2/52, https://ui.adsabs.harvard.edu/abs/2011AJ....142...52P

Prša A, Conroy KE, Horvat M, Pablo H, Kochoska A, Bloemen S, Giammarco J, Hambleton KM, Degroote P (2016) Physics of eclipsing binaries. II. Toward the increased model fidelity. Astrophys J Suppl Ser 227:29, https://doi.org/10.3847/1538-4365/227/2/29, http://adsabs.harvard.edu/abs/2016ApJS..227...29P

Puls J (2017) FASTWIND reloaded: Complete comoving frame transfer for "all" contributing lines. In: Proceedings IAU symposium, vol 329, pp 435–435, https://doi.org/10.1017/S1743921317000229, http://adsabs.harvard.edu/abs/2017IAUS..329..435P

Puls J, Urbaneja MA, Venero R, Repolust T, Springmann U, Jokuthy A, Mokiem MR (2005) Atmospheric NLTE-models for the spectroscopic analysis of blue stars with winds. II. Line-blanketed models. Astron Astrophys 435:669–698, https://doi.org/10.1051/0004-6361:20042365

Pylyser E, Savonije GJ (1988) Evolution of low-mass close binary systems with a compact mass accreting component. Astron Astrophys 191:57–70, https://ui.adsabs.harvard.edu/abs/1988A&A...191...57P

Quintero EA, Eenens P, Rauw G (2020) The massive binary system 9 Sgr revisited: New insights into disentangling methods. Astron Nachr 341(6–7):628–637, https://doi.org/10.1002/asna.202013696, https://onlinelibrary.wiley.com/doi/abs/10.1002/asna.202013696, 2010.01181

Raghavan D, McAlister HA, Henry TJ, Latham DW, Marcy GW, Mason BD, Gies DR, White RJ, ten Brummelaar TA (2010) A survey of stellar families: multiplicity of solar-type stars. Astrophys J Suppl Ser 190(1):1, https://doi.org/10.1088/0067-0049/190/1/1

Ramírez-Tannus MC, Backs F, de Koter A, Sana H, Beuther H, Bik A, Brandner W, Kaper L, Linz H, Henning Th, Poorta J (2021) A relation between the radial velocity dispersion of young clusters and their age. Evidence for hardening as the formation scenario of massive close binaries. Astron Astrophys 645:L10, https://doi.org/10.1051/0004-6361/202039673

Rappaport S, Verbunt F, Joss PC (1983) A new technique for calculations of binary stellar evolution application to magnetic braking. Astrophys J 275:713–731, https://doi.org/10.1086/161569, https://ui.adsabs.harvard.edu/abs/1983ApJ...275..713R

Raskin G, van Winckel H, Hensberge H, Jorissen A, Lehmann H, Waelkens C, Avila G, de Cuyper JP, Degroote P, Dubosson R, Dumortier L, Frémat Y, Laux U, Michaud B, Morren J, Perez Padilla J, Pessemier W, Prins S, Smolders K, van Eck S, Winkler J (2011) HERMES: A high-resolution fibre-fed spectrograph for the Mercator telescope. Astron Astrophys 526:A69, https://doi.org/10.1051/0004-6361/201015435

Raucq F, Gosset E, Rauw G, Manfroid J, Mahy L, Mennekens N, Vanbeveren D (2017) Observational signatures of past mass-exchange episodes in massive binaries: The case of LSS 3074. Astron Astrophys 601:A133, https://doi.org/10.1051/0004-6361/201630330, https://www.aanda.org/articles/aa/abs/2017/05/aa30330-16/aa30330-16.html

Rauw G, Nazé Y (2016) X-ray emission from interacting wind massive binaries: A review of 15 years of progress. Adv Space Res 58:761–781, https://doi.org/10.1016/j.asr.2015.09.026, http://adsabs.harvard.edu/abs/2016AdSpR..58..761R

Rauw G, Blomme R, Waldron WL, Corcoran MF, Pittard JM, Pollock AMT, Runacres MC, Sana H, Stevens IR, Loo SV (2002a) A multi-wavelength investigation of the non-thermal radio emitting O-star 9 Sgr. Astron Astrophys 394(3):993–1008, https://doi.org/10.1051/0004-6361:20020926, https://www.aanda.org/articles/aa/abs/2002/42/aa2716/aa2716.html

Rauw G, Nazé Y, Gosset E, Stevens IR, Blomme R, Corcoran MF, Pittard JM, Runacres MC (2002b) An XMM-Newton observation of the Lagoon Nebula and the very young open cluster NGC 6530. Astron Astrophys 395:499–513, https://doi.org/10.1051/0004-6361:20021230

Rauw G, Vreux JM, Stevens IR, Gosset E, Sana H, Jamar C, Mason KO (2002c) Phase-resolved X-ray and optical spectroscopy of the massive binary HD 93403. Astron Astrophys 388:552–562, https://doi.org/10.1051/0004-6361:20020523

Rauw G, Sana H, Spano M, Gosset E, Mahy L, De Becker M, Eenens P (2012) 9 Sagittarii: uncovering an O-type spectroscopic binary with an 8.6 year period. Astron Astrophys 542:A95, https://doi.org/10.1051/0004-6361/201219254

Rauw G, Blomme R, Nazé Y, Spano M, Mahy L, Gosset E, Volpi D, van Winckel H, Raskin G, Waelkens C (2016) Testing the theory of colliding winds: The periastron passage of 9 Sagittarii. I. X-ray and optical spectroscopy. Astron Astrophys 589:A121, https://doi.org/10.1051/0004-6361/201526871

Reitze D, Adhikari RX, Ballmer S, Barish B, Barsotti L, Billingsley G, Brown DA, Chen Y, Coyne D, Eisenstein R, Evans M, Fritschel P, Hall ED, Lazzarini A, Lovelace G, Read J, Sathyaprakash BS, Shoemaker D, Smith J, Torrie C, Vitale S, Weiss R, Wipf C, Zucker M (2019) Cosmic explorer: The U.S. contribution to gravitational-wave astronomy beyond LIGO. In: Bulletin of the American Astronomical Society, eprint: arXiv:1907.04833, vol 51, p 35, https://doi.org/10.48550/arXiv.1907.04833, https://ui.adsabs.harvard.edu/abs/2019BAAS...51g..35R

Retter A, Marom A (2003) A model of an expanding giant that swallowed planets for the eruption of V838 Monocerotis. Mon Not R Astron Soc 345:L25–L28, https://doi.org/10.1046/j.1365-8711.2003.07190.x, https://ui.adsabs.harvard.edu/abs/2003MNRAS.345L..25R

Ricker GR, Winn JN, Vanderspek R et al (2015) Transiting exoplanet survey satellite (TESS). J Astron Telesc Instrum Syst 1:014003, https://doi.org/10.1117/1.JATIS.1.1.014003, https://ui.adsabs.harvard.edu/abs/2015JATIS...1a4003R

Ritter A (1898) On the constitution of gaseous celestial bodies. Astrophys J 8:293, https://doi.org/10.1086/140534, https://ui.adsabs.harvard.edu/abs/1898ApJ.....8..293R

Rogers FJ, Nayfonov A (2002) Updated and expanded OPAL equation-of-state tables: implications for helioseismology. Astrophys J 576:1064–1074, https://doi.org/10.1086/341894, https://ui.adsabs.harvard.edu/abs/2002ApJ...576.1064R

Röpke FK, De Marco O (2023) Simulations of common-envelope evolution in binary stellar systems: Physical models and numerical techniques. Living Rev Comput Astrophys 9:2, https://doi.org/10.1007/s41115-023-00017-x, https://ui.adsabs.harvard.edu/abs/2023LRCA....9....2R

Rosseland S (1924) Electrical state of a star. Mon Not R Astron Soc 84:720–728, https://doi.org/10.1093/mnras/84.9.720, https://ui.adsabs.harvard.edu/abs/1924MNRAS..84..720R

Rossiter MW (1980) “Women’s work” in science, 1880–1910. Isis 71(3):381–398, https://www.jstor.org/stable/230118, 230118

Roxburgh IW (2004) 2-dimensional models of rapidly rotating stars I. Uniformly rotating zero age main sequence stars. Astron Astrophys 428:171–179, https://doi.org/10.1051/0004-6361:20041202, https://ui.adsabs.harvard.edu/abs/2004A&A...428..171R

Rucinski SM, Pribulla T, van Kerkwijk MH (2007) Contact binaries with additional components. III. A search using adaptive optics. Astron J 134(6):2353–2365, https://doi.org/10.1086/523353, https://ui.adsabs.harvard.edu/abs/2007AJ....134.2353R/abstract

Russell HN (1902) An improved method of calculating the orbit of a spectroscopic binary. Astrophys J 15:252, https://doi.org/10.1086/140914, https://ui.adsabs.harvard.edu/abs/1902ApJ....15..252R

Russell HN (1914) Relations between the spectra and other characteristics of the stars. Popular Astron 22:275–294, https://ui.adsabs.harvard.edu/abs/1914PA.....22..275R

Saliba G (1994) A history of Arabic astronomy: planetary theories during the golden age of Islam. New York University Press, New York, http://archive.org/details/historyofarabica0000sali

Salpeter EE (1955) The luminosity function and stellar evolution. Astrophys J 121:161, https://doi.org/10.1086/145971, https://ui.adsabs.harvard.edu/abs/1955ApJ...121..161S

Sana H, Stevens IR, Gosset E, Rauw G, Vreux JM (2004) A phase-resolved XMM-Newton campaign on the colliding-wind binary HD 152248. Mon Not R Astron Soc 350:809–828, https://doi.org/10.1111/j.1365-2966.2004.07719.x, http://adsabs.harvard.edu/abs/2004MNRAS.350..809S

Sana H, Rauw G, Nazé Y, Gosset E, Vreux JM (2006) An XMM-Newton view of the young open cluster NGC 6231 - II. The OB star population. Mon Not R Astron Soc 372:661–678, https://doi.org/10.1111/j.1365-2966.2006.10847.x, http://adsabs.harvard.edu/abs/2006MNRAS.372..661S

Sana H, James G, Gosset E (2011a) The massive star binary fraction in young open clusters - III. IC 2944 and the Cen OB2 association. Mon Not R Astron Soc 416:817–831, https://doi.org/10.1111/j.1365-2966.2011.18698.x, https://ui.adsabs.harvard.edu/abs/2011MNRAS.416..817S

Sana H, Le Bouquin JB, De Becker M, Berger JP, de Koter A, Mérand A (2011b) The non-thermal radio emitter HD 93250 resolved by long baseline interferometry. Astrophys J Lett 740:L43, https://doi.org/10.1088/2041-8205/740/2/L43, http://adsabs.harvard.edu/abs/2011ApJ...740L..43S

Sana H, de Mink SE, de Koter A, Langer N, Evans CJ, Gieles M, Gosset E, Izzard RG, Bouquin JBL, Schneider FRN (2012) Binary interaction dominates the evolution of massive stars. Science 337(6093):444–446, https://doi.org/10.1126/science.1223344, https://science.sciencemag.org/content/337/6093/444

Sana H, de Koter A, de Mink SE, Dunstall PR, Evans CJ, Hénault-Brunet V, Maíz Apellániz J, Ramírez-Agudelo OH, Taylor WD, Walborn NR, Clark JS, Crowther PA, Herrero A, Gieles M, Langer N, Lennon DJ, Vink JS (2013a) The VLT-FLAMES tarantula survey. VIII. Multiplicity properties of the O-type star population. Astron Astrophys 550:A107, https://doi.org/10.1051/0004-6361/201219621

Sana H, Le Bouquin JB, Mahy L, Absil O, De Becker M, Gosset E (2013b) Three-dimensional orbits of the triple-O stellar system HD 150136. Astron Astrophys 553:A131, https://doi.org/10.1051/0004-6361/201321189

Sana H, Le Bouquin JB, Lacour S, Berger JP, Duvert G, Gauchet L, Norris B, Olofsson J, Pickel D, Zins G, Absil O, de Koter A, Kratter K, Schnurr O, Zinnecker H (2014) Southern massive stars at high angular resolution: observational campaign and companion detection. Astrophys J Suppl Ser 215:15, https://doi.org/10.1088/0067-0049/215/1/15, https://ui.adsabs.harvard.edu/abs/2014ApJS..215...15S

Sana H, Ramírez-Tannus MC, de Koter A, Kaper L, Tramper F, Bik A (2017) A dearth of short-period massive binaries in the young massive star forming region M 17. Evidence for a large orbital separation at birth? Astron Astrophys 599:L9, https://doi.org/10.1051/0004-6361/201630087

Sana H, Shenar T, Bodensteiner J et al (2025) A high fraction of close massive binary stars at low metallicity. Nat Astron 9(9):1337–1346, https://doi.org/10.1038/s41550-025-02610-x, https://www.nature.com/articles/s41550-025-02610-x

Sandage AR (1953) The color-magnitude diagram for the globular cluster M 3. Astron J 58:61–75, https://doi.org/10.1086/106822, https://ui.adsabs.harvard.edu/abs/1953AJ.....58...61S

Santolaya-Rey AE, Puls J, Herrero A (1997) Atmospheric NLTE-models for the spectroscopic analysis of luminous blue stars with winds. Astron Astrophys 323:488–512

Sanyal D, Grassitelli L, Langer N, Bestenlehner JM (2015) Massive main-sequence stars evolving at the Eddington limit. Astron Astrophys 580:A20, https://doi.org/10.1051/0004-6361/201525945, https://ui.adsabs.harvard.edu/abs/2015A&A...580A..20S

Saumon D, Chabrier G, van Horn HM (1995) An equation of state for low-mass stars and giant planets. Astrophys J Suppl Ser 99:713, https://doi.org/10.1086/192204, https://ui.adsabs.harvard.edu/abs/1995ApJS...99..713S

Schleicher DRG, Hidalgo JP, Galli D (2023) Survival of fossil fields during the pre-main sequence evolution of intermediate-mass stars. Astron Astrophys 678:A204, https://doi.org/10.1051/0004-6361/202346809

Schneider FRN, Izzard RG, de Mink SE, Langer N, Stolte A, de Koter A, Gvaramadze VV, Hußmann B, Liermann A, Sana H (2013) Ages of young star clusters, massive blue stragglers, and the upper mass limit of stars: analyzing age-dependent stellar mass functions. Astrophys J 780(2):117, https://doi.org/10.1088/0004-637X/780/2/117, https://dx.doi.org/10.1088/0004-637X/780/2/117

Schneider FRN, Langer N, de Koter A, Brott I, Izzard RG, Lau HHB (2014) Bonnsai: A Bayesian tool for comparing stars with stellar evolution models. Astron Astrophys 570:A66, https://doi.org/10.1051/0004-6361/201424286

Schneider FRN, Ohlmann ST, Podsiadlowski P, Röpke FK, Balbus SA, Pakmor R, Springel V (2019) Stellar mergers as the origin of magnetic massive stars. Nature 574(7777):211–214, https://doi.org/10.1038/s41586-019-1621-5, https://www.nature.com/articles/s41586-019-1621-5

Schootemeijer A, Langer N, Grin NJ, Wang C (2019) Constraining mixing in massive stars in the Small Magellanic Cloud. Astron Astrophys 625:A132, https://doi.org/10.1051/0004-6361/201935046, https://ui.adsabs.harvard.edu/abs/2019A%26A...625A.132S/abstract

Schroder KP, Pols OR, Eggleton PP (1997) A critical test of stellar evolution and convective core 'overshooting' by means of zeta Aurigae systems. Mon Not R Astron Soc 285:696–710, https://doi.org/10.1093/mnras/285.4.696, https://ui.adsabs.harvard.edu/abs/1997MNRAS.285..696S

Schrödinger E (1926) Quantisierung als Eigenwertproblem. Ann Phys 384:361–376, https://doi.org/10.1002/andp.19263840404, https://ui.adsabs.harvard.edu/abs/1926AnP...384..361S

Schwarzschild K (1900) Ein Verfahren der Bahnbestimmung bei spectroskopischen Doppelsternen. Astron Nachr 152:65, https://doi.org/10.1002/asna.19001520502, https://ui.adsabs.harvard.edu/abs/1900AN....152...65S

Schwarzschild K (1906) On the equilibrium of the Sun's atmosphere. Nachr K Gesellschaft Wiss Göttingen Math-phys Klasse 195:41–53

Schwarzschild K (1916) On the gravitational field of a mass point according to Einstein's theory. Sitzungsberichte K Preuss Akad Wiss:189–196, https://ui.adsabs.harvard.edu/abs/1916AbhKP1916..189S

Schwarzschild M (1958) Structure and evolution of the stars. Princeton University Press, https://ui.adsabs.harvard.edu/abs/1958ses..book.....S

Sen K, Langer N, Marchant P, Menon A, de Mink SE, Schootemeijer A, Schürmann C, Mahy L, Hastings B, Nathaniel K, Sana H, Wang C, Xu XT (2022) Detailed models of interacting short-period massive binary stars. Astron Astrophys 659:A98, https://doi.org/10.1051/0004-6361/202142574, https://ui.adsabs.harvard.edu/abs/2022A&A...659A..98S/abstract

Shenar T, Hainich R, Todt H, Moffat AFJ, Sander A, Oskinova LM, Ramachandran V, Munoz M, Pablo H, Sana H, Hamann WR (2018) The shortest-period Wolf-Rayet binary in the Small Magellanic Cloud: Part of a high-order multiple system. Spectral and orbital analysis of SMC AB 6. Astron Astrophys 616:A103, https://doi.org/10.1051/0004-6361/201833006, https://ui.adsabs.harvard.edu/abs/2018A&A...616A.103S

Shenar T, Bodensteiner J, Abdul-Masih M, Fabry M, Mahy L, Marchant P, Banyard G, Bowman DM, Dsilva K, Hawcroft C, Reggiani M, Sana H (2020) The "Hidden" companion in LB-1 unveiled by spectral disentangling. Astron Astrophys 639:L6, https://doi.org/10.1051/0004-6361/202038275

Shenar T, Sana H, Mahy L, El-Badry K, Marchant P, Langer N, Hawcroft C, Fabry M, Sen K, Almeida LA, Abdul-Masih M, Bodensteiner J, Crowther PA, Gieles M, Gromadzki M, Hénault-Brunet V, Herrero A, de Koter A, Iwanek P, Kozłowski S, Lennon DJ, Maíz Apellániz J, Mróz P, Moffat AFJ, Picco A, Pietrukowicz P, Poleski R, Rybicki K, Schneider FRN, Skowron DM, Skowron J, Soszyński I, Szymański MK, Toonen S, Udalski A, Ulaczyk K, Vink JS, Wrona M (2022) An X-ray-quiet black hole born with a negligible kick in a massive binary within the Large Magellanic Cloud. Nat Astron 6:1085–1092, https://doi.org/10.1038/s41550-022-01730-y, https://ui.adsabs.harvard.edu/abs/2022NatAs...6.1085S

Shu FH, Lubow SH, Anderson L (1976) On the structure of contact binaries. I - The contact discontinuity. Astrophys J 209:536–546, https://doi.org/10.1086/154748, http://adsabs.harvard.edu/abs/1976ApJ...209..536S

Shu FH, Lubow SH, Anderson L (1979) On the structure of contact binaries. III - Mass and energy flow. Astrophys J 229:223–241, https://doi.org/10.1086/156948, http://adsabs.harvard.edu/abs/1979ApJ...229..223S

Shultz ME, Wade GA, Rivinius T, Neiner C, Alecian E, Bohlender D, Monin D, Sikora J, MiMeS Collaboration, BinaMIcS Collaboration (2018) The magnetic early B-type stars I: Magnetometry and rotation. Mon Not R Astron Soc 475:5144–5178, https://doi.org/10.1093/mnras/sty103, https://ui.adsabs.harvard.edu/abs/2018MNRAS.475.5144S

Sills A, Pinsonneault MH, Terndrup DM (2000) The angular momentum evolution of very low mass stars. Astrophys J 534(1):335, https://doi.org/10.1086/308739, https://iopscience.iop.org/article/10.1086/308739/meta

Simon KP, Sturm E (1994) Disentangling of composite spectra. Astron Astrophys 281:286–291

Simón-Díaz S, Herrero A (2014) The IACOB project. I. Rotational velocities in northern Galactic O- and early B-type stars revisited. The impact of other sources of line-broadening. Astron Astrophys 562:A135, https://doi.org/10.1051/0004-6361/201322758

Skrutskie MF, Cutri RM, Stiening R, Weinberg MD, Schneider S, Carpenter JM, Beichman C, Capps R, Chester T, Elias J, Huchra J, Liebert J, Lonsdale C, Monet DG, Price S, Seitzer P, Jarrett T, Kirkpatrick JD, Gizis JE, Howard E, Evans T, Fowler J, Fullmer L, Hurt R, Light R, Kopan EL, Marsh KA, McCallon HL, Tam R, Van Dyk S, Wheelock S (2006) The two micron all sky survey (2MASS). Astron J 131:1163–1183, https://doi.org/10.1086/498708, http://adsabs.harvard.edu/abs/2006AJ....131.1163S

Skumanich A (1972) Time scales for Ca II emission decay, rotational braking, and lithium depletion. Astrophys J 171:565, https://doi.org/10.1086/151310, https://ui.adsabs.harvard.edu/abs/1972ApJ...171..565S

Smith MA (1979) Rotational studies of lower main-sequence stars. Publ Astron Soc Pac 91(544):737, https://doi.org/10.1086/130579, https://iopscience.iop.org/article/10.1086/130579/meta

Smith N, Andrews JE, Van Dyk SD, Mauerhan JC, Kasliwal MM, Bond HE, Filippenko AV, Clubb KI, Graham ML, Perley DA, Jencson J, Bally J, Ubeda L, Sabbi E (2016) Massive star mergers and the recent transient in NGC 4490: A more massive cousin of V838 Mon and V1309 Sco. Mon Not R Astron Soc 458(1):950–962, https://doi.org/10.1093/mnras/stw219

Soker N, Tylenda R (2003) Main-sequence stellar eruption model for V838 monocerotis. Astrophys J 582:L105–L108, https://doi.org/10.1086/367759, https://ui.adsabs.harvard.edu/abs/2003ApJ...582L.105S

Soter S, deGrasse Tyson N (eds) (2001) Cosmic horizons: astronomy at the cutting edge. New Press, New York

Southworth J (2020) Binary stars: A cheat sheet. arXiv:191213400 [astro-ph] http://arxiv.org/abs/1912.13400, 1912.13400

Stahler SW, Palla F (2004) The formation of stars. Wiley-VCH, https://ui.adsabs.harvard.edu/abs/2004fost.book.....S

Stauffer JR, Hartmann LW (1987) The distribution of rotational velocities for low-mass stars in the pleiades. Astrophys J 318:337, https://doi.org/10.1086/165371, https://ui.adsabs.harvard.edu/abs/1987ApJ...318..337S

Stephenson FR, Green DA (2003) Was the supernova of AD 1054 reported in European history? J Astron Hist Heritage 6:46–52, https://ui.adsabs.harvard.edu/abs/2003JAHH....6...46S

Stępień K (2009) Large-scale circulations and energy transport in contact binaries. Mon Not R Astron Soc 397:857–867, https://doi.org/10.1111/j.1365-2966.2009.14981.x, https://ui.adsabs.harvard.edu/abs/2009MNRAS.397..857S

Stępień K (2011) Evolution of the progenitor binary of V1309 Scorpii before merger. Astron Astrophys 531:A18, https://doi.org/10.1051/0004-6361/201116689

Stevens IR, Blondin JM, Pollock AMT (1992) Colliding winds from early-type stars in binary systems. Astrophys J 386:265–287, https://doi.org/10.1086/171013, http://adsabs.harvard.edu/abs/1992ApJ...386..265S

Stevenson S, Vigna-Gómez A, Mandel I, Barrett JW, Neijssel CJ, Perkins D, de Mink SE (2017) Formation of the first three gravitational-wave observations through isolated binary evolution. Nat Commun 8:14906, https://doi.org/10.1038/ncomms14906, https://ui.adsabs.harvard.edu/abs/2017NatCo...814906S

Sundqvist JO, Puls J (2018) Atmospheric NLTE models for the spectroscopic analysis of blue stars with winds. IV. Porosity in physical and velocity space. Astron Astrophys 619:A59, https://doi.org/10.1051/0004-6361/201832993

Sundqvist JO, Björklund R, Puls J, Najarro F (2019) New predictions for radiation-driven, steady-state mass-loss and wind-momentum from hot, massive stars. I. Method and first results. Astron Astrophys 632:A126, https://doi.org/10.1051/0004-6361/201936580, https://ui.adsabs.harvard.edu/abs/2019A&A...632A.126S

Sung H, Chun MY, Bessell MS (2000) UBVRI and Hα photometry of the young open cluster NGC 6530. Astron J 120:333–348, https://doi.org/10.1086/301450, http://adsabs.harvard.edu/abs/2000AJ....120..333S

Sweet PA (1950) The importance of rotation in stellar evolution. Mon Not R Astron Soc 110:548, https://doi.org/10.1093/mnras/110.6.548, https://ui.adsabs.harvard.edu/abs/1950MNRAS.110..548S

Szymanski M, Kubiak M, Udalski A (2001) Contact binaries in OGLE-I database. Acta Astron 51:259–277, https://ui.adsabs.harvard.edu/abs/2001AcA....51..259S

Tassoul JL (2000) Stellar rotation. Cambridge University Press, https://ui.adsabs.harvard.edu/abs/2000stro.book.....T

ten Brummelaar TA, McAlister HA, Ridgway ST, Bagnuolo WG Jr, Turner NH, Sturmann L, Sturmann J, Berger DH, Ogden CE, Cadman R, Hartkopf WI, Hopper CH, Shure MA (2005) First results from the CHARA array. II. A description of the instrument. Astrophys J 628:453–465, https://doi.org/10.1086/430729, https://ui.adsabs.harvard.edu/abs/2005ApJ...628..453T

Thiele TN (1883) Neue Methode zur Berechung von Doppelsternbahnen. Astron Nachr 104:245, https://ui.adsabs.harvard.edu/abs/1883AN....104..245T

Timmes FX, Swesty FD (2000) The accuracy, consistency, and speed of an electron-positron equation of state based on table interpolation of the Helmholtz free energy. Astrophys J Suppl Ser 126:501–516, https://doi.org/10.1086/313304, https://ui.adsabs.harvard.edu/abs/2000ApJS..126..501T

Tkachenko A, Pavlovski K, Johnston C, Pedersen MG, Michielsen M, Bowman DM, Southworth J, Tsymbal V, Aerts C (2020) The mass discrepancy in intermediate- and high-mass eclipsing binaries: The need for higher convective core masses. Astron Astrophys 637:A60, https://doi.org/10.1051/0004-6361/202037452, https://ui.adsabs.harvard.edu/abs/2020A&A...637A..60T

Tokovinin A, Moe M (2020) Formation of close binaries by disc fragmentation and migration, and its statistical modelling. Mon Not R Astron Soc 491:5158–5171, https://doi.org/10.1093/mnras/stz3299, https://ui.adsabs.harvard.edu/abs/2020MNRAS.491.5158T

Tokovinin A, Thomas S, Sterzik M, Udry S (2006) Tertiary companions to close spectroscopic binaries. Astron Astrophys 450:681–693, https://doi.org/10.1051/0004-6361:20054427

Toonen S, Hamers A, Portegies Zwart S (2016) The evolution of hierarchical triple star-systems. Comput Astrophys Cosmol 3:6, https://doi.org/10.1186/s40668-016-0019-0, http://adsabs.harvard.edu/abs/2016ComAC...3....6T

Toonen S, Portegies Zwart S, Hamers AS, Bandopadhyay D (2020) The evolution of stellar triples. The most common evolutionary pathways. Astron Astrophys 640:A16, https://doi.org/10.1051/0004-6361/201936835, https://ui.adsabs.harvard.edu/abs/2020A%26A...640A..16T/abstract, 2004.07848

Torres G, Andersen J, Giménez A (2010) Accurate masses and radii of normal stars: Modern results and applications. Astron Astrophys Rev 18(1–2):67, https://doi.org/10.1007/s00159-009-0025-1, https://ui.adsabs.harvard.edu/abs/2010A%26ARv..18...67T/abstract

Turcotte DL, Schubert G (2014) Geodynamics, 3rd edn. Cambridge University Press, Cambridge

Tutukov AV, Yungelson LR (1993) The merger rate of neutron star and black hole binaries. Mon Not R Astron Soc 260:675–678, https://doi.org/10.1093/mnras/260.3.675, https://ui.adsabs.harvard.edu/abs/1993MNRAS.260..675T

Tylenda R (2004) On the light echo in V838 Mon. Astron Astrophys 414:223–233, https://doi.org/10.1051/0004-6361:20034015

Tylenda R, Soker N (2006) Eruptions of the V838 Mon type: Stellar merger versus nuclear outburst models. Astron Astrophys 451:223–236, https://doi.org/10.1051/0004-6361:20054201, https://ui.adsabs.harvard.edu/abs/2006A&A...451..223T

Tylenda R, Soker N, Szczerba R (2005) On the progenitor of V838 monocerotis. Astron Astrophys 441:1099–1109, https://doi.org/10.1051/0004-6361:20042485

Tylenda R, Hajduk M, Kamiński T, Udalski A, Soszyński I, Szymański MK, Kubiak M, Pietrzyński G, Poleski R, Wyrzykowski Ł, Ulaczyk K (2011) V1309 Scorpii: Merger of a contact binary. Astron Astrophys 528:A114, https://doi.org/10.1051/0004-6361/201016221, https://ui.adsabs.harvard.edu/abs/2011A&A...528A.114T/abstract

van den Ancker ME, The PS, Feinstein A, Vazquez RA, de Winter D, Perez MR (1997) A multiwavelength study of star formation in the very young open cluster NGC 6530. Astron Astrophys Suppl Ser 123:63–82, https://doi.org/10.1051/aas:1997306, https://ui.adsabs.harvard.edu/abs/1997A&AS..123...63V

van den Bos WH (1926) Orbital elements of binary stars. Circ Union Observ Johannesburg 68:352–359, https://ui.adsabs.harvard.edu/abs/1926CiUO...68..352V

van den Heuvel EPJ, Portegies Zwart SF, de Mink SE (2017) Forming short-period Wolf-Rayet X-ray binaries and double black holes through stable mass transfer. Mon Not R Astron Soc 471:4256–4264, https://doi.org/10.1093/mnras/stx1430, https://ui.adsabs.harvard.edu/abs/2017MNRAS.471.4256V

Vanbeveren D, De Donder E, Van Bever J, Van Rensbergen W, De Loore C (1998) The WR and O-type star population predicted by massive star evolutionary synthesis. New Astron 3(7):443–492, https://doi.org/10.1016/S1384-1076(98)00020-7, https://www.sciencedirect.com/science/article/pii/S1384107698000207

Vanon R, Edelmann PVF, Ratnasingam RP, Varghese A, Rogers TM (2023) Three-dimensional simulations of massive stars. II. Age dependence. Astrophys J 954:171, https://doi.org/10.3847/1538-4357/ace9db, https://ui.adsabs.harvard.edu/abs/2023ApJ...954..171V

Verbunt F, Zwaan C (1981) Magnetic braking in low-mass X-ray binaries. Astron Astrophys 100:L7–L9

Vilhu O (1973) Evolving models for contact binaries. Astron Astrophys 26:267

Villaseñor JI, Taylor WD, Evans CJ, Ramírez-Agudelo OH, Sana H, Almeida LA, de Mink SE, Dufton PL, Langer N (2021) The B-type binaries characterization programme I. Orbital solutions for the 30 Doradus population. Mon Not R Astron Soc 507:5348–5375, https://doi.org/10.1093/mnras/stab2197, https://ui.adsabs.harvard.edu/abs/2021MNRAS.507.5348V

Villaseñor JI, Sana H, Mahy L, Shenar T, Bodensteiner J, Britavskiy N, Lennon DJ, Moe M, Patrick LR, Pawlak M, Bowman DM, Crowther PA, de Mink SE, Deshmukh K, Evans CJ, Fabry M, Fouesneau M, Holgado G, Langer N, Maíz Apellániz J, Mandel I, Oskinova LM, Pauli D, Ramachandran V, Renzo M, Rix HW, Rocha DF, Sander AAC, Schneider FRN, Sen K, Simón-Díaz S, van Loon JTh, Toonen S, Vink JS (2025) Binarity at LOw Metallicity (BLOeM): Enhanced multiplicity of early B-type dwarfs and giants at $Z = 0.2\ Z_\odot$. Astron Astrophys 698:A41, https://doi.org/10.1051/0004-6361/202453166, https://ui.adsabs.harvard.edu/abs/2025A&A...698A..41V

Vink JS, de Koter A, Lamers HJGLM (2001) Mass-loss predictions for O and B stars as a function of metallicity. Astron Astrophys 369:574–588, https://doi.org/10.1051/0004-6361:20010127, https://ui.adsabs.harvard.edu/abs/2001A&A...369..574V

Virtanen P, Gommers R, Oliphant TE, Haberland M, Reddy T, Cournapeau D, Burovski E, Peterson P, Weckesser W, Bright J, van der Walt SJ, Brett M, Wilson J, Millman KJ, Mayorov N, Nelson ARJ, Jones E, Kern R, Larson E, Carey CJ, Polat i, Feng Y, Moore EW, VanderPlas J, Laxalde D, Perktold J, Cimrman R, Henriksen I, Quintero EA, Harris CR, Archibald AM, Ribeiro AH, Pedregosa F, van Mulbregt P, SciPy 1 0 Contributors (2020) SciPy 1.0: Fundamental algorithms for scientific computing in Python. Nat Methods 17:261–272, https://doi.org/10.1038/s41592-019-0686-2, https://ui.adsabs.harvard.edu/abs/2020NatMe..17..261V

Vitense E (1953) Die Wasserstoffkonvektionszone der Sonne. Mit 11 Textabbildungen. Z Astrophys 32:135, https://ui.adsabs.harvard.edu/abs/1953ZA.....32..135V

Vogel HC (1890) Spectrographische Beobachtungen an Algol. Astron Nachr 123(19):289–292, https://doi.org/10.1002/asna.18901231902, https://onlinelibrary.wiley.com/doi/abs/10.1002/asna.18901231902

Vogt H (1925) Zum Strahlungsgleichgewicht der Sterne. Astron Nachr 223:229, https://doi.org/10.1002/asna.19242231403, https://ui.adsabs.harvard.edu/abs/1925AN....223..229V

Von Mertens A (2024) Attention is discovery: the life and legacy of astronomer Henrietta Leavitt. MIT Press

von Zeipel H (1910) Sur l'application des séries de M. Lindstedt à l'étude du mouvement des comètes périodiques. Astron Nachr 183:345, https://doi.org/10.1002/asna.19091832202, https://ui.adsabs.harvard.edu/abs/1910AN....183..345V

von Zeipel H (1924) The radiative equilibrium of a rotating system of gaseous masses. Mon Not R Astron Soc 84:665–683, https://doi.org/10.1093/mnras/84.9.665, https://ui.adsabs.harvard.edu/abs/1924MNRAS..84..665V

Vrancken J, Abdul-Masih M, Escorza A, Menon A, Mahy L, Marchant P (2024) Constraining the overcontact phase in massive binary evolution: III. Period stability of known B+B and O+B overcontact systems. Astron Astrophys 691:A150, https://doi.org/10.1051/0004-6361/202450849, https://ui.adsabs.harvard.edu/abs/2024A&A...691A.150V

Wade GA, Neiner C, Alecian E et al (2016) The MiMeS survey of magnetism in massive stars: Introduction and overview. Mon Not R Astron Soc 456:2–22, https://doi.org/10.1093/mnras/stv2568, https://ui.adsabs.harvard.edu/abs/2016MNRAS.456....2W

Wang M, Huang WJ, Kondev FG, Audi G, Naimi S (2021) The AME 2020 atomic mass evaluation (II). Tables, graphs and references. Chin Phys C 45:030003, https://doi.org/10.1088/1674-1137/abddaf, https://ui.adsabs.harvard.edu/abs/2021ChPhC..45c0003W

Webbink RF (1977) The evolution of low-mass close binary systems. V - Transport processes in the envelopes of contact components. Astrophys J 215:851–863, https://doi.org/10.1086/155422, http://adsabs.harvard.edu/abs/1977ApJ...215..851W

Webster BL, Murdin P (1972) Cygnus X-1—a spectroscopic binary with a heavy companion? Nature 235(5332):37–38, https://doi.org/10.1038/235037a0, https://www.nature.com/articles/235037a0

Wellstein S, Langer N, Braun H (2001) Formation of contact in massive close binaries. Astron Astrophys 369:939–959, https://doi.org/10.1051/0004-6361:20010151, https://ui.adsabs.harvard.edu/abs/2001A&A...369..939W

Wheeler JC (1979) Blue stragglers as long-lived stars. Astrophys J 234:569–578, https://doi.org/10.1086/157531, https://ui.adsabs.harvard.edu/abs/1979ApJ...234..569W

Wickramasinghe DT, Tout CA, Ferrario L (2014) The most magnetic stars. Mon Not R Astron Soc 437:675–681, https://doi.org/10.1093/mnras/stt1910, https://ui.adsabs.harvard.edu/abs/2014MNRAS.437..675W

Wilson RE (2001) Binary star morphology and the name overcontact. Inf Bull Variable Stars 5076:1, https://ui.adsabs.harvard.edu/abs/2001IBVS.5076....1W

Wilson RE, Devinney EJ (1971) Realization of accurate close-binary light curves: application to MR Cygni. Astrophys J 166:605, https://doi.org/10.1086/150986, http://adsabs.harvard.edu/abs/1971ApJ...166..605W

Wright NJ, Drake JJ, Mamajek EE, Henry GW (2011) The stellar-activity–rotation relationship and the evolution of stellar dynamos. Astrophys J 743(1):48, https://doi.org/10.1088/0004-637X/743/1/48

Yakut K, Eggleton PP (2005) Evolution of close binary systems. Astrophys J 629:1055–1074, https://doi.org/10.1086/431300, https://ui.adsabs.harvard.edu/abs/2005ApJ...629.1055Y

Yang Y, Yuan H, Dai H (2019) A comprehensive study of three early-type contact twin binaries: CT Tau, GU Mon, and V701 Sco. Astron J 157:111, https://doi.org/10.3847/1538-3881/aafee1, https://ui.adsabs.harvard.edu/abs/2019AJ....157..111Y

Yaşarsoy B, Yakut K (2014) An interacting O + O supergiant close binary system: Cygnus OB2-5 (V729 Cyg). New Astron 31:32–36, https://doi.org/10.1016/j.newast.2014.02.005, https://ui.adsabs.harvard.edu/abs/2014NewA...31...32Y

Zahn JP (1975) The dynamical tide in close binaries. Astron Astrophys 41:329–344, https://ui.adsabs.harvard.edu/abs/1975A&A....41..329Z

Zahn JP (1977) Tidal friction in close binary stars. Astron Astrophys 500:121–132, https://ui.adsabs.harvard.edu/abs/1977A&A....57..383Z/abstract

Zahn JP (1992) Circulation and turbulence in rotating stars. Astron Astrophys 265:115–132, https://ui.adsabs.harvard.edu/abs/1992A&A...265..115Z

Zotti G, Hoffmann SM, Wolf A, Chéreau F, Chéreau G (2020) The simulated sky: stellarium for cultural astronomy research. J Skyscape Archaeol 6(2):221–258, https://doi.org/10.1558/jsa.17822, https://journal.equinoxpub.com/JSA/article/view/17822

Index

M. Fabry, *Detailed Structure and Evolution Modeling of the Tightest Massive Binary Stars*, Springer Theses, https://doi.org/10.1007/978-3-032-28788-5

GPSR Compliance
The European Union's (EU) General Product Safety Regulation (GPSR) is a set of rules that requires consumer products to be safe and our obligations to ensure this.

If you have any concerns about our products, you can contact us on

ProductSafety@springernature.com

In case Publisher is established outside the EU, the EU authorized representative is:

Springer Nature Customer Service Center GmbH
Europaplatz 3
69115 Heidelberg, Germany

www.ingramcontent.com/pod-product-compliance
Ingram Content Group UK Ltd.
Pitfield, Milton Keynes, MK11 3LW, UK
UKHW020911180926
475648UK00002B/10
9783032287878